T0321366

PHOTONIC CRYSTALS

Qihuang Gong | Xiaoyong Hu

PHOTONIC CRYSTALS

PRINCIPLES AND APPLICATIONS

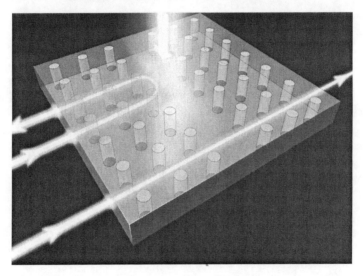

PAN STANFORD PUBLISHING

Published by

Pan Stanford Publishing Pte. Ltd.
Penthouse Level, Suntec Tower 3
8 Temasek Boulevard
Singapore 038988

Email: editorial@panstanford.com
Web: www.panstanford.com

British Library Cataloguing-in-Publication Data
A catalogue record for this book is available from the British Library.

Photonic Crystals: Principles and Applications

ISBN 978-981-4267-30-4 (Hardcover)
ISBN 978-981-4364-83-6 (eBook)

Printed in the USA

Contents

Preface

Photonic crystals are artificial microstructures that have spatially periodic distribution of dielectric constant and unique abilities of manipulating and controlling propagation states of photons. Therefore, they provide a perfect basis for the construction of integrated photonic circuits and chips and various other photonic devices. Moreover, the research of novel optical phenomena and newfangled devices based on photonic crystals is conducive to the construction and realization of optical computing systems, optical interconnection networks, and ultrahigh-speed information-processing systems in the future.

In this book, we introduce some important principles and research results in device applications of photonic crystals. Along with explaining the primary principles of photonic crystals, we interpret in detail the realization approach, the fabrication method, measurement systems, and potential applications of diversified optical devices based on photonic crystals. The areas of focus include photonic crystal all-optical switching, tunable photonic crystal filter, photonic crystal laser, photonic crystal logic devices, and photonic crystal sensors. The book provides extensive references, important principles, and experimental results that may be indispensible for students and researchers who are interested in photonic crystals and want to enter this research field quickly. We have authored this book to present the wonderful world of photonic crystals and sincerely hope that students and researchers will find it very useful.

We are grateful to Ms. Yanan Song for her meticulous work in spelling and grammar correction.

Qihuang Gong
Xiaoyong Hu
December 2013

Chapter 1

Introduction

One of the development goals of the integrated photonic technology is to realize ultrahigh-speed information processing and all-optical logic computing based on micro/nano-scale integrated photonic devices. Microstructure photonic materials having the ability of modulating propagation states of photons are very important basis for the development of the integrated photonic technology. The appearance of photonic crystal offers great hope for the realization of these goals. As the counterpart of semiconductor materials in photonics, photonic crystals possess unique photonic band structures, which contain pass bands, stop bands, and defect states [1,2]. This makes it possible to control the behaviors of photons at will. Photonic crystals, also called photonic bandgap materials, have been considered an important basis for the realization of micro/ nano-scale integrated photonic devices due to their unique photonic bandgap properties. Great efforts have been made to study the physical properties and applications of photonic crystals.

1.1 Brief Research History of Photonic Crystal

The concept of photonic crystal was proposed by Yablonovitch and John in 1987 [1,2]. Yablonovitch reported that a three-dimensional

Photonic Crystals: Principles and Applications
Qihuang Gong and Xiaoyong Hu
Copyright © 2014 Pan Stanford Publishing Pte. Ltd.
ISBN 978-981-4267-30-4 (Hardcover), 978-981-4364-83-6 (eBook)
www.panstanford.com

periodic structure has an electromagnetic bandgap, and the spontaneous emission of atoms in the periodic microstructure can be rigorously forbidden when the electronic band edge overlaps the electromagnetic bandgap [1]. John reported that the strong localization of photons could occur in a frequency range in certain disordered superlattice microstructures with sufficiently high dielectric contrast [2]. Owing to their unique ability to control the propagation states of photons, photonic crystals have great potential applications in the fields of optical computing, integrated photonic circuits, and ultrahigh-speed information processing. Inspired by the pioneering works by Yablonovitch and John, many scientists devote to the research field of photonic crystals. Photonic crystals have gradually become one of the research foci of physics and nano-photonics.

From 1987 to the early 1990s, majority of research works focused on the mechanism of forming a photonic bandgap, the photonic band structures of various crystal lattice structures, the localized states of photons, and waveguiding properties of photonic crystals [3–9]. In fact, only a few photonic crystal samples can be found in nature. Photonic crystals have to be fabricated artificially for the purpose of study and practical applications. Therefore, it is imperative that various techniques be developed for the fabrication of photonic crystals according to people's actual needs. Then, in the latter stage of the 1990s, great efforts were made to develop various techniques of fabricating photonic crystals, modulate the spontaneous emission of atoms by using the photonic bandgap effect, and study the novel nonlinear optical effects of photonic crystals [10–15]. During this period, the self-assembly method and the laser direct writing techniques through the two-photon-absorption of resin were developed to fabricate high-quality photonic crystals [16,17]. At present, these two methods have become very important techniques for the fabrication of three-dimensional photonic crystals. The approach to obtaining the complete photonic bandgap was also a research focus during this period [18,19].

In the beginning of the 21st century, the research of photonic crystals focused on the fabrication of high-quality photonic crystals, study of physical phenomena in photonic crystals, and the realization of optical devices based on photonic crystals [20–25].

High-quality three-dimensional photonic crystals were fabricated by use of the vertical deposition method, the laser direct writing method, and the laser holographic lithography method [26,27]. The novel phenomena of negative refraction effect and slow light effect in photonic crystals were also confirmed during this period [28–30]. Various integrated photonic devices based on photonic crystals, such as photonic crystal filter, photonic crystal optical switching, and photonic crystal laser, were realized experimentally [31–33]. In recent years, great attention has been paid to the realization of integrated photonic devices based on the new physical effects and phenomena in photonic crystals [34–36]. Moreover, photonic metamaterials, propagation and localization properties of surface plasmon polariton in metal photonic crystals, and the quantum electrodynamics of high-quality photonic crystal microcavity coupled with quantum dots have been studied extensively [37–40].

From the brief discussion of the research history, we can see that the fabrication of high-quality photonic crystal samples and the applications of photonic crystal in micro/nano-scale integrated photonic devices have been always the most pursued research topics, which will be discussed in detail in the following chapters.

1.2   Fundamental Principles of Photonic Crystal

Photonic crystals are a kind of microstructure photonic materials with a spatially periodic distribution of dielectric constant. The unique properties of photonic crystal are its photonic bandgap effect, which originates from the modulation of light by the spatially periodic distribution of dielectric constant. The photonic bandgap, also called the stop band, corresponds to the frequency range where the density of states of photons is zero in the photonic band structures. An incident electromagnetic wave with a resonant frequency dropping into the photonic bandgap will be reflected completely by the photonic crystal. The reason lies in that there do not exist corresponding Bloch modes that can propagate through the photonic crystal [1,2].

When a lattice defect is introduced in a perfect photonic crystal structure, the lattice defect will support an electromagnetic wave mode with a certain resonant frequency. Accordingly, defect states will appear in the photonic bandgap. The electric-field distribution of the defect mode will be confined in the defect site [41]. By use of the lattice defect, high-quality microcavity can be formed in photonic crystals. There is a drastic increase of the density of states of photons in the center of the defect mode compared with that of the stop band. Owing to the strong photon confinement effect of photonic crystal microcavity, the interactions of light and matter is enhanced greatly in the photonic crystal microcavity. So, not only novel nonlinear optical phenomena can be observed in the photonic crystal microcavity, but also the thresholdless laser emission is promising to be reached in high-quality photonic crystal microcavity structure [42,43].

The propagation and the confinement of light in a photonic crystal can be calculated based on the Maxwell's electromagnetic theory and the Bloch theory. Various numerical calculation methods have been developed to study the photonic band structure, the photon confinement effect, and nonlinear optical effects of photonic crystals, such as the finite-different time-domain (FDTD) method, the plane-wave expansion method, the transfer matrix method, and so on [44–46]. These numerical calculation methods provide a great help for people to understand the unique properties of photonic crystal and to design suitable photonic crystal structure for the realization of integrated photonic devices.

The thorough understanding of the principles of the photonic crystal is of great help for the research and applications of photonic crystals. The fundamental principles of photonic crystal, including the origination of the photonic bandgap, the configuration of the photonic crystal, and the measurement and characterization method of the photonic band structures, along with various numerical calculation methods, will be discussed in detail in the following chapters.

1.3 Fabrication of Photonic Crystals

High-quality photonic crystal samples are the important basis for the fundamental researches and practical applications of photonic crystals. Much attention has been paid to the development of novel techniques to fabricate high-quality photonic crystals.

The traditional methods to fabricate one-dimensional photonic crystals are based on expensive film growth techniques, including the molecular beam epitaxy (MBE) method, the chemical vapor deposition (CVD) method, and the pulsed laser deposition (PLD) method. These methods are very complicated and time consuming. Recently, various convenient methods have been developed to fabricate one-dimensional photonic crystals, such as the spin-coating method, the sol-gel method, and so on [47]. This makes it possible to fabricate one-dimensional photonic crystal by use of not only inorganic materials but also organic polymer materials very quickly.

The fabrication of two-dimensional photonic crystals is more complicated than that of one-dimensional photonic crystals. The expensive microfabrication technology of the microelectronic industry can be adopted to fabricate high-quality two-dimensional semiconductor photonic crystals [48]. The laser holography lithography method is also developed to fabricate two-dimensional organic photonic crystals and photonic quasicrystals [49]. Two-dimensional organic photonic crystals with various complicated lattice structures can be achieved by use of the laser holography lithography method.

To achieve large-size and single-crystal samples of three-dimensional photonic crystals is still a long-term goal and dream. The traditional self-assembling method can be adopted to obtain large-size samples of three-dimensional photonic crystals made of polystyrene nanoparticles or silicon dioxide nanoparticles. However, there exist a large number of structural defects in the three-dimensional colloidal photonic crystal [50]. Recently, the direct laser writing method and the laser holography lithography method were developed to fabricate high-quality three-dimensional photonic crystals with various lattice structures [51]. However, up to now, it is still a great challenge to realize large-size single-crystal photonic crystals with the complete photonic bandgap.

The fabrication methods play an important role in the study and practical applications of photonic crystals. Great achievements have been made for the fabrication of high-quality photonic crystal samples. The mechanism and preparation process will be discussed in detail for various fabrication methods in the following chapters.

1.4   Photonic Crystal Optical Devices

The major application of photonic crystals is to construct and realize micro/nano-scale integrated photonic devices. In recent years, along with the great achievement obtained in the fabrication of high-quality photonic crystal samples, various integrated photonic devices have been realized based on photonic crystals, such as photonic crystal all-optical switching, photonic crystal laser, photonic crystal sensors, and so on [52].

All-optical switching and logic device are essential components of optical computing and ultrahigh-speed information processing systems. The concept of photonic crystal all-optical switching was proposed by Scalora *et al.* in 1994 [53]. However, limited by small nonlinear optical coefficient of conventional materials, the research progress of the photonic crystal all-optical switching was very slow. Until recently, the limitation of the nonlinear material was gradually overcome. Ultralow-power and ultrafast photonic crystal all-optical switching has been reported lately [54]. This has greatly promoted the research process of photonic crystal all-optical switching. Logic devices can also be realized based on photonic bandgap effect and strong photon confinement effect of photonic crystal microcavity. Up to now, there are few experimental reports of photonic crystal logic devices.

Micro/nano-scale photonic crystal laser plays an important role in the construction of integrated photonic circuits and optical communication systems. Owing to the steep change of the density of states of photons in photonic band edges, and the strong photon confinement effect of photonic crystal microcavity, low threshold laser emission can be achieved when the gain medium is introduced in the photonic crystal. Recently, it has been possible to obtain a photonic crystal microcavity with an ultrahigh quality factor of 10^6 order by the microfabrication etching technology [55]. Low-threshold photonic crystal laser has been reported recently [56].

Photonic crystal filter has great potential applications in the wavelength division multiplex (WDM) systems. Optical channel of the filter can be formed by the photonic crystal microcavity mode. Moreover, multichannel and narrow-band filter can also be realized based on the coupling of several identical photonic crystal microcavities [57]. When a nonlinear optical material is adopted to construct photonic crystal filter, ultrafast tunability of the filtering channel can be reached [58]. This provides more flexible applications of photonic crystal filter.

The photonic crystal sensor is a kind of micro/nano-scale integrated photonic devices widely used in fields of biochemical sensing and environment detection. Various photonic crystal sensors, including fluid sensor, gas sensor, and biochemical sensor, can be realized based on high quality-factor photonic crystal microcavity structures [59]. The real-time, labelless, and quick detection can be reached on photonic crystal sensors.

The realization mechanism, the characterization method, the research progress, and potential applications of these micro/nano-scale integrated photonic crystal devices will be discussed in detail in the following chapters. Although it will take some time for these photonic crystal devices to meet the stringent criteria of practical applications, broad applications can be expected confidently.

References

1. E. Yablonovitch, "Inhibited spontaneous emission in solid-state physics and electronics," *Phys. Rev. Lett.* **58**, 2059–2062 (1987).

2. S. John, "Strong localization of photons in certain disordered dielectric superlattices," *Phys. Rev. Lett.* **58**, 2486–2489 (1987).

3. E. Yablonovitch and T. J. Gmitter, "Photonic band structure: the face-centered-cubic case," *Phys. Rev. Lett.* **63**, 1950–1953 (1989).

4. E. Yablonovitch, T. J. Gmitter, and K. M. Leung, "Photonic band structure: the face-centered-cubic case employing nonspherical atoms," *Phys. Rev. Lett.* **67**, 2295–2298 (1991).

5. E. Yablonovitch, T. J. Gmitter, R. D. Meade, A. M. Rappe, K. D. Brommer, and J. D. Joannopoulos, "Donor and acceptor modes in photonic band structure," *Phys. Rev. Lett.* **67**, 3380–3383 (1991).

6. S. John and N. Akozbek, "Nonlinear optical solitary in a photonic band gap," *Phys. Rev. Lett.* **71**, 1168–1171 (1993).

7. D. F. Sievenpiper, M. E. Sickmiller, and E. Yablonovitch, "3D wire mesh photonic crystals," *Phys. Rev. Lett.* **76**, 2480–2483 (1996).

8. I. Tarhan and G. H. Watson, "Photonic band structure of fcc colloidal crystal," *Phys. Rev. Lett.* **76**, 315–318 (1996).

9. C. M. Anderson and K. P. Giapis, "Larger two-dimensional photonic band gaps," *Phys. Rev. Lett.* **77**, 2949–2952 (1996).

10. S. Fan, P. R. Villeneuve, J. D. Joannopoulos, and E. F. Schubert, "High extraction efficiency of spontaneous emission from slabs of photonic crystals," *Phys. Rev. Lett.* **78**, 3294–3297 (1997).

11. D. F. Sievenpiper, E. Yablonovitch, S. Fan, P. R. Villeneuve, J. D. Joannopoulos, and E. F. Schubert, "3D metallo-dielectric photonic crystals with strong capacitive coupling between metallic islands," *Phys. Rev. Lett.* **80**, 2829–2832 (1998).

12. S. Fan, P. R. Villeneuve, J. D. Joannopoulos, and E. F. Schubert, "High extraction efficiency of spontaneous emission from slabs of photonic crystals," *Phys. Rev. Lett.* **78**, 3294–3297 (1997).

13. E. P. Petrov, V. N. Bogomolo, I. I. Kalosha, and S. V. Gaponenko "Spontaneous emission of organic molecules embedded in a photonic crystal," *Phys. Rev. Lett.* **81**, 77–80 (1998).

14. V. Berger, "Nonlinear photonic crystals," *Phys. Rev. Lett.* **81**, 4136–4139 (1998).

15. J. F. Bertone, P. Jiang, K. S. Hwang, D. M. Mittleman, and V. L. Colvin, "Thickness dependence of the optical properties of ordered silica-air and air-polymer photonic crystal," *Phys. Rev. Lett.* **83**, 300–303 (1999).

16. Y. A. Vlasov, N. Yao, and D. J. Norris, "Synthesis of photonic crystals for optical wavelengths from semiconductor quantum dots," *Adv. Mater.* **11**, 165–169 (1999).

17. H. B. Sun, S. Matsuo, and H. Misawa, "Three-dimensional photonic crystal structures achieved with two-photon-absorption photopolymerization of resin," *Appl. Phys. Lett.* **74**, 786–788 (1999).

18. K. Busch and S. John, "Liquid-crystal photonic-band-gap materials: the tunable electromagnetic vacuum," *Phys. Rev. Lett.* **83**, 967–970 (1999).

19. Z. Y. Li, B. Y. Gu, and G. Z. Yang, "Large absolute band gap in 2D anisotropic photonic crystals," *Phys. Rev. Lett.* **81**, 2574–2577 (1998).

20. J. D. Joannopoulos, "Self-assembly lights up," *Nature* **414**, 257–258 (2001).

21. M. C. Netti, A. Harris, J. J. Baumberg, D. M. Whittaker, M. B. D. Charlton, M. E. Zoorob, and G. J. Parker, "Optical trirefringence in photonic crystal waveguides," *Phys. Rev. Lett.* **86**, 1526–1529 (2001).

22. V. I. Kopp, A. Z. Genack, and Z. Q. Zhang, "Large coherence area thin-film photonic stop-band lasers," *Phys. Rev. Lett.* **86**, 1753–1756 (2001).

23. M. Notomi, K. Yamada, A. Shinya, J. Takahashi, C. Takahashi, and I. Yokohama, "Extremely large group-velocity dispersion of line-defect waveguides in photonic crystal slabs," *Phys. Rev. Lett.* **87**, 253902 (2001).

24. A. L. Pokrovsky and A. L. Efros, "Electrodynamics of metallic photonic crystals and the problem of left-handed materials," *Phys. Rev. Lett.* **89**, 093901 (2002).

25. A. Chutinan, S. John, and O. Toader, "Diffractionless flow of light in all-optical microchips," *Phys. Rev. Lett.* **90**, 123901 (2003).

26. Y. A. Vlasov, X. Z. Bo, J. C. Sturm, and D. J. Norris, "On-chip natural assembly of silicon photonic bandgap crystals," *Nature* **414**, 289–293 (2001).

27. M. Campbell, D. N. Sharp, M. T. Harrison, R. G. Denning, and A. J. Turberfield, "Fabrication of photonic crystals for the visible spectrum by holographic lithography," *Nature* **404**, 53–56 (2000).

28. S. Foteinopoulou, E. N. Economou, and C. M. Soukoulis, "Refraction in media with a negative refractive index," *Phys. Rev. Lett.* **90**, 107402 (2003).

29. E. Cubukcu, K. Aydin, and E. Ozbay, "Subwavelength resolution in a two-dimensional photonic-crystal-based superlens," *Phys. Rev. Lett.* **91**, 207401 (2003).

30. M. F. Yanik, W. Suh, Z. Wang, and S. Fan, "Stopping light in a waveguide with an all-optical analog of electromagnetically induced transparency," *Phys. Rev. Lett.* **93**, 233903 (2004).

31. W. Jiang and R. T. Chen, "Multichannel optical add-drop processes in symmetrical waveguide-resonator systems," *Phys. Rev. Lett.* **91**, 213901 (2001).

32. D. A. Mazurenko, R. Kerst, J. I. Dijkhuis, A. V. Akimov, V. G. Golubev, D. A. Kurdyukov, A. B. Pevtsov, and A. V. Selkin, "Ultrafast optical switching in three-dimensional photonic crystals," *Phys. Rev. Lett.* **91**, 213903 (2001).

33. M. Notomi, H. Suzuki, T. Tamamura, and K. Edagawa, "Lasing action due to the two-dimensional quasiperiodicity of photonic quasicrystal with a penrose lattice," *Phys. Rev. Lett.* **92**, 123906 (2004).

34. Y. Chassagneux, R. Colombelli, W. Maineult, S. Barbieri, H. E. Beere, D. A. Ritchie, S. P. Khanna, E. H. Linfield, and A. G. Davies, "Electrically pumped photonic-crystal terahertz lasers controlled by boundary conditions," *Nature* **457**, 174–178 (2009).

35. X. Y. Hu, P. Jiang, C. Y. Ding, H. Yang, and Q. H. Gong, "Picosecond and low-power all-optical switching based on an organic photonic-bandgap microcavity," *Nat. Photon.* **2**, 185–189 (2008).

36. S. E. Baker, M. D. Pocha, A. S. P. Chang, D. J. Sirbuly, S. Cabrini, S. D. Dhuey, T. C. Bond, and S. E. Letant, "Detection of bio-organism simulants using random binding on a defect-free photonic crystal," *Appl. Phys. Lett.* **97**, 113701 (2010).

37. Z. Wang, Y. D. Chong, J. D. Joannopoulos, and M. Soljacic, "Reflection-free one-way edge modes in a Gyromanetic photonic crystal," *Phys. Rev. Lett.* **100**, 013905 (2008).

38. B. I. Popa and S. A. Cummer, "Compact dielectric particles as a building block for low-loss magnetic metamaterials," *Phys. Rev. Lett.* **100**, 207401 (2008).

39. Z. Yu, G. Veronis, Z. Wang, and S. Fan, "One-way electromagnetic waveguide formed at the interface between a plasmonic metal under a static magnetic field and a photonic crystal," *Phys. Rev. Lett.* **100**, 023902 (2008).

40. D. Englund, A. Majumdar, A. Faraon, M. Toishi, N. Stoltz, P. Petroff, and J. Vuckovic, "Resonant excitation of a quantum dot strongly coupled to a photonic crystal nanocavity," *Phys. Rev. Lett.* **104**, 073904 (2010).

41. K. Guven, and E. Ozbay, "Coupling and phase analysis of cavity structures in two-dimensional photonic crystals," *Phys. Rev. B* **71**, 085108 (2005).

42. M. Fatih, S. Fan, and M. Soljacic, "High-contrast all-optical bistable switching in photonic crystal microcavities," *Appl. Phys. Rev.* **83**, 2739–2741 (2003).

43. R. Colombelli, K. Srinivasan, M. Troccoli, O. Painter, C. F. Gmachl, D. M. Tennant, A. M. Sergent, D. L. Sivco, A. Y. Cho, and F. Capasso, "Quantum cascade surface-emitting photonic crystal laser," *Science* **302**, 1374–1377 (2003).

44. M. Okano, and S. Noda, "Analysis of multimode point-defect cavities in three-dimensional photonic crystals using group theory in frequency and time domains," *Phys. Rev. B* **70**, 125105 (2004).

45. Y. C. Hsue, A. J. Freeman, and B. Y. Gu, "Extended plane-wave expansion method in three-dimensional anisotropic photonic crystals," *Phys. Rev. B* **72**, 195118 (2005).

46. Z. Y. Li, and K. M. Ho, "Application of structural symmetries in the plane-wave-based transfer-matrix method for three-dimensional photonic crystal waveguides," *Phys. Rev. B* **68**, 245117 (2003).

47. T. Komikado, S. Yoshida, and S. Umegaki, "Surface-emitting distributed-feedback dye laser of a polymeric multiplayer fabricated by spin coating," *Appl. Phys. Lett.* **89**, 061123 (2006).

48. N. Courjal, S. Benchabane, J. Dahdah, G. Ulliac, Y. Gruson, and V. Laude, "Acousto-optically tunable lithium niobate photonic crystal," *Appl. Phys. Lett.* **96**, 131103 (2010).

49. G. Q. Liang, W. D. Mao, Y. Y. Pu, H. Zou, H. Z. Wang, and Z. H. Zeng, "Fabrication of two-dimensional coupled photonic crystal resonator arrays by holographic lithography," *Appl. Phys. Lett.* **89**, 041902 (2006).

50. W. Lee, A. Chan, M. A. Bevan, J. A. Lewis, and P. V. B, "Nanoparticle-mediated assembly of colloidal crystals on patterned substrates," *Langmuir* **20**, 5262–5270 (2004).

51. D. Xia, J. Zhang, X. He, and S. R. J. Brueck, "Fabrication of three-dimensional photonic crystal structures by interferometric lithography and nanoparticle self-assembly," *Appl. Phys. Lett.* **93**, 071105 (2008).

52. Y. Chassagneux, R. Colombelli, W. Maineult, S. Barbieri, H. E. Beere, D. A. Ritchie, S. P. Khanna, E. H. Linfield, and A. G. Davies, "Electrically pumped photonic-crystal terahertz lasers controlled by boundary conditions," *Nature* **457**, 174–178 (2009).

53. M. Scalora, J. P. Dowling, C. M. Bowden, and M. J. Bloemer, "Optical limiting and switching of ultrashort pulses in nonlinear photonic band gap materials," *Appl. Phys. Lett.* **73**, 1368–1371 (1994).

54. X. Y. Hu, P. Jiang, C. Y. Ding, H. Yang, and Q. H. Gong, "Picosecond and low-power all-optical switching based on an organic photonic-bandgap microcavity," *Nat. Photon.* **2**, 185–189 (2008).

55. P. B. Deotare, M. W. Mccutcheon, I. W. Frank, M. Khan, and M. Loncar, "High quality factor photonic crystal nanobeam cavities," *Appl. Phys. Lett.* **94**, 121106 (2009).

56. Y. Chassagneux, R. Colombelli, W. Maineult, S. Barbieri, H. E. Beere, D. A. Ritchie, S. P. Khanna, E. H. Linfield, and A. G. Davies, "Electrically pumped photonic-crystal terahertz lasers controlled by boundary conditions," *Nature* **457**, 174–178 (2009).

57. P. Kohli, C. Christensen, J. Muehlmeier, R. Biswas, G. Tuttle, and K. M. Ho, "Add-drop filters in three-dimensional layer-by-layer photonic crystals using waveguides and resonant cavities," *Appl. Phys. Lett.* **89**, 231103 (2006).

58. C. Schuller, J. P. Reithmaier, J. Zimmermann, M. Kamp, A. Forchel, and S. Anand, "Polarization-dependent optical properties of planar photonic crystals infiltrated with liquid crystals," *Appl. Phys. Lett.* **87**, 121105 (2005).

59. H. Yang, and P. Jiang, "Macroporous photonic crystal-based vapor detectors created by doctor blade coating," *Appl. Phys. Lett.* **98**, 011104 (2011).

Chapter 2

Fundamental Properties of Photonic Crystals

Photons can be used as an excellent information carrier. The advantages of photons lie in the large information capacity and the high information transmission speed. To achieve ultrahigh-speed information processing and all-optical computing based on micro/nano-photonic devices is one target of the integrated photonic technology. Proper photonic materials are the essential basis for the realization of micro/nano-photonic devices. Are there photonic materials that can control the propagation states of photons? Photonic crystals are a kind of materials that can perform such functions. The propagation states of photons can be tailored by use of photonic crystals in much the same way as electrons controlled by semiconductor materials. Photonic crystals are artificial photonic materials made of two (or even more) kinds of dielectric materials periodically placed in space. The dielectric distribution of a photonic crystal varies periodically in space, which provides a "periodic potential" for electromagnetic waves entering in the photonic crystal. The electromagnetic waves encounter strong coherent scattering in the interfaces of different dielectric components. Owing to the strong modulation of the electromagnetic wave by the spatially periodic distribution of dielectric constant, photonic bandgaps appear in the dispersion

Photonic Crystals: Principles and Applications
Qihuang Gong and Xiaoyong Hu
Copyright © 2014 Pan Stanford Publishing Pte. Ltd.
ISBN 978-981-4267-30-4 (Hardcover), 978-981-4364-83-6 (eBook)
www.panstanford.com

relation curves of photonic crystal. The incident light whose frequencies drop in the photonic bandgap cannot propagate in the photonic crystal. The propagation states of electromagnetic waves can be tailored at will by use of photonic crystals due to their unique photonic bandgap effect. Just like semiconductor materials, when a structure defect is introduced in a perfect photonic crystal, defect states will appear in the photonic bandgap. Electromagnetic waves with a certain resonant frequency will be strongly confined around the defect structure site. The defect states can possess very high transmittance with elaborately designed defect structures. An incident light whose frequency is resonant with that of the defect mode could transmit through the photonic crystal based on the photon tunneling effect. The role of photonic crystals in the photonic technology is just like semiconductor materials in the microelectronic technology. Various novel integrated photonic devices can be realized based on photonic crystals.

2.1 Configuration of Photonic Crystals

As a kind of artificial photonic materials, photonic crystals are constructed by two (or even more) kinds of dielectric materials arranged periodically in space. Accordingly, the distribution of dielectric function has a spatial periodicity. According to different spatial periodicities of dielectric materials, photonic crystal can be categorized into one-dimensional (1D) photonic crystal, two-dimensional (2D) photonic crystal, and three-dimensional (3D) photonic crystal.

2.1.1 One-Dimensional Photonic Crystal

One-dimensional photonic crystals possess the dielectric periodicity in only one direction, for example, in the Z axle direction. One-dimensional photonic crystals can be constructed by alternatively placing a dielectric layer with a high refractive index and a dielectric layer with a low refractive index. The schematic structure of a one-dimensional photonic crystal is shown in Fig. 2.1. One-dimensional photonic crystals possess the configuration of multiple dielectric layers. The photonic bandgap effect of a one-dimensional photonic crystal arises from the strong multiple scattering in the interfaces between high-refractive-index layers and low-refractive-index layers, and the subsequent destructive interference. When an electromagnetic wave is incident along the direction perpendicular to the plane of the dielectric layers, the propagation states of the electromagnetic wave will be controlled by the photonic bandgap effect. The electromagnetic wave will see a homogeneous distribution of dielectric function when it is incident in the direction parallel to the plane of the dielectric layers. There exists no photonic bandgap effect that can influence the propagation process of the electromagnetic wave in this direction. Therefore, one-dimensional photonic crystals cannot provide the complete confinement and controlling of photons in all three dimensions and all the directions. The major properties of the photonic bandgap of one-dimensional photonic crystals are mainly determined by the following parameters: the thickness of two kinds of dielectric layers, the lattice constant (corresponding to the dielectric periodicity), and the refractive index contrast of high and low dielectric layers. The dispersion properties of dielectric materials also have a very important influence on the photonic bandgap effect of one-dimensional photonic crystals.

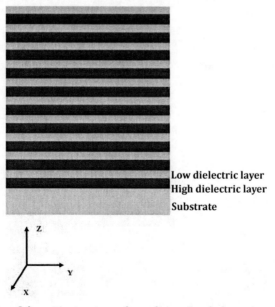

Figure 2.1   Schematic structure of one-dimensional photonic crystal.

From the point of view of practical applications, the frequency range of the photonic bandgap and pass bands of one-dimensional photonic crystals should be far away from the linear absorption bands of materials constructing the photonic crystals. Or else, most of the energy of the electromagnetic waves propagating in the photonic crystals will be absorbed strongly by the dielectric materials, so that the functions of the photonic bandgap effect cannot be brought into play.

2.1.2 Two-Dimensional Photonic Crystals

Two-dimensional photonic crystals possess the dielectric periodicity in two independent directions, for example, the X and Y axle directions. Two-dimensional photonic crystals have two kinds of configurations: the dielectric-rod-type and the air-hole-type photonic crystal. The schematic structure of a two-dimensional photonic crystal is shown in Fig. 2.2a. Dielectric-rode-type two-dimensional photonic crystals are constructed by periodic arrays of dielectric rods with a high refractive index embedded in the background material with a low refractive index. The dielectric function of the photonic crystal is distributed periodically in the direction perpendicular to axis of dielectric rods. When an electromagnetic wave is incident along the plane perpendicular to the dielectric-rods, the propagation states of the electromagnetic wave will be controlled by the photonic bandgap effect originating from strong multiple scattering and interference in the interface between dielectric-rods and low-refractive-index background material. The incident electromagnetic wave will see a homogeneous distribution of dielectric function when it is incident in the direction parallel to dielectric-rods. So, there exists no photonic bandgap effect that can influence the propagation process of the electromagnetic wave in this direction. Air-hole-type two-dimensional photonic crystals can be constructed by periodic arrays of cylindrical air holes embedded in a background material with a high refractive index, as shown in Fig. 2.2b. The dielectric function of the photonic crystal is distributed periodically in the direction perpendicular to the axis direction of air holes. When an electromagnetic wave is incident in the direction perpendicular to the air holes, the propagation states of the electromagnetic wave will be controlled by the photonic bandgap effect due to the strong multiple scattering and interference in the interface between air holes and the high-refractive-index background material. The incident electromagnetic wave will see a homogeneous distribution of dielectric function when it is incident in the direction parallel to the axis of air holes. There exists no photonic bandgap effect that can influence the propagation process of the electromagnetic wave in this direction. The properties of the photonic bandgap of a two-dimensional photonic crystal are mainly determined by the structural parameters of photonic crystal, the contrast of the refractive index between high dielectric material and low dielectric material, the defect structure, and the dispersion properties of dielectric materials. For the purpose of practical applications, the dielectric material should be transparent for an incident electromagnetic wave. Or else, most of the electromagnetic waves propagating in the photonic crystals will be absorbed strongly by the dielectric materials.

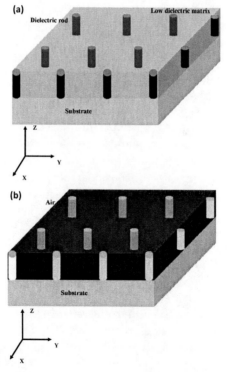

Figure 2.2   Schematic structure of two-dimensional photonic crystal. (a) For dielectric-rod type. (b) For air-hole type.

If a two-dimensional photonic crystal possesses a finite thickness in the Z direction, it is also called a two-dimensional photonic crystal slab. Two-dimensional photonic crystal slab can control the propagation states of an incident electromagnetic wave in the X–Y plane by the photonic bandgap effect due to its periodic distribution of dielectric function.

The electromagnetic wave is confined in the photonic crystal slab in the Z direction by the effect of totally internal reflection in the upside interface between photonic crystal slab and air, and in the bottom interface between the photonic crystal slab and the substrate. Two-dimensional photonic crystal slabs are a very important basis for the practical application of photonic crystals in the fields of integrated photonic devices and integrated photonic circuits.

2.1.3 Three-Dimensional Photonic Crystals

Three-dimensional photonic crystals have the dielectric periodicity in all the three independent directions, i.e., the X, Y, and Z axle directions. Since the dielectric function of the three-dimensional photonic crystal is distributed periodically in three independent directions, it is possible that an electromagnetic wave incident in any direction will suffer from the controlling of the photonic bandgap effect. It is found that when the dielectric constant contrast is high enough, a complete photonic bandgap can be reached in a three-dimensional photonic crystal. An electromagnetic wave propagating in the three-dimensional photonic crystal will be controlled by the photonic bandgap no matter toward which direction it propagates. Therefore, the propagation states of the electromagnetic wave can be fully tailored in all the three dimensions by use of a three-dimensional photonic crystal. Unlike one- and two-dimensional photonic crystals, the three-dimensional photonic crystal has much diverse crystalline configurations, such as the face-centered-cubic (fcc) structure, the body-centered-cubic (bcc) structure, the diamond structure, and so on. The crystal cell can have various shapes, including sphere, rode, cube, and filament. The schematic structure of a three-dimensional photonic crystal with a layer-by-layer structure is shown in Fig. 2.3.

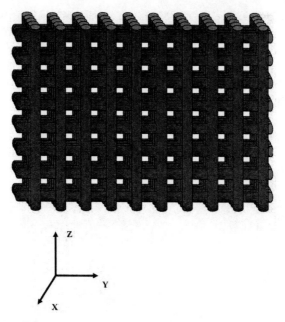

Figure 2.3   Schematic structure of a three-dimensional photonic crystal constructed by stacking dielectric rods.

In order to realize photonic bandgap effect, the periodicity of dielectric distribution of a photonic crystal should be in the same order as that of the wavelength of the incident signal light. For a photonic crystal whose photonic bandgap is located in the microwave range, the lattice constant is usually in the order of several millimeters or centimeters. Accordingly, the total size of a sample of photonic crystals in the microwave range is relatively large, often in the order of several hundred millimeters. For a photonic crystal in the visible and near-infrared range, the lattice constant should be in the submicrometer order. The total size of the near-infrared photonic crystal is usually small. Therefore, precision fabrication techniques are needed.

2.2   Origination of Photonic Bandgap

Photonic crystals consist of different dielectric materials that were placed periodically in space. The period of the distribution of dielectric function is in the same order as that of the wavelength of the incident electromagnetic wave. The physical origination of the generation of the photonic bandgap lies in the strong multiple scattering of incident electromagnetic wave in the interfaces between different dielectric materials and the subsequently destructive interference. The propagation process of the electromagnetic wave in a photonic crystal can be described by use of the basic Maxwell equations [1]:

$$\nabla \times \vec{E} = -\frac{\partial \vec{B}}{\partial t} \tag{2.1}$$

$$\nabla \cdot \vec{D} = \rho \tag{2.2}$$

$$\nabla \times \vec{H} = \vec{J} + \frac{\partial \vec{D}}{\partial t} \tag{2.3}$$

$$\nabla \cdot \vec{B} = 0 \tag{2.4}$$

where $\vec{E}$ is electric field, $\vec{B}$ is magnetic induction field, $\vec{D}$ is electric displacement, ρ is electric charges density, $\vec{H}$ is magnetic field, and $\vec{J}$ is current density. The parameters $\vec{E}$, $\vec{D}$, $\vec{B}$, and $\vec{H}$ meet the following relations:

$$\vec{D} = \varepsilon \vec{E} \tag{2.5}$$

$$\vec{B} = \mu \vec{H} \tag{2.6}$$

where ε is the electric permittivity, μ is the magnetic permeability. A photonic crystal can be regarded as a composite dielectric material. Originally there is no free electric charge and current in the dielectric materials. Therefore, the photonic crystal does not have the resources for electromagnetic wave. Accordingly, the parameters ρ and $\vec{J}$ are set as $\rho = 0$ and $\vec{J} = 0$, respectively. Then the Maxwell equations turn into the following form [1]:

$$\nabla \times \vec{E} = -\frac{\partial \vec{B}}{\partial t} \tag{2.7}$$

$$\nabla \cdot \vec{D} = 0 \tag{2.8}$$

$$\nabla \times \vec{H} = \frac{\partial \vec{D}}{\partial t} \tag{2.9}$$

$$\nabla \cdot \vec{B} = 0 \tag{2.10}$$

The components of the electric field and the magnetic field of an incident electromagnetic wave can be written as follows:

$$\vec{E}(\vec{r},t) = \vec{E}(\vec{r})e^{i\omega t}$$

(2.11)

$$\vec{H}(\vec{r},t) = \vec{H}(\vec{r})e^{i\omega t}$$

(2.12)

From equations (2.5) and (2.11), it can be obtained that

$$\vec{D}(\vec{r},t) = \varepsilon(\vec{r})\vec{E}(\vec{r})e^{i\omega t}$$

(2.13)

Then take equations (2.12) and (2.13) into equation (2.9), $\vec{E}(\vec{r})$ can be replaced by

$$\vec{E}(\vec{r}) = -\frac{i}{\omega\varepsilon(\vec{r})}\nabla \times \vec{H}(\vec{r})$$

(2.14)

From equations (2.6) and (2.12), it can be obtained that

$$\vec{B}(\vec{r},t) = \mu\vec{H}(\vec{r})e^{i\omega t}$$

(2.15)

Taking (2.14) and (2.15) into (2.7), we can obtain [1]

$$\nabla \times \left(\frac{1}{\varepsilon(\vec{r})}\nabla \times \vec{H}(\vec{r})\right) = (\mu\omega^2)\vec{H}(\vec{r})$$

(2.16)

Equation (2.16) is the master equation describing the propagation properties of an electromagnetic wave in a photonic crystal. This equation is determined by the field component $\vec{H}(\vec{r})$. The dielectric function has a periodic distribution

$$\varepsilon(\vec{r}) = \varepsilon(\vec{r} + \vec{R})$$

(2.17)

where $\vec{R}$ is the lattice vector of the photonic crystal.

Similarly, we can also write the master equation in terms of $\vec{E}(\vec{r})$ as [1]

$$\nabla \times \nabla \times \vec{E}(\vec{r}) = (\mu\omega^2)\varepsilon(\vec{r})\vec{E}(\vec{r})$$

(2.18)

To conveniently understand the physical origination of photonic bandgap, let us contrast the master equation of photons with the

Schrödinger equation of electrons. The propagation states of an electron in a semiconductor crystal can be described by the wave function of electrons $\Psi(\vec{r},t)$

$$\Psi(\vec{r},t)=\Psi(\vec{r})e^{(iE/h)t} \qquad (2.19)$$

where E is the electron energy. The Schrödinger equation is used to describe the propagation of an electron in a semiconductor crystal

$$\left(\frac{p^2}{2m}+V(\vec{r})\right)\Psi(\vec{r})=E\Psi(\vec{r}) \qquad (2.20)$$

where p is the momentum of the electron, m is the mass of the electron. $V(\vec{r})$ is the periodic potential formed by the periodically placed atoms, which meets the following relation:

$$V(\vec{r})=V(\vec{r}+\vec{R}) \qquad (2.21)$$

It is very clear that there is a similar corresponding relation between the master equation of photonic crystal and the Schrödinger equation of semiconductor crystal, as shown in Table 2.1 [1].

Table 2.1   Corresponding relation between the master equation and the Schrödinger equation

	Photons	**Electrons**
Wave functions	$\vec{H}(\vec{r},t)=\vec{H}(\vec{r})e^{i\omega t}$	$\Psi(\vec{r},t)=\Psi(\vec{r})e^{(iE/h)t}$
The periodicity	$\varepsilon(\vec{r})=\varepsilon(\vec{r}+\vec{R})$	$V(\vec{r})=V(\vec{r}+\vec{R})$
Hermitian operator	$\nabla\times\left(\frac{1}{\varepsilon(\vec{r})}\nabla\times\right)$	$\left(\frac{p^2}{2m}+V(\vec{r})\right)$
Eigenvalue	$\mu\omega^2$	E

Owing to the modulation of periodic dielectric function $\varepsilon(\vec{r})=\varepsilon(\vec{r}+\vec{R})$, the electromagnetic wave is scattered in the interface of different potential regions and subsequently the scattered electromagnetic waves interference. As a result, bandgap appears in the energy band structure of photonic crystal. The photonic bandgap structure of a photonic crystal can be obtained by solving the master equations (2.16) and (2.18), which are complex differential equations.

2.3 Characterization of Photonic Bandgap

The photonic bandgap includes a certain frequency range. When the frequency of an incident electromagnetic wave drops in the photonic bandgap, the electromagnetic wave will be reflected completely by the photonic crystal due to the strong Bragg scattering and subsequent destructive interference. Usually, the properties of the photonic bandgap can be characterized by use of dispersion relations, transmittance spectrum, and reflectance spectrum of the photonic crystal. The photonic band structure of a photonic crystal can be conveniently obtained from the diagram of dispersion relation of the photonic crystal. The transmittance spectrum (or the reflectance spectrum) not only provides the position, the frequency range, and the width of the photonic bandgap but also gives the data of the transmittance (or reflectance) of the pass band and the photonic bandgap, and the slope of the photonic bandedge. Therefore, the transmittance and the reflectance spectrum are of great importance to the characterization of the photonic bandgap in experiment. The diagrams of dispersion relation, the transmittance spectrum, the reflectance spectrum, and the electric-field distribution are often used to characterize the properties of a defect mode located in the photonic bandgap. The quality factor (Q value) and the line width are also very important factors for a defect mode.

2.3.1 Dispersion Relation

The dispersion relation of a photonic crystal can be obtained by solving the Maxwell equation. All possible eigenvalues of the master equations (2.16) and (2.18) can be obtained by letting wave vector k change in the first Brillouin zone. Usually, the first Brillouin zone is also called the irreducible Brillouin zone. For a one-dimensional photonic crystal, the first Brillouin zone can be defined as a region of wave vector k [1]

$$-\frac{\pi}{a} \leq k < \frac{\pi}{a} \qquad (2.22)$$

where a is the lattice constant. For each wave vector k, there exists a series of electromagnetic wave modes, with discrete resonant frequencies, which construct the pass band of the photonic crystal. The frequency range between two neighboring pass bands, where no eigenvalue exists for the master equations (2.16) and (2.18), is the photonic bandgap. According to the electromagnetic variational theory, the electromagnetic modes with high resonant frequency will concentrate their energy on the low dielectric material region of a photonic crystal. While the electromagnetic modes with low resonant frequency will concentrate their energy on the high dielectric material region of the photonic crystal. Therefore, the pass band under the fundamental photonic bandgap is also called the dielectric band. The pass band above the fundamental photonic bandgap is also called the air band.

2.3.2 Transmittance Spectrum

In the ideal case, the transmittance of the frequency range corresponding to a photonic bandgap is zero, while the transmittance of the pass bands is 100%. Therefore, a deep dip will appear in the transmittance spectrum of a photonic crystal. If we neglect the influence of the intrinsic material absorption on the photonic bandgap structure, the position and the width of photonic bandgap can be determined through the transmittance spectrum of a photonic crystal by use of deep dips.

2.3.3 Reflectance Spectrum

In the ideal case, the reflectance of the frequency range corresponding to a photonic bandgap is 100%, while the reflectance of the pass bands is zero. Therefore, a high peak will appear in the reflectance spectrum of a photonic crystal. This corresponds to the position of the photonic bandgap. If we neglect the influence of the intrinsic material absorption on the photonic bandgap structure, the position and the width of photonic bandgap can be determined through the reflectance spectrum of a photonic crystal by use of high reflection peaks.

To obtain the photonic bandgap structure of a photonic crystal, various simulation methods have been developed to solve Maxwell equations, such as multiple scattering method, plane wave expanding method (PWE), Green function method (GF), finite-difference time-domain (FDTD) method, and transfer matrix method (TMM). Essentially, the problem of solving the Maxwell equations (2.16) and (2.18) is converted into the problem of solving the eigenvalues of eigen-equations.

2.4 One-Dimensional Photonic Crystal

One-dimensional photonic crystals have a configuration of multi-layers, constructed by alternatively placing a high-refractive-index layer and a low-refractive-index layer. According to the Bragg law, the central wavelength λ_c of the Bragg reflection peak for a multilayer structure meets the relation [1]

$$d_1 = \frac{\lambda_c}{4n_1} \qquad (2.23)$$

$$d_2 = \frac{\lambda_c}{4n_2} \qquad (2.24)$$

where d_1 and d_2 are the thickness of high dielectric layers and low dielectric layers, respectively. n_1 and n_2 are the refractive index of high dielectric layers and low dielectric layers, respectively. The lattice constant a of a one-dimensional photonic crystal can be calculated by

$$a = d_1 + d_2 \qquad (2.25)$$

The optical thickness of d_1 and d_2 is ($\lambda_c/4$). Therefore, one-dimensional photonic crystal is also called ($\lambda/4$) Bragg reflector. Due to the destructive interference in the interface of difference dielectric layers, the photonic bandgap centered at λ_c is generated. This offers an approach to constructing a one-dimensional photonic crystal whose photonic bandgap is in the required frequency range.

The transfer matrix method (TMM) can be adopted to conveniently calculate the photonic bandgap structure of one-dimensional photonic crystals. When an electromagnetic wave propagates through a dielectric multilayer system, the incident, reflected and transmitted electric fields are connected via a transfer matrix T [2–6]

$$\begin{pmatrix} E^{in} \\ E^{re} \end{pmatrix} = T \begin{pmatrix} E^{tr} \\ 0 \end{pmatrix} \tag{2.26}$$

where E^{in}, E^{re} and E^{tr} are the incident, reflected, and transmitted electric fields, respectively. The transfer matrix T can be expressed as

$$T = \begin{pmatrix} e_{11} & e_{12} \\ e_{21} & e_{22} \end{pmatrix} \tag{2.27}$$

where e_{ij} $(i,j = 1,2)$ is the element of the transfer matrix. For a one-dimensional photonic crystal having a simple configuration of $(AB)^m$, T can be calculated by using a sequential product of the transfer matrices for every successive interface:

$$T = T_{airA} T_A T_{AB} T_B T_{BA} T_A \cdots\cdots T_{AB} T_B T_{BA} \cdots\cdots T_A T_{AB} T_B T_{Bair} \tag{2.28}$$

where T_{airA} and T_{Bair} represent the propagation of light through the interfaces air $\rightarrow$ A and B $\rightarrow$ air, respectively. T_A and T_B represent the propagation of light within layer A and B, respectively. T_{AB}, and T_{BA} represent the propagation of light through the interfaces A $\rightarrow$ B, B $\rightarrow$ A, respectively. The transmittance of the one-dimensional photonic crystal system can be calculated by the relation [5,6]

$$t = \left| \frac{1}{e_{11}} \right|^2 \tag{2.29}$$

Then the reflectance of the one-dimensional photonic crystal can be calculated by the relation [5,6]

$$r = \left| \frac{e_{21}}{e_{11}} \right|^2 \tag{2.30}$$

2.4.1 One-Dimensional All-Dielectric Photonic Crystals

One-dimensional all-dielectric photonic crystals are composed of alternatively placed two (or even more) different dielectric materials. The position and width of the photonic bandgap are determined by the structural parameter and the refractive index contrast of the photonic crystal. The larger the refractive index contrast is, the wider the photonic bandgap becomes. The photonic bandgap properties can be described by analyzing a practical sample of one-dimensional photonic crystal. Let us consider a one-dimensional photonic crystal constructed by alternatively placing a silicon dioxide (SiO_2) layer and a magnesium fluoride (MgF_2) layer. The photonic crystal has a configuration of $(AB)^m$, where A and B represent SiO_2 layer and MgF_2 layer, respectively; m is the number of repetition period. The refractive index is set as 1.45 for SiO_2 and 1.38 for MgF_2, respectively. The thickness is 103.5 nm for SiO_2 layers and 108.7 nm for MgF_2 layers, respectively. The transmittance and reflectance spectra of the one-dimensional photonic crystal calculated by the transfer matrix method are shown in Fig. 2.4. Figure 2.4a,b shows the transmittance and the reflectance spectra for the transverse-electric (TE)-polarized incident electromagnetic wave. Figure 2.4c,d shows the transmittance and the reflectance spectra for the transverse-magnetic (TM)-polarized incident electromagnetic wave. There exists remarkable photonic bandgap for both the TE- and TM-polarized light. In fact, both the TE and TM incident light can see periodic distribution of dielectric function. The central wavelength of the fundamental photonic bandgap (the first order photonic bandgap) is located at 600 nm both for the TE and TM incident light. The period of the dielectric distribution of the photonic crystal with a $(AB)^m$ configuration is the same one for both the TE- and TM-polarized light. The width of the fundamental bandgap is 33.8 nm. Even though there are 30 repetition periods in the photonic crystal, the transmittance of the center of the photonic bandgap still reaches 18.6%. The reason lies in that the refractive index contrast between SiO_2 and MgF_2 is relatively small, only achieving 0.07. Therefore, the Bragg scattering effect and the subsequent interference in the interface of SiO_2 layers and MgF_2 layers is relatively weak. Perfect and ideal photonic bandgap effect is very difficult to obtain with such a low refractive index contrast.

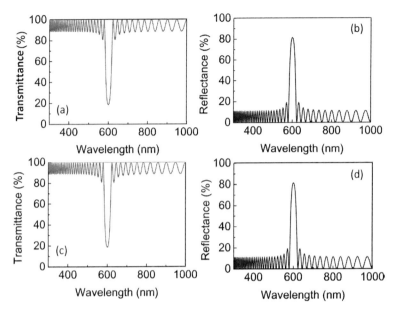

Figure 2.4 Transmittance and reflectance spectra of the one-dimensional photonic crystal composed of SiO_2 and MgF_2 calculated by the transfer matrix method. (a) Transmittance spectrum for incident TE electromagnetic wave. (b) Reflectance spectrum for incident TE electromagnetic wave. (c) Transmittance spectrum for incident TM electromagnetic wave. (d) Reflectance spectrum for incident TM electromagnetic wave.

In order to achieve a better photonic bandgap effect, let us increase the refractive index contrast. Let us consider a one-dimensional photonic crystal constructed by alternatively placing titanium dioxide (TiO_2) layer and SiO_2 layer, also presenting a configuration of $(AB)^m$. The transmittance spectrum of the one-dimensional TiO_2/SiO_2 was calculated by use of the transfer matrix method and the results are shown in Fig. 2.5b. The transmittance spectrum is for the TE-polarized electromagnetic wave. The refractive index of TiO_2 is 2.3. The refractive index contrast reaches 0.85 for the photonic crystal. The thickness is 103.5 nm for TiO_2 layers and 108.7 nm for SiO_2 layers. There are still 30 repetition periods in the photonic crystal. According to equations (2.23) and (2.24), the position of the photonic bandgap will shift in the direction of long wavelength with the increase of the effective refractive index of the photonic crystal. The center wavelength of the photonic bandgap shifts to 790 nm. The width of the photonic bandgap increases with the increment of the refractive index contrast. The width of the photonic bandgap increases to 227.7 nm. The transmittance of the photonic bandgap is zero. The transmittance contrast between the pass bands and the photonic bandgap is almost 100%. This indicates that the photonic crystal possesses perfect photonic bandgap effect, which is of great importance for the practical applications of one-dimensional photonic crystal.

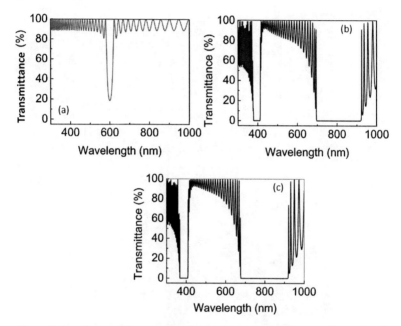

Figure 2.5 Transmittance and reflectance spectra of the one-dimensional photonic crystal and TE electromagnetic wave calculated by the transfer matrix method. (a) For photonic crystal composed of SiO_2 and MgF_2. (b) For photonic crystal composed of TiO_2 and SiO_2. (c) For photonic crystal composed of TiO_2 and MgF_2.

To further increase the refractive index contrast, Let us consider a one-dimensional photonic crystal constructed by alternatively placing titanium dioxide (TiO_2) layer and MgF_2 layer, also having a configuration of $(AB)^m$. The transmittance spectrum of the one-dimensional TiO_2/MgF_2 photonic crystal was calculated by use of the transfer matrix method and the results are shown in Fig. 2.5c. The transmittance spectrum is for the TE-polarized electromagnetic wave. The thickness is 103.5 nm for TiO_2 layers and 108.7 nm for MgF_2 layers. The refractive index contrast increases to 0.92. The number of the repetition period is still 30. The center wavelength of the photonic bandgap shifts to 800 nm. The width of the photonic bandgap increases to 246.1 nm. The transmittance contrast between the pass bands and the photonic bandgap is 100%. High-quality one-dimensional photonic crystals can be obtained easily with high refractive index contrast.

Not only can the refractive index contrast greatly influence the photonic bandgap, but the position of the photonic bandgap is different for an electromagnetic wave incident in different directions. The angle response properties of the one-dimensional TiO_2/MgF_2 photonic crystal are calculated by the transfer matrix method and the results are plotted in Fig. 2.6. It is very clear that the position of the photonic bandgap gradually shifts in the direction of short wavelength with the increase of the incident angle. The lattice constant encountered by the incident electromagnetic wave will be shortened with the increase of the incident angle. This makes the photonic bandgap shift in the direction of short wavelength.

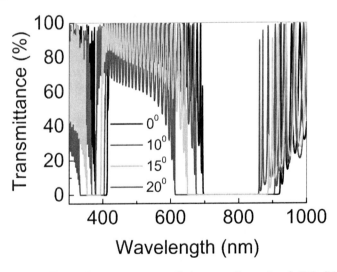

Figure 2.6 Transmittance spectra of the one-dimensional TiO_2/MgF_2 photonic crystal with different incident angle. The incident light is a TE polarized electromagnetic wave.

2.4.2 One-Dimensional Metallodielectric Photonic Crystals

One-dimensional metallodielectric photonic crystals are constructed by alternatively placing a metal layer and a dielectric layer. Just like all-dielectric photonic crystal, the metallodielectric photonic crystal also possesses remarkable photonic bandgap effect. Owing to the strong Bragg scattering in the interface of the dielectric and metal layers, obvious photonic bandgap can also be generated in one-dimensional metallodielectric photonic crystals. The metal materials have very distinctive properties compared with the dielectric material, such as the negative dielectric constant, surface plasmon polariton, and excellent electrical conductivity. There exist great differences between metallic and dielectric materials. Therefore, many interesting properties can be obtained for one-dimensional metallodielectric photonic crystals. For example, the available refractive index contrast of dielectric materials is greatly restricted in the visible range, which leads to a relatively narrow photonic bandgap for one-dimensional all-dielectric photonic crystals. By introducing the metallic component into a one-dimensional dielectric photonic crystal, the width of the photonic bandgap can be enlarged considerably [7]. Moreover, a very broad omnidirectional stop band covering almost all the visible range, from 450 to 750 nm, can be obtained for a one-dimensional metallic-organic photonic crystal [8].

2.4.2.1 One-dimensional metallodielectric periodic photonic crystal structures

2.4.2.1.1 *Basic characteristics*

A one-dimensional metallodielectric periodic photonic crystal is composed of alternatively placing a metal layer and a dielectric material layer. The material absorptions are neglected in the discussion of all-dielectric one-dimensional photonic crystal. However, the material dispersion of metal must be taken into account when metal materials are adopted as a component of a photonic crystal. Noble metals, including silver, gold, and copper, possess strong dispersion in the visible and near-infrared range. The Drude model can be used to characterize the dielectric function of metal. The dielectric function $\varepsilon(\omega)$ of a metal can be written as [9–11]

$$\varepsilon(\omega) = 1 - \frac{\omega_p^2}{\omega^2 + i\gamma\omega} \tag{2.31}$$

where ω_p is the plasma frequency of the bulk metal, γ is the damping coefficient, which is related to the absorption and thermal losses. The plasma frequency ω_p can be calculated by [11]

$$\omega_p = \left(\frac{ne^2}{\varepsilon_0 m}\right)^{\frac{1}{2}} \tag{2.32}$$

where ε_0 is the free space permittivity, n is the electron density, e is the electronic charge, and m is the mass of electrons. The plasma frequency of silver is 3.8 eV, which corresponds to the wavelength of 326 nm [10]. An electromagnetic wave with a resonant frequency larger than the plasma frequency ω_p will be absorbed by silver, and cannot transmit through it. The electromagnetic waves with resonant frequencies much less than the plasma frequency ω_p can penetrate a very small distance in the metal away from the surface, which is also called the skin depth. The skin depth is determined by the imaginary part of the refractive index of metal, which can be calculated by $(\lambda/4\pi n'')$, where λ is the wavelength of the incident light, n'' is the imaginary part of the refractive index of metal [9–11]. The skin depth of silver is about 12 nm in the visible and near-infrared range. Therefore, the transmittance of an electromagnetic wave with a resonant frequency less than the plasma frequency propagating through a 20 nm-thick silver film is very small. However, it has been pointed out that the absorption losses of a one-dimensional metallodielectric photonic crystal can be reduced by the resonant tunneling effect [11]. An incident electromagnetic wave can resonantly tunnel through the photonic crystal with minimum absorption losses. Obvious photonic bandgap effect can also be achieved in a one-dimensional metallo-dielectric photonic crystal. It has also been found that the width of the photonic bandgap of a one-dimensional metallodielectric photonic crystal is much larger than that of a one-dimensional all-dielectric photonic crystal. The reason is that an electromagnetic wave will be reflected stronger in the interface of metal and dielectric materials than in the interface of two different dielectric materials. Moreover, a one-dimensional metallodielectric photonic crystal can act as a perfect reflection mirror, whose reflectivity can be even larger than that of the bulk metal [12].

2.4.2.1.2 Effective plasma frequency theory

The effective plasma frequency theory can be adopted to study the photonic band structures of one-dimensional metallodielectric photonic crystals [13]. A one-dimensional metallodielectric photonic crystal could be considered an effective metallic medium with a well-defined effective plasma frequency Ω_p, which is lower than the inherent plasma frequency ω_p of the metal. The effective dielectric function $\varepsilon_{eff}(\omega)$ of the one-dimensional metallodielectric photonic crystal can be written as [13]

$$\varepsilon_{eff}(\omega) = \tilde{\varepsilon}_0 \left(1 - \frac{\Omega_p^2}{\omega^2} \right) \qquad (2.33)$$

where $\tilde{\varepsilon}_0$ is the effective static dielectric constant of the one-dimensional metallodielectric photonic crystal, which can be simply taken as the geometric mean of the static dielectric constant of the metal and dielectric layers. The first photonic bandgap starts from the effective plasma frequency Ω_p. The effective plasma frequency can be written as [13]

$$\Omega_p = \frac{\pi c}{nd} \qquad (2.34)$$

where c is the velocity of light in vacuum, n and d are the refractive index and the thickness of the dielectric layer, respectively. The effective plasma frequency is inversely proportional to the optical thickness of the dielectric layer and is independent of either the constituent metal or the thickness of the metallic layer. By increasing the optical thickness of the dielectric layer, the effective plasma frequency of the metallodielectric photonic crystal can be depressed into extremely low frequencies. Accordingly, the position of the first photonic bandgap can be adjusted by changing the thickness of the dielectric layers. The electric-field distributions of the three lowest photonic bands of a one-dimensional metallo-dielectric photonic crystal with thick dielectric layers are cavity-mode-like. As a result, the one-dimensional metallodielectric photonic crystal can be regarded as a metallic Fabry–Perot cavity for frequencies much lower than the bulk plasma frequency ω_p of the constituent metal. The cavity modes can be written as [13]

$$\omega_m = m\frac{\pi c}{n(d + 2\delta)} \tag{2.35}$$

where m is the cavity mode index, n and d are the refractive index and the thickness of the dielectric layer, respectively. δ is the skin depth of the metal. The value of the first order cavity mode is equal to that of the effective plasma frequency Ω_p [13].

2.4.2.1.3 *Properties of photonic bandgap*

Just like all-dielectric photonic crystals, the position and width of the metallodielectric photonic bandgap are also determined by the structural parameter and the refractive index contrast of the photonic crystal. The photonic bandgap properties can be described by analyzing a practical sample of one-dimensional metallodielectric photonic crystal. Let us consider a one-dimensional metallodielectric photonic crystal constructed by alternatively placing TiO_2 layers and silver layers. The photonic crystal has a configuration of $(AB)^m$, where A is TiO_2 layers and B is silver layers, respectively. The thickness of TiO_2 layers is 250 nm. The number of the repletion period m is 15. The transmittance spectrum of the one-dimensional metallodielectric photonic crystal with different thicknesses of silver layers was calculated by the transfer matrix method and the results are shown in Fig. 2.7. The incident light is a TM-polarized electromagnetic wave. The number of the repetition period is 15. Like a thick silver film, the transmittance of the metallodielectric photonic crystal is very low in the frequency range of larger than the plasma frequency ω_p due to the strong absorption induced by the interband transitions in silver. However, high transmittance can be obtained in the frequency range of less than the plasma frequency ω_p. The reason lies in the strong photon resonantly tunneling effect. A photonic bandgap centered at 560 nm can still be obtained due to strong scattering and interference in the interface of metal and dielectric material. The long-wavelength pass band is shut off at about 900 nm due to the large dispersion with the extinction coefficient of silver [14]. The width of the photonic bandgap is gradually enlarged with the increase of the thickness of silver layers. The width of the photonic bandgap is 10 nm for 1 nm-thick silver layers. The width of the photonic bandgap increases to 27.7 nm for 3 nm-thick silver layers.

The width is enlarged to 58.5 nm when the thickness of silver layers increases to 7 nm. This is related to the changes of the average refractive index of the metallodielectric photonic crystal. It is very clear that the transmittance of the dielectric bandedge of the pass band is larger than that of the air bandedge. For the metallodielectric photonic crystal with 7 nm-thick silver layers, the transmittance of the dielectric bandedge and air bandedge of the pass band is 60% and 43%, respectively. In the dielectric bandedge of the pass band, the solution to the Maxwell equations is a standing wave. The electric field of the incident light is mainly confined in the dielectric layers. While in the air bandedge of the pass band, the electric field of the incident light is mainly confined in the metal layers, which leads to very large absorption losses [14]. With the increase of the thickness of silver layers, the average transmittance of the pass band decreases greatly. The average transmittance of the pass band is 91% for the metallodielectric

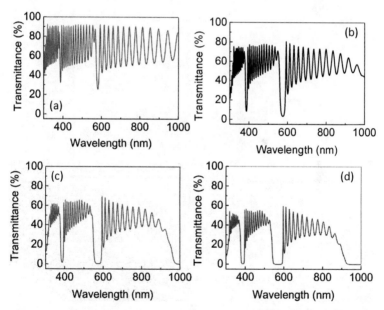

Figure 2.7   Transmittance spectra of one-dimensional TiO$_2$/Ag photonic crystal calculated by the transfer matrix method. The incident light is a TM polarized electromagnetic wave. The thickness of TiO$_2$ layers is 250 nm. (a) For 1 nm-thick silver layers. (b) For 3 nm-thick silver layers. (c) For 5 nm-thick silver layers. (d) For 7 nm-thick silver layers.

photonic crystal with 1 nm-thick silver layers. The average transmittance of the pass band decreases to 75% for 3 nm-thick silver layers. When the thickness of silver layers increases to 7 nm, the average transmittance of the pass band decreases to 55%. The reason is that the absorption losses of total silver layers increase with the increment of the thickness of silver layers. The number of the repetition period also has a great influence on the average transmittance of the pass band of the metallodielectric photonic crystal. The transmittance spectra of the one-dimensional TiO_2/ silver photonic crystal as a function of repetition period number are shown in Fig. 2.8. The average transmittance of the pass band decreases greatly with the increase of the repetition period number. The average transmittance changes from 55% to 25% when the number of the repetition period increases from 15 to 25. The absorption losses of total silver layers increase with the increment of the repetition period. This makes the average transmittance of the pass band decrease.

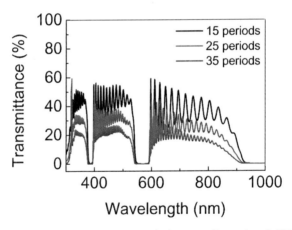

Figure 2.8 Transmittance spectra of the one-dimensional TiO_2/silver photonic crystal as a function of repetition period number. The incident light is a TM polarized electromagnetic wave.

Like one-dimensional all-dielectric photonic crystals, the position of the photonic bandgap of one-dimensional metallodielectric photonic crystals is different for an electromagnetic wave incident in different directions. The angle response properties of the one-dimensional TiO_2/silver photonic crystal are calculated by the transfer matrix method and the results are plotted in Fig. 2.9.

The position of the photonic bandgap gradually shifts in the direction of short wavelength with the increase of the incident angle. The lattice constant encountered by the incident electromagnetic wave will be shortened with the increase of the incident angle. This makes the photonic bandgap shift in the direction of short wavelength [15].

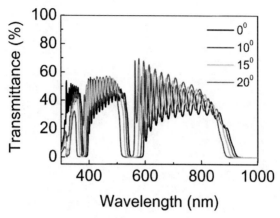

Figure 2.9   Transmittance spectra of the one-dimensional TiO_2/silver photonic crystal as a function of incident angle. The incident light is a TM polarized electromagnetic wave.

To reduce the absorption losses of the one-dimensional metallodielectric photonic crystal, a unique structural cell of dielectric/dielectric-metal-dielectric can be adopted. In this photonic crystal structure, the number of the metal layers will be reduced greatly, which can result in a remarkable increase in the transmittance of the pass band. Ye *et al.* [15] reported a low-loss one-dimensional metallodielectric photonic crystal fabricated by inserting Ag layers into ZnS/MgF_2 quarter-wave multilayers. The metallodielectric photonic crystal, fabricated by thermal evaporation of ZnS, MgF_2, Ag, and MgF_2 layers alternatively onto a glass substrate, has a configuration of 3.5 period of ZnS/MgF_2-Ag-MgF_2. The thickness of ZnS, MgF_2, and Ag layers are 61.7 nm, 52.5 nm, and 19 nm, respectively. The transmittance spectrum at normal incidence shows a wide and deep photonic bandgap between 420 and 790 nm, which is about 100 nm wider than that of the ZnS/MgF_2 photonic crystal. The rejection of the photonic bandgap is 10 dB per lattice constant, which is much better than the ZnS/MgF$_2$ photonic crystal. Moreover, the photonic bandedge is very sharp for the metallodielectric photonic crystal. The extension of the photonic bandgap of this metallodielectric photonic crystal is caused by a redistribution of the displacement field within the photonic crystal when a third component is inserted into the photonic crystal [15]. When a metal layer is inserted in the center of each unit cell, the change of the distribution of displacement field results in the change of the properties of the photonic bandgap [15]. Different insertion materials and positions will have different influences on the properties of the photonic bandgap [15].

2.4.2.1.4 *Transparent photonic band* [16]

Traditionally, the number of the periodic units will greatly influence the transmittance of the pass bands. However, under certain circumstances a transparent photonic band can be formed in a one-dimensional metallodielectric photonic crystal with a configuration of $(M_{1/2}DM_{1/2})^m$ or $(D_{1/2}MD_{1/2})^m$, where M and D are metal and dielectric layers, respectively [16]. An incident light at a frequency within the transparent band exhibits 100% transmittance, which is independent of the number of the periodic units. The transparent photonic band corresponds to the excitation of pure eigenstate modes across the entire Bloch band in structures possessing mirror symmetry. Each frequency in the transparent band is in an eigenstate common to translation and surface-wave operators. The existence of the transparent band depends on the surface plasmon excitation and the manner in which the first and last layers are truncated. In the dispersion relations of a one-dimensional metallodielectric photonic crystal possessing the transparent photonic bands, the pass bands are below the light line of the dielectric material so that the fields are evanescent not only in the metal layers but also in the dielectric layers. These modes are evanescent and bounded to the metal/dielectric interfaces. The thickness of the metal layers is thin enough to allow coupled eigenstates throughout the entire structure.

2.4.2.2 One-dimensional metallodielectric quasiperiodic photonic crystals

For a one-dimensional metallodielectric photonic crystal with a periodic arrangement of the metal layers, the transmittance will gradually decrease with the increase of the periodic number of the metallodielectric photonic crystal. While it is possible to arrange the layers in either a chirped or quasiperiodic geometry in order to maintain high levels of transparency due to their unique photon localization properties, even when more metal is added into the one-dimensional metallodielectric photonic crystal [17]. Cantor-like structures are fractal nonperiodic multilayers in a way similar to the Cantor set configuration [18]. Two characteristic parameters of a Cantor-like multiplayer are the generator $G = 3,5,7,...$ and the generation number $N = 1,2,3,....$ A Cantor-like multilayer can be generated by the method depicted as follows [18]: Set a metal layer (labeled B) as a seed. Replace certain parts of the seed (determined by the value of G) with a dielectric layer (labeled A). Then repeat the same procedure over all the remaining inclusions of the initial materials, as if they are seeds. After repeating these steps for N times, a Cantor multilayer characterized by a pair of numbers (G, N) is generated. The schematic structure of a Cantor multilayer with $G = 3$ and $N = 3$ is shown in Fig. 2.10a. Fibonacci multilayer, a self-similar fractal sequences, can be constructed recursively as $S_{j+1} = \{S_{j-1}, S_j\}$ for $j \geq 1$; with $S_0 = \{B\}$(the metal layer), $S_1 = \{A\}$ (dielectric layer) [17,19]. In this sequence one has $S_2 = \{BA\}$, $S_3 = \{ABA\}$, $S_4 = \{BAABA\}$, $S_5 = \{ABABAABA\}$, $S_6 = \{BAABAABABAABA\}$, ... The schematic structure of a Fibonacci multilayer S_6 is shown in Fig. 2.10b. In order to compare the transmittance properties, the total thickness of the metal layers of the Cantor-like multilayer, the Fibonacci multilayer and the periodic multilayer is the same value. A is MgF$_2$ and B is Ag. More than 50% transmittance can be achieved in the pass band of the (3,3) Cantor-like multiplayer and the S_6 Fibonacci multiplayer, and this pass band is much larger than that of the periodic multiplayer.

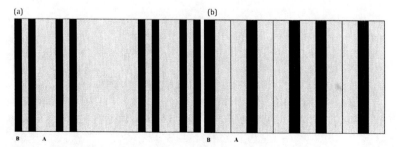

(a)   (b)

B   A   B   A

Figure 2.10   Schematic structure of a Cantor multilayer with $G = 3$ and $N = 3$ (a) and S_6 Fibonacci multilayer (b).

2.4.2.3 Applications of one-dimensional metallodielectric photonic crystals

2.4.2.3.1 *Modifying emission*

Metallodielectric gratings, a kind of one-dimensional metallo-dielectric photonic crystals, can be used to modify the emission of atoms, molecules, and ions. For example, Kalkman *et al.* found that the spectral shape and bandwidth of the emission of Er^{3+} ions in silica glass around 1.5 μm can be strongly modified by the presence of a metallodielectric grating [20]. The metallodielectric grating could be fabricated by use of a sequence process of etching a grating structure in the Er^{3+} doped silica glass, and subsequently depositing a silver film. The depth and the pitch of the metallodielectric grating were 230 nm and 1070 nm, respectively. The intensity of the photoluminescence peak could be enhanced by more than two orders. The remarkable PL enhancement originates from the coupling of the PL of Er^{3+} ions to surface plasmon polaritons (SPPs) of silver grating, which radiates at a resonance condition [20]

$$\vec{K}_0 \sin\theta_{int} = \pm\vec{K}_{SPP} \pm m\vec{G} \tag{2.36}$$

where $\vec{K}_0$ is the wave vector of the radiated light, θ_{int} is the angle between the radiated light and the normal direction of the grating plane, $\vec{K}_{SPP}$ is the wave vector of surface plasmon polaritons, $\vec{G}$ is the reciprocal lattice vector of the silver grating, and m is an integer. One or more reciprocal lattice vectors of the silver grating have to be added into (or subtracted from) the SPP wave vector in order to meet the resonance condition because the SPP wave vector is much larger than the photon wave vector. When the resonance condition is satisfied, the PL intensity of Er^{3+} ions will be enhanced greatly. Wu *et al.* [21] fabricated a one-dimensional metallodielectric heterostructural photonic crystal that had a configuration of $(ABCB)^2(CD)^2$, where A is alumina (Al_2O_3) layer, B and D is MgF_2 layer, C is aluminum (Al) layer. The thickness is 48 nm for A, 31 nm for B, 10 nm for C, and 47 nm for D, respectively. The metallodielectric heterostructural photonic crystal possesses large pass band in short wavelength, broad and deep photonic bandgap (also called the forbidden band) in long wavelength, and sharp cutoff edge between the pass and stop bands. In the emission spectrum of a pure BaF_2 crystal, the slow component drops in the stop band of the metallodielectric photonic crystal, while the fast component is in the pass band of the metallodielectric photonic crystal. They found that when the pure BaF_2 crystal is coated with the metallodielectric photonic crystal, the slow component of the scintillator light is suppressed greatly. The suppression ratio of the slow to the fast component reaches 28 dB [21].

2.4.2.3.2 *Negative refraction lens*

One-dimensional metallodielectric photonic crystal can also be adopted to reach a superlens. Shin and Fan [22] proposed a negative refraction lens by use of a one-dimensional metallodielectric photonic crystal, in which each unit cell consists of a metal and a dielectric layer. The thickness of both the metal and dielectric layers is $0.2\lambda_p$, where λ_p is related to the bulk plasma frequency ω_p of the constituent metal [22]

$$\lambda_p = \frac{2\pi c}{\omega_p} \tag{2.37}$$

where c is the velocity of light in vacuum. The negative refraction effect can be attracted from the constant frequency contour (CFC) based on the conservation of the parallel wave vector. The propagation direction of the electromagnetic energy flow is determined by the gradient $\nabla_k\omega$. They found that the refraction direction of an incident light shows negative refraction behavior. The radius of the air CFC is smaller than the maximum wave vector allowed in the photonic crystal. As a result, the negative refraction occurs for all angles of incidence. In the visible range where the plasmonic properties of metal become prominent, this metallodielectric photonic crystal could provide all-angle negative refraction when light is incident from a positive index medium. They constructed a one-dimensional photonic crystal composed of five layers of Ag and four layers of Si_3N_4. The thickness of both Ag and Si_3N_4 layers is 40 nm. When a point source is placed at 48 nm away from the slab edge of the left-hand side, the image of the point source is formed at the symmetric point on the right-hand side of the metallodielectric photonic crystal. The full width at half maximum (FWHM) of the electric-field distribution curve of the image is 0.31λ. Subwavelength focusing can be achieved.

2.4.2.3.3   *Optical limiting*

The optical limiting effect can be reached by use of the strong optical nonlinearity of the metal or dielectric materials of a metallo-dielectric photonic crystal. Larciprete *et al.* [23] demonstrated the optical limiting properties of a one-dimensional metallodielectric photonic crystal, fabricated by alternatively depositing 109 nm-thick zinc oxide (ZnO) layer and 17 nm-thick silver layer. The metallodielectric photonic crystal only has four periodic units. The transmittance resonance of the low frequency band edge is located at 532 nm, which corresponds to the frequency doubled Nd:yttrium-aluminum-garnet (Nd:YAG) laser beam. The high-refractive-index material of the photonic crystal, ZnO, possesses low linear absorption and strong two-photon absorption (TPA) at a wavelength of 532 nm. The TPA coefficient of ZnO is 2.5 cm/GW [23]. Bendickson *et al.* and Centini *et al.* have pointed out that near the photonic bandedge the field intensity of an incident light may be enhanced by several orders of magnitude due to strong field localization effect [24,25]. Silver has a relatively low refractive index value in the visible range. The electromagnetic field at a frequency below the metal plasma frequency will be strongly confined in the dielectric layers in a metallodielectric photonic crystal due to the high reflective properties of the metal layers [11]. Therefore, the electric-field distribution of a high power laser at a wavelength of 532 nm will be strongly confined in the nonlinear medium region, which leads to a very large change of the complex refractive index of ZnO. As a result, a self-limiting effect could be achieved very easily. The second harmonic of a Q-switched Nd:YAG laser (with a pulse width of 5 ns and a repetition rate of 14 Hz) is used as the excitation laser. The transmittance is about 57% at a low input intensity. When the intensity of the incident light increases to 2 GW/cm^2, a transmittance decrease of 50% is achieved due to the strong two-photon absorption in ZnO layers. Scalora *et al.* also [26] reported an ultra-wide bandwidth optical limiting effect in a one-dimensional metallodielectric photonic crystal made of Cu and ZnO. High third-order nonlinear optical susceptibility of Cu is adopted to achieve optical limiting effect across the entire visible range. Chirping ZnO layer thickness dramatically improves the linear transmittance through the photonic crystal and achieves large field distribution inside Cu to obtain large nonlinearity. They found that at high intensity, the transmittance can be reduced by nearly two orders of magnitude compared with the weak intensity in a single pass through the metallodielectric photonic crystal [26].

2.4.2.3.4 *Optical absorber* [27]

One-dimensional metallo-dielectric photonic crystals can also find great applications in the fields of photothermal technology, thermophotovoltaics, and blackbody emission [27]. Yu *et al.* found that the absorption of a bulk metal in the visible and infrared range could be enhanced greatly by periodically inserting dielectric layers in the bulk metal to form a one-dimensional metallo-dielectric photonic crystal, which originates from the peculiar photonic band structures of metallodielectric photonic crystals [27]. No propagation electromagnetic modes could exist in the photonic bandgap of the metallodielectric photonic crystal. For electromagnetic modes at frequencies within the pass band, an absorption enhancement still occurs due to strong reflection and multiple Bragg scattering in the interface of the dielectric and metal layers. Yu *et al.* calculated the absorption spectra of a one-dimensional metallodielectric photonic crystal made of 12 nm-thick Ag lays and 120 nm-thick MgF_2 layers. The visible range (from 400 to 700 nm) drops in the pass band of the metallodielectric photonic crystal. They found that this metallodielectric photonic crystal could render a large absorption enhancement in the visible range. The absorption of an 80 nm-thick Ag film is about 4% in the visible range. While for the metallodielectric photonic crystal with only eight repetition periods the average absorption increases to 45%, in the visible range, which is more than one order of magnitude larger than that of the bulk Ag film.

2.5　Two-Dimensional Photonic Crystals

Two-dimensional photonic crystal can realize totally two-dimensional photon confinement in a plane where the dielectric function is distributed periodically in space. The basic structure configuration of the two-dimensional photonic crystal consists of periodic dielectric rods embedded in air, also called the dielectric rod type, and periodic air holes embedded in high-refractive-index material, also called the air hole type. According to the practical requirement, the lattice structure of a two-dimensional photonic crystal can be adopted as a periodic lattice structure, such as the square lattice, the triangular lattice, and so on, or a quasiperiodic lattice structure, such as the octagonal quasicrystal, dodecagonal quasicrystal, and so on. For a practical two-dimensional photonic crystal with a finite thickness, various optical devices can be realized in a photonic crystal slab. This makes that two-dimensional photonic crystals are of great importance for the realization of integrated photonic circuits and integrated photonic chip. Many photonic devices have been studied experimentally and theoretically based on two-dimensional photonic crystals. The photonic bandgap structure of two-dimensional photonic crystals can be calculated by the multiple-scattering method, plane-wave expansion method (PWE), Green function method (GF), and so on.

2.5.1 Theoretical Calculation Models

2.5.1.1 The multiple-scattering method

The multiple-scattering method treats two-dimensional photonic crystals with a finite size perpendicular to the propagation direction of an incident light as scattering objects in an open geometry, where the radiation boundary condition is naturally imposed [28,29]. The total light field seen by a specific scattering object includes the field coming from the external excitation source and the scattered fields by all the other scattering objects. The generalized transmittance coefficient, the dispersion relations, and the decay length insider the photonic bandgap can be retrieved from the far-field total scattering amplitude. Consider a two-dimensional photonic crystal composed of N identical cylinders with a radius R and a dielectric constant ε under an external light source $u_{\text{inc}}(\vec{\rho})$ at a frequency ω. $\vec{\rho} = (\rho, \theta)$ indicates the position in the two-dimensional plane. u represents the electric field for the TM modes and the magnetic field for the TE mode. The total field seen by the cylinder j, located at $\vec{\rho} = (\rho_j, \theta_j)$, can be written as the sum of the incident and the scattered fields [28]:

$$u(\vec{\rho}) = u_{\text{inc}}(\vec{\rho}) + u_{\text{scatt}}(\vec{\rho}) = \sum_{m=-\infty}^{\infty} \left[a_m(j) J_m(k_0 \rho_{\rho j}) + \tilde{B}_m(j) H_m(k_0 \rho_{\rho j}) \right] e^{im\theta_{\rho j}}$$

$$(2.38)$$

where $k_0 = \omega/c$ is the wave vector, $\vec{\rho}_{pj} = \vec{\rho} - \vec{\rho}_j$, J_m, and H_m are the first kind Bessel function and Hankel function, respectively. The near-field radiation pattern can be obtained from the Poynting vector [28]

$$\vec{S}(\vec{\rho}) = \frac{-c}{8\pi k_0}\text{Im}[u(\vec{\rho})\nabla u^*(\vec{\rho})] \tag{2.39}$$

In the far field, the total scattering field amplitude is [28]

$$f_s(\theta) = \sqrt{\frac{2}{\pi k_0}}e^{-i\frac{\pi}{4}}\sum_{i=1}^{N}\sum_{m=-\infty}^{\infty}e^{-ik_0\rho_i\cos(\theta_i-\theta)}\times(-i)^m e^{im\theta}\tilde{B}_m(i) \tag{2.40}$$

The far-field energy flux can be written as [28]

$$\vec{S}(\rho,\theta) = \frac{c[|a_0(\theta)+f_s(\theta)|^2]}{8\pi\rho}\hat{\rho} \tag{2.41}$$

The generalized transmittance coefficient T can be calculated by dividing the far-field energy by that of the incident light at $\theta = 0$ [28]

$$T = \left|1 + \frac{f_s(0)}{a_0(0)}\right|^2 \tag{2.42}$$

The transmittance spectrum of a two-dimensional photonic crystal with a finite size can be obtained very conveniently, which is of great importance for comparison with the experimental measurement results. The transmittance properties of the two-dimensional photonic crystal with a finite size are mainly focused upon in the experimental study.

2.5.1.2   The plane-wave expansion method

In the plane-wave expansion method, the fields are expanded in a set of harmonic modes [30]. Since the permittivity $\varepsilon(\vec{r})$ and the permeability $\mu(\vec{r})$ are spatially periodical modulation functions in a two-dimensional photonic crystal, the permittivity $\varepsilon(\vec{r})$, the permeability $\mu(\vec{r})$, and the magnetic field $\vec{H}_\omega(\vec{r})$ can be expanded in terms of Fourier series [31]:

$$\varepsilon(\vec{r}) = \sum_{\vec{G}}E^{i\vec{G}\cdot\vec{r}}\varepsilon_{\vec{G}} \tag{2.43}$$

$$\mu(\vec{r}) = \sum_{\vec{G}}E^{i\vec{G}\cdot\vec{r}}\mu_{\vec{G}} \tag{2.44}$$

$$\vec{H}_\omega(\vec{r}) = \sum_{\vec{G}} E^{i(\vec{k}+\vec{G})\vec{r}} \vec{H}_{\vec{k},\vec{G}} \tag{2.45}$$

where $\vec{G}$ is the reciprocal lattice vector. Then the master equation (2.16) can be written as [31]:

$$-\sum_{\vec{G}'} (\vec{k}+\vec{G}) \times \varepsilon^{-1}_{\vec{G}-\vec{G}'}(\vec{k}+\vec{G}') \times \vec{H}_{\vec{k},\vec{G}'} = \omega^2 \sum_{\vec{G}'} \mu_{\vec{G}-\vec{G}'} \vec{H}_{\vec{k},\vec{G}'} \tag{2.46}$$

For TM-polarized electromagnetic modes, equation (2.45) can be changed into the form of [31]

$$\sum_{\vec{G}'} \varepsilon^{-1}_{\vec{G}-\vec{G}'} (\vec{k}+\vec{G}) \cdot (\vec{k}+\vec{G}') H_{\vec{k},\vec{G}'} = \omega^2 \sum_{\vec{G}'} \mu_{\vec{G}-\vec{G}'} H_{\vec{k},\vec{G}'} \tag{2.47}$$

Equation (2.46) is an eigenvalue equation. The photonic band structure (or dispersion relations) of a two-dimensional photo-nic crystal can be obtained by solving equation (2.45) very conveniently [32–34].

2.5.1.3   The Green function method

The Green function method is based on Green's dyadic technique. The introduction of a dyadic Dyson's equation enables the straight-forward construction of Green's dyadics associated with arbitrarily complex geometries. The total dielectric tensor of a scattering system embedded in an infinite homogeneous reference medium can be described by [35]

$$\varepsilon(\vec{r}, \omega) = \varepsilon_r(\omega) + \varepsilon_s(\vec{r}, \omega) \tag{2.48}$$

where $\varepsilon_r(\omega)$ is the dielectric tensor of the reference medium, $\varepsilon_s(\vec{r}, \omega)$ is the dielectric tensor of the scattering medium. When an incident monochromatic field $\vec{E}^0(\vec{r})e^{-i\omega t}$ propagates in the scattering system, the scattered field $\vec{E}(\vec{r})$ can be obtained by solving the vectorial wave equation [35]

$$-\nabla \times \nabla \times \vec{E}(\vec{r}) + k^2 \varepsilon_r(\omega)\vec{E}(\vec{r}) + k^2 \varepsilon_s(\vec{r},\omega)\vec{E}(\vec{r}) = 0 \tag{2.49}$$

where k is the wave number in vacuum. Introducing the operators $\bar{L}$, $\bar{e}_r$, and $\bar{e}_s$ for $-\nabla \times \nabla \times$, $k^2\varepsilon_r(\omega)$, $k^2\varepsilon_s(\vec{r},\omega)$, respectively. The

Green's operator, also called the field propagator, associated with the total system can be written as [37]

$$(\vec{L} + \vec{e}_r + \vec{e}_s)\vec{G} = 1 \qquad (2.50)$$

where 1 is the unit operator. To calculate Green's tensor $\vec{G}$, Dyson's equation is adopted [36–39]:

$$\vec{G} = \vec{G}^0 - \vec{G}^0 \vec{e}_s \vec{G} \qquad (2.51)$$

where $\vec{G}^0$ is the Green operator associated with the reference system. The photonic band structure of a two-dimensional photonic crystal can be obtained by solving equation (2.51).

2.5.2 Negative Refraction of All-Dielectric Photonic Crystal

Recently, the negative refraction effect has attracted great attention due to its potential applications in the fields of integrated photonic devices and nanophotonics [40]. Materials with negative index can be hardly found in nature. For an all-dielectric two-dimensional photonic crystal, the dielectric constants of component materials are positive. Negative refraction can still be achieved in an all-dielectric two-dimensional photonic crystal due to the strong dispersion properties of the photonic crystal structure.

2.5.2.1 Mechanism of negative refraction

In 1998, Kosaka and Notomi found that the negative refraction phenomenon can exist in photonic crystals in regimes of negative group velocity and negative effective index above the first band near the Brillouin zone center [41,42]. Subsequently, Luo *et al.* found that the negative refraction effect can be achieved in a frequency range for any incident angle, also called all-angle negative refraction, in the lowest photonic band near a corner of the Brillouin zone in a two-dimensional photonic crystal [43]. The outstanding advantages of the negative refraction in the lowest photonic band are single mode and high transmittance, which are of great importance for the realization of subwavelength focusing and imaging. Constant-frequency contours of a two-dimensional photonic crystal structure are often used to analyze refraction phenomena. The gradient vectors of constant-frequency contours indicate the group velocities of the photonic modes. The propagation direction of the refracted electromagnetic modes is determined by the conservation of the frequency and the wave-vector component parallel to the interface of air and photonic crystal. Therefore, if a constant-frequency contour has a convex configuration due to the negative-definite photonic effective mass, defined as $(\partial^2\omega)/(\partial k_i \partial k_j)$, negative refraction will occur for an incident electromagnetic mode with a frequency corresponding to the constant-frequency contour [43]. Luo *et al.* also pointed out that two key criteria should be satisfied in order to obtain single-beam all-angle negative refraction [43]:

(1) The constant-frequency contour of the photonic crystal is all convex with a negative photonic effective mass $(\partial^2\omega)/(\partial k_i \partial k_j)$.

(2) All incoming wave vectors at such a frequency are included within the constant-frequency contour of the photonic crystal.

Gupta and Luo also proposed the concept of photonic crystal slab superlens based on negative refraction effect [43,44]. The photonic crystal slab superlens can focus a point source on one side of the photonic crystal slab into a real image on the other side of the photonic crystal slab due to the strong dispersion effect of the photonic crystal structure. Subsequently, Cubukcu *et al.* experimentally demonstrated photonic crystal slab superlens in the microwave regime [45]. Li and Lin found that [46] only a point source placed in the vicinity of the photonic crystal slab could form a good quality image on the opposite side of the photonic crystal slab. The image was strongly confined in the near-field region of the photonic crystal and gradually degraded and disappeared beyond the near-field domain. Moreover, the image–slab distance has little dependence on the source-slab distance and the slab thickness [46]. Luo *et al.* found that the amplification of near-field waves originates from the coupling between the incident evanescent field and surface photon bound states of the photonic crystal slab [47]. Belov *et al.* gave a general description of the physical mechanism of photonic crystal slab superlens: the spatial harmonics of propagating and evanescent modes produced by a source refract into the eigenmodes of the photonic crystal slab at the front interface. These eigenmodes propagate normally to the interface and deliver the distribution of near-field electric field from the front interface to the back interface without disturbances. The incoming waves refract at the back interface and form an image. In this way the incident field with subwavelength details is transported from one interface to the other one [48]. This is also called the canalization mechanism because the subwavelength spatial spectrum of a source is canalized by the eigenmodes of the photonic crystal slab having the same longitudinal components of wave vector and group velocities directed across the photonic crystal slab [48]. Following these works, many researchers have dedicated themselves to the study of photonic crystal slab superlens whose effective refractive index is equal to the ideal value of −1 for all incident angles [49–51].

2.5.2.2 Tunable negative refraction

Tunable negative refraction is of great importance for the practical applications of all-dielectric photonic crystal in the fields of near-field optics and integrated photonic devices. The all-angle negative refraction effect does not appear in all two-dimensional photonic crystals due to its relatively harsh requirement for the photonic crystal. Zhang *et al.* presented an approach that can not only create all-angle negative refraction but also conveniently adjust the frequency region of the all-angle negative refraction in a two-dimensional all-dielectric photonic crystal [52]. For two-dimensional all-dielectric photonic crystals without all-angle negative refraction properties formerly, the frequency region of the all-angle negative refraction can be created by adding a fraction of a metal component to the center of each dielectric cylinder. The insertion of a metallic cylinder in the center of each dielectric cylinder leads to the change of the field distribution in the unit cell, which then leads to a change in the photonic band structure and the shape of constant-frequency contours. Therefore, when taking metal cylinders with a certain size to insert in the center of every dielectric cylinder, the convex shape of constant-frequency contours could appear and the all-angle refraction region could be created [53]. Accordingly, the frequency range of the all-angle negative refraction can be enlarged or decreased by adjusting the size of the center metal cylinders. This means that people can tune the all-angle negative refraction at will by adjusting the size of the metal core. More often than not, the negative refraction and the superlens effect exist in a two-dimensional all-dielectric photonic crystal only for a certain polarized wave, S wave (TE-polarized wave) or P wave (TM-polarized wave). Zhang also proposed a simple and efficient method to engineer absolute negative refraction for both polarizations of electromagnetic wave in the above metal cores doped dielectric cylinder photonic crystals by adding an additional dielectric component to the existing photonic crystals [54]. The dielectric properties of the additional dielectric component and the insertion position in a unit cell are chosen according to the field distribution of Bloch states at photonic band edges. For a triangular photonic crystal at 2Γ point, the additional dielectric material can be inserted at the six symmetric points of Wigner unit cells, where the field distribution is strong for both TE- and TM-polarized waves. This leads to the changes of the field distribution in the unit cell and the shift of the photonic band edge and the shape of constant-frequency contours [54]. Therefore, it is possible to achieve the all-angle negative refraction for both polarized waves through inserting suitable dielectric component and proper position.

Huang *et al.* proposed an approach to create negative refraction and all-angle negative refraction in frequency windows by use of surface modification in two-dimensional all-dielectric photonic crystals [55]. When introducing proper surface grating, negative refraction and all-angle negative refraction can be achieved along the Γ–X direction in the first band of a square lattice all-dielectric photonic crystal. Huang *et al.* found that the surface periodicity of two-dimensional photonic crystal influences the constant-frequency contours and the folding of the photonic band structure [56]. Consider a photonic crystal slab with surface normal along the Γ–X direction, the incident electromagnetic wave with a frequency larger than ω_x will be reflected by the photonic crystal because the frequency of the incident light will drop in the photonic bandgap. ω_x is the frequency at X point. The surface grating will provide a momentum boost along the surface to the incident plane wave and that it will be coupled to the Bloch waves inside the photonic crystal. The refracted light will propagate on the opposite side of the surface normal with respect to the incident light. So, negative refraction is achieved. The effect of the surface grating is equivalent to bringing down constant-frequency contours around the M point to the X point in a certain range and making the convex shape of constant-frequency contours. The all-angle negative refraction can be generated when the wave vector k_a of constant-frequency contours meets $(\pi/a) - k_a \geq (\omega/c)$, where $2a$ is the grating period [55]. By adjusting the structure parameters of the surface grating, the frequency range of the all-angle negative refraction can be tuned.

Feng *et al.* proposed a method to achieve tunable negative refraction in a two-dimensional active magneto-optical photonic crystal [57]. The Voigt effect is a well-known magneto-optical effect where the permittivity is changed by an applied magnetic field, whose direction is perpendicular to the propagating direction of the electromagnetic wave [58]. In the Voigt effect, the permittivity for the electromagnetic wave with its electric field perpendicular to the external magnetic field is different from that parallel to the external field. When the electric field of the electromagnetic wave is parallel to the external magnetic field, the permittivity keeps unchanged. When the electric field of the electromagnetic wave is perpendicular to the external magnetic field, the permittivity depends on the magnetic field [59]. The anisotropy of constant-frequency contours determines the propagation and refraction of electromagnetic waves in two-dimensional photonic crystals. The refraction of incident light varies with the magnitude of external magnetic field. It was found that the refraction of the incident light in the magneto-optical photonic crystal varies from positive refraction, to totally internal reflection in bandgap, to negative refraction, and to positive refraction in the lowest photonic band with the increment of the magnitude of the external magnetic field. The refraction angle of an incident light could be changed from 90° to −90° [57]. Accordingly, a frequency range of all-angle negative refraction could be created and tuned by changing the magnitude of the external magnetic field.

2.5.2.3 Disorder influences on subwavelength focusing properties

The structure disorders introduced into the fabrication process will affect the image quality and may restrict the practical applications of all-dielectric photonic crystal slab superlens [44,51]. Feng *et al.* found that a rotational randomness of the unit cells beyond 30° could destroy the focusing effect of a two-dimensional dielectric-rod photonic crystal slab [60,61]. To systematically study the influences of structure disorders on the subwavelength focusing properties of all-dielectric photonic crystal slab superlens, let us consider a two-dimensional all-dielectric photonic crystal superlens previously used by Luo *et al.* The photonic crystal slab superlens is constructed by arrays of square lattice of cylindrical air holes with infinitely length embedded in a dielectric matrix. The dielectric constant of the dielectric material and air is set at 12 and 1, respectively. The refractive index of the dielectric material corresponds to that of silicon in 1550 nm. The lattice constant is a. The diameter of air holes is $0.7a$. The width of the photonic crystal slab is about $11.6637a$. The air holes at the edge of the left- and right-hand side of the photonic crystal slab are cut off by $0.175a$, which provides an intermediate medium between the interface of the photonic crystal slab and the surrounding air to decrease the reflection [62]. The incident light is a transverse electric polarized wave with the electric field vector perpendicular to the cylindrical air holes. The frequency of the incident light is $0.192 \times (2\pi c/a)$, which is within the frequency range of all-angle negative refraction of the photonic crystal slab. The light is incident in the Γ–M direction. The structure disorders are formed by randomly perturbing the position or diameter of air holes of the photonic crystal slab. To introduce a positional disorder, the positions of the air holes are moved away from their lattice sites by a distance $\Delta x^i = \gamma_x^i nL$ for the i-th air hole in the x direction, and $\Delta y^i = \gamma_x^i nL$ for the i-th air hole in the y direction. Here $\gamma_{x,y}^i \in [-1,1]$ is a random number; n is the disorder degree. $L = 0.15a$. L is the maximum disorder amplitude, which will make two nearest-neighbor air holes overlapped with each other. The magnetic-field H_z distributions of the photonic crystal slab superlens with different degrees of positional disorder for two continuous-wave point sources are plotted in Fig. 2.11. The distance between two point sources is the value of one wavelength λ of the incident light. The distance between the left-side edge of the photonic crystal slab and two point sources is $0.5a$. Two real images are formed in the right-hand side of the photonic crystal slab at a distance of $0.3189a$ away from the slab edge. With the increase of the disorder degree, the perfect quality of the image is gradually damaged. But the photonic crystal slab superlens can tolerate up to a 10% degree of positional disorder. By introducing a diameter disorder, the diameter of the air holes is changed to $D^i = D_0 + \gamma^i nL$ for the

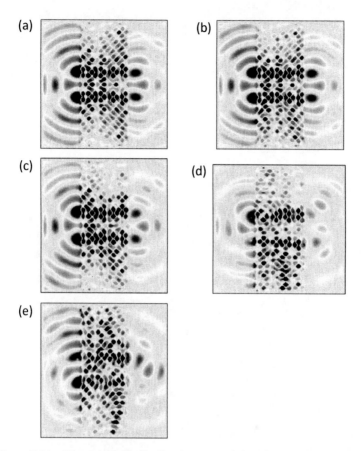

Figure 2.11   Magnetic-field H_z distributions of the photonic crystal slab superlens with different degrees of positional disorder for two point sources. (a) For zero disorder degree. (b) For the disorder degree of 5%. (c) For the disorder degree of 10%. (d) For the disorder degree of 15%. (e) For the disorder degree of 20%. Reprinted from *Phys. Lett. A*, **373**(17), X. Hu, *et al.*, Disorders influences on the focusing effect of all-dielectric photonic crystal slab superlens, 1588–1594, Copyright (2009), with permission from Elsevier.

i-th air hole. Here D_0 is the normal diameter, $D_0 = 0.7a$. $\gamma^i \in [-1,1]$. is a random number. n is the disorder degree. $L = 0.5a$. The magnetic-field H_z distributions of the photonic crystal slab superlens with different degrees of diameter disorder are shown in Fig. 2.12. When the disorder degree is less than 15%, the photonic crystal slab

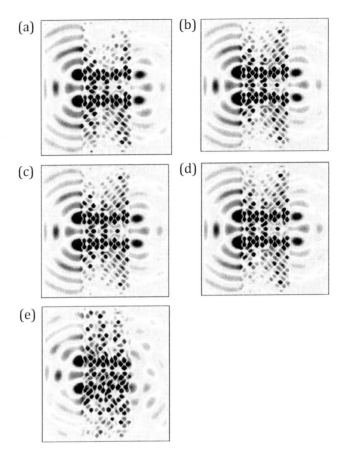

Figure 2.12 Magnetic-field H_z distributions of the photonic crystal slab superlens with different degrees of diameter disorder for two point sources. (a) For zero disorder degree. (b) For the disorder degree of 5%. (c) For the disorder degree of 10%. (d) For the disorder degree of 15%. (e) For the disorder degree of 20%. Reprinted from *Phys. Lett. A*, **373**(17), X. Hu, *et al.*, Disorders influences on the focusing effect of all-dielectric photonic crystal slab superlens, 1588–1594, Copyright (2009), with permission from Elsevier.

superlens possesses excellent focusing function. Therefore, it can be confirmed that the all-dielectric photonic crystal slab superlens can tolerate less than 10% degree of structure disorder. This means that the photonic crystal slab superlens can perform very well even if the photonic crystal possesses a certain range of disorder in the position or diameter of the air holes. This is of great importance for the practical applications of photonic crystal slab superlens. The field intensity distributions across the image center along the direction of parallel to the edge of the photonic crystal slab are shown in Fig. 2.13. The spatial resolution R of the imaging, defined as the ratio of the full width at half maximum to the wavelength λ of the incident light, is $R = 0.32$. This indicates that the photonic crystal slab possesses the excellent subwavelength resolution. It is very clear that the photonic crystal slab superlens can maintain excellent focusing function when less than a 10% degree of positional disorder or a 15% degree of diameter disorder is introduced in the photonic crystal slab. Therefore, less than 10% of structure disorder is tolerable for the photonic crystal slab superlens. This also implies that the photonic crystal slab superlens is more sensitive to the positional disorder than to the diameter disorder. The reason may be that the positional disorder can destroy the spatially periodic structure of the photonic crystal slab, while the spatially periodic lattice of the photonic crystal slab is maintained for the diameter disorder. As a result, the photonic bandgap structure and the dispersion properties of the photonic crystal slab can be damaged more seriously for positional disorder than it is the case for diameter disorder [62].

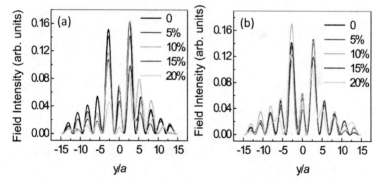

Figure 2.13 Field intensity distributions crossing the image center along the direction of parallel to the edge of the photonic crystal slab. (a) For positional disorder and two point sources. (b) For diameter disorder and two point sources. Reprinted from *Phys. Lett. A*, **373**(17), X. Hu, *et al.*, Disorders influences on the focusing effect of all-dielectric photonic crystal slab superlens, 1588–1594, Copyright (2009), with permission from Elsevier.

2.5.3 Two-Dimensional Metal Photonic Crystal

2.5.3.1 Basic properties

Two-dimensional metallic photonic crystals represent a specific class of artificial photonic materials where the photonic bandgap evolves from zero frequency [63]. This may imply potential applications in the far-infrared and microwave range. Two important characteristics of metallic photonic crystals are plasmonic response and plasmon damping. The plasmon damping determines the losses of the metallic photonic crystal. The basic difference between metallic and dielectric photonic crystals lies in that at small wave vector $\vec{k}$ the lowest mode has a form of $\omega = \tilde{c}k$ for dielectric photonic crystals, where $\tilde{c}$ is determined by the average dielectric constant, while in metallic photonic crystals the lowest mode has a finite frequency in the Γ point [64]. The physical reason of this difference is very simple: at a very small ω, for example, $\omega < 1/\tau$, where τ is the relaxation time of electrons in the metal, the average dielectric constant of the metallic photonic crystal is imaginary and the propagation of electromagnetic wave with small value of k is forbidden [64]. If the wavelength of an incident light is of the order of the lattice constant a, the incident light would be modulated in such a way that the electric field would be small in the metal cylinders. So the incident light could propagate in the photonic crystal but have a cutoff frequency $\omega_c = \chi c/a$, where χ is a small factor dependent of the filling fraction of metal f. At a high frequency, for example, $\omega > 1/\tau$, the average dielectric function $\bar{\varepsilon}(\omega)$ can be written as [64]

$$\bar{\varepsilon}(\omega) = \varepsilon_0 - f\frac{\varepsilon_0\omega_p^2}{\omega^2} \tag{2.52}$$

where ε_0 is the free space permittivity and ω_p is the plasma frequency of the metal. The incident light at a frequency larger than the cutoff frequency $\sqrt{f}\omega_p$ can propagate in the metallic photonic crystal. There exists a very large photonic bandgap, in the frequency range from zero to the cutoff frequency ω_p, which originates from the metallicity of the photonic crystal structure [65]. While the photonic bandgaps appearing in the frequency range of larger than the cutoff frequency originate from the periodicity of the photonic crystal structure [66]. Just like the all-dielectric photonic crystal microcavity, when a structure defect is introduced in the metallic photonic crystal, defect modes with high transmittance can be formed in the photonic bandgap [67,68].

2.5.3.2 Tunable metallic photonic crystals

The position and width of the photonic bandgap of a tunable metallic photonic crystal can be tuned by adjusting the external parameters. Third-order nonlinear optical effects can also be adopted to achieve tunable metallic photonic crystal. Kee *et al.* proposed an approach to achieve tunable photonic bandgaps of two-dimensional photonic crystal composed of arrays of metallic rods by infiltrating with liquid crystal [69]. Liquid crystal is a promising material for constructing tunable photonic crystals because the refractive index of liquid crystal varies with the external electric field or temperature [70]. Liquid crystal can be infiltrated in the void region of two-dimensional metallic photonic crystals. It was found that the infiltration of liquid crystal enlarges the width of the photonic bandgap between the first and second pass bands and creates another photonic bandgap between the third and fourth bands. The reason lies in that the infiltration of liquid crystal increases the refractive index contrast between the characteristic impedance of metallic material and the background medium [69]. As a result, the width of the first photonic bandgap is enlarged and the second photonic bandgap appears. Moreover, the center frequency of the first photonic bandgap decreases after the liquid crystal is infiltrated in the metallic photonic crystal due to the increase of the average refractive index of the photonic crystal structure. Through adjusting the external electric field or temperature, the refractive index of liquid crystal changes and leads to the variation of the refractive index contrast and the average refractive index of the metallic photonic crystal. Then the width and position of the photonic bandgap can be tuned. Contrary to metallic photonic crystals infiltrated with liquid crystal, the infiltration of liquid crystal into dielectric photonic crystal can decrease the width of the photonic bandgaps due to the reduction of the refractive index contrast.

Serebryannikov *et al.* proposed a method to control the locations of the cutoff frequency and higher opaque ranges in transmission spectrum of s-polarized waves in two-dimensional metallic photonic crystals by adjusting the incident direction [71]. The cutoff frequency ω_{cut} could be estimated by [71]

$$\omega_{cut}(\theta) \approx \omega_p / \cos\theta \tag{2.53}$$

where θ is the incident angle and ω_p is the plasma frequency. It was found that strong sensitivity to the variation of the incident angle could be realized by a proper choice of parameters of the rods and host medium. In particular, the difference in cutoff values can vary from a few percent in the case of a high ε host medium to several tens percent in the case of an air medium, when the incident angle varies from 0 to $\pi/3$.

2.5.3.3 Applications

2.5.3.3.1 *Enhancing emission efficiency*

Metallic photonic crystal can be used to control the emission of infrared light. Kim *et al.* proposed a two-dimensional metallic photonic crystal bandpass filter used in the photon recycling system to improve the emission efficiency of an incandescent light source [72]. The inefficiency of incandescent light sources lies in that a large amount of infrared light emission is wasted as heat rather than converted into useful visible light. If the wasted infrared light can be recycled to generate useful visible light, the efficiency of incandescent light sources can be improved greatly. A metallic photonic crystal bandpass filter enclosing the filament of the incandescent light source can be used to recycle the infrared light. The filter transmits the useful visible light and reflects the undesired infrared light back to the filament. The returned infrared light is reused to heat the filament and it consequently reduces the required input energy to maintain the temperature of the filament. A luminous efficiency of 125 lm/W can be obtained at a temperature of 2800 K, which is 8 times larger than that of a traditional blackbody filament and comparable to the current most efficient lighting devices [72].

2.5.3.3.2 *Thermal emitter applications*

Photonic crystals possess the unique properties to modify spontaneous emission. Similarly, photonic crystals could also be adopted to tailor thermal radiation. There are two research areas with regard to the tailoring of thermal radiation by use of photonic crystals. One area is to design highly selective narrow-band thermal emitters, exhibiting wavelength, directional and polarization selectivity [73]. The other area is to design wide-band thermal emitters, exhibiting near-blackbody thermal emission within a given wavelength range and largely attenuated emission outside the given frequency range [74].

Celanovic *et al.* reported a two-dimensional tungsten photonic crystal as a selective infrared thermal emitter [75]. The selective infrared thermal radiation source exhibited the emission properties close to the blackbody for frequency range near the photonic bandgap and relatively sharp cutoff for wavelength above the photonic bandgap. The tungsten photonic crystal is composed of square lattice of deep cylindrical holes embedded in the tungsten film. The mechanism of coupling resonant cavity electromagnetic modes has been adopted to enhance the thermal emission over a relatively broad wavelength range [75]. It was found that large emission enhancement for wavelengths shorter than 2 micron could be achieved when the lattice constant is 1 micron [75]. Not only high emission below the cutoff wavelength but also low emission at long wavelengths could be achieved. The cutoff frequency could be tailored by adjusting the geometry of the photonic crystal, which provides the freedom to design a selective thermal emitter with a precisely positioned spectral cutoff.

Pralle *et al.* reported that a metallic photonic crystal could be used to enhance the narrow-band infrared emission [76]. A 150 nm-thick Au film was deposited on the surface of silicon. Periodic arrays of square lattice air holes were patterned in the Au film. A narrow-band thermal emitter needs a material with a sharp absorption feature because the emission efficiency must equal the absorption efficiency according to Kirchhoff's radiation law [76]. This could be achieved by use of surface plasmon modes in patterned metal films. Since the wave vector of incident light is less than that of the surface plasmon, the incoming radiation cannot directly generate surface plasmon modes on a smooth metal surface. A periodically patterned metal surface provides additional reciprocal lattice vectors. A narrow-band incident radiation can be coupled to surface plasmon modes with the help of reciprocal lattice vectors. The frequency range of the incident radiation coupled to surface plasmon modes can be tuned by adjusting the lattice period of the photonic crystal. The infrared emission is restricted in the narrow-band where strong absorption appears. Moreover, the emission wavelengths are defined by the periodicity of the metal photonic crystal because the lattice constant scales linearly with the wavelength of surface plasmon modes [76]. Therefore, the infrared emission of the emitter can be tuned by adjusting the lattice periodicity.

Sai *et al.* proposed a high temperature resistive infrared emitter by use of two-dimensional tungsten photonic crystal [77]. The two-dimensional tungsten photonic crystal was composed of rectangular microcavities with the period of 1 μm fabricated on single crystalline and polycrystalline tungsten substrates. The aperture size was a and the thickness of the photonic crystal was d. When the aspect ratio a/d was set at 1, the light emission of the photonic crystal increased significantly with increasing a in the wavelength range from 300 nm to 2 μm [77]. Moreover, the high emission region broadened simultaneously. This behavior corresponds to the lengthening of the cutoff wavelength of the microcavities that are basically represented by $2a$ [77]. This indicates that the emission enhancement is related to the resonance between electro-magnetic waves and standing wave modes inside the microcavities. The metal photonic crystal with 200 nm wall thickness made from a single crystalline tungsten shows very high thermal stability over 1400 K, while the polycrystalline photonic crystal is deformed at a high temperature because of the grain growth.

2.6 Three-Dimensional Photonic Crystal

Three-dimensional photonic crystals have the periodic distribution of the dielectric function in all the three dimensions. This makes it possible to achieve complete photonic bandgap and three-dimensional strong photon confinement in three-dimensional photonic crystals. Unfortunately, it is still a great challenge to achieve the complete photonic bandgap in the visible and infrared range in three-dimensional photonic crystals due to the relatively low dielectric constant of conventional materials. Exploring the approaches to realize large and even complete photonic bandgap and creating the efficient and rapid fabrication methods of single-crystal photonic crystal samples are two important research orientations in the field of three-dimensional photonic crystals.

2.6.1 Theoretical Calculation Models

Owing to the complicated spatial structures of three-dimensional photonic crystals, the theoretical calculation of the photonic bandgap is very complex. The traditional plane-wave expansion method, solving the Maxwell equations in the frequency domain, and the three-dimensional finite-difference time-domain method, solving the Maxwell equations in the frequency domain, are inefficient and time consuming when calculating the photonic band structures of three-dimensional photonic crystals [78,79]. People have thus made great efforts to develop many efficient simulation methods to calculate the photonic band structure of three-dimensional photonic crystals.

2.6.1.1 The plane-wave-based transfer-matrix method

The plane-wave-based transfer-matrix method was firstly proposed by Li *et al.* in 2003 [80,81]. The transfer-matrix method based on the plane-wave expansion of electromagnetic fields was combined with the Bloch theory to calculate the photonic band structure of the three-dimensional layer-by-layer photonic crystal. The column vector of the electric fields in the left- and right-hand of the unit cell of the photonic crystal can be written as (Ω_0^+, Ω_0^-) and (Ω_1^+, Ω_1^-), and the propagation of the electromagnetic waves can be described by the transfer matrix [81]

$$\begin{pmatrix} \Omega_1^+ \\ \Omega_1^- \end{pmatrix} = T \begin{pmatrix} \Omega_0^+ \\ \Omega_0^- \end{pmatrix} \tag{2.54}$$

According to the Bloch theory $u(\vec{r} + \vec{R}) = e^{i\vec{k}\cdot\vec{R}} u(\vec{r})$, where $\vec{R}$ is the lattice vector, $\vec{k}$ is the Bloch wave vector, equation (2.53) can be written as [81]

$$T \begin{pmatrix} \Omega_0^+ \\ \Omega_0^- \end{pmatrix} = e^{i\vec{k}\cdot\vec{a}_3} \begin{pmatrix} \Omega_0^+ \\ \Omega_0^- \end{pmatrix} \tag{2.55}$$

where $\vec{a}_3$ is one of the primitive lattice vectors of the three-dimensional photonic crystal. Equation (2.55) can be converted into an S matrix algorithm [81]

$$\begin{pmatrix} S_{11} & 0 \\ S_{21} & -I \end{pmatrix} \begin{pmatrix} \Omega_0^+ \\ \Omega_0^- \end{pmatrix} = e^{i\vec{k}\cdot\vec{a}_3} \begin{pmatrix} I & -S_{12} \\ 0 & -S_{22} \end{pmatrix} \begin{pmatrix} \Omega_0^+ \\ \Omega_0^- \end{pmatrix} \tag{2.56}$$

Equation (2.56) is a standard eigenvalue function. Through calculating the eigenvalue, the photonic band structure of the three-dimensional layer-by-layer photonic crystal can be obtained [82].

2.6.1.2 The group-theory-based plane-wave expansion method

The group-theory-based plane-wave expansion method, proposed by Okano and Noda in 2004, is an efficient approach for studying the defect states of three-dimensional photonic crystals [83]. This method adopts the projection operator for the electromagnetic fields to classify the defect modes into a basis of the irreducible representation of the point group. According to the Maxwell equation, the wave equation of the electric flux density $\vec{D}(\vec{r},\omega,t)$ can be written as [83]

$$\Theta\vec{D}(\vec{r},\omega,t)=\nabla\times\nabla\times\frac{\vec{D}(\vec{r},\omega,t)}{\varepsilon(\vec{r})}=\frac{\omega^2}{c^2}\vec{D}(\vec{r},\omega,t) \qquad (2.57)$$

R is a symmetry operator of $\varepsilon(\vec{r})$, and equation (2.57) turns into the form [83]

$$\Theta R\vec{D}(\vec{r},\omega,t)=\nabla\times\nabla\times\frac{R\vec{D}(\vec{r},\omega,t)}{\varepsilon(\vec{r})}=\frac{\omega^2}{c^2}R\vec{D}(\vec{r},\omega,t)=R\Theta\vec{D}(\vec{r},\omega,t) \qquad (2.58)$$

The symmetry point group of $\varepsilon(\vec{r})$ corresponds to that of Θ. By introducing a projection operator P, the resonant modes of the defect structure can be classified into a basis of the irreducible representation of the point group of $\varepsilon(\vec{r})$. It is very convenient to calculate the resonant modes and electric field distribution of the defect states.

2.7 Reaching a Complete Photonic Bandgap

The complete photonic band can prohibit the propagation of light for any polarization and incident angle. High refractive index contrast and proper lattice structure are two essential factors to reach a complete photonic bandgap [84]. For example, high refractive index of more than 2.7 is needed to achieve a complete three-dimensional photonic bandgap for a three-dimensional photonic crystal with a close-packed or non-close-packed face-centered-cubic lattice structure [85,86]. For the conventional dielectric material, the dielectric constant is relatively small in the visible and infrared range [87]. This makes it quite difficult to reach a complete photonic bandgap in practice. There are two methods widely used to achieve a complete photonic bandgap. One method is to design special lattice structures to tailor the photonic band structure of photonic crystals [88–90]. The other one is to adopt novel materials, such as left-handed materials and semiconductor quantum dots, to adjust the dispersion relations of photonic crystals [91].

2.7.1 The Designing Lattice Structure Method

Constructing special lattice structure is an effective approach to realize the complete three-dimensional photonic bandgap. The diamond lattice structure is a promising candidate for the realization of complete photonic bandgap. When dielectric microspheres are arranged in the diamond lattice structure, a complete photonic bandgap can be obtained even when the refractive index contrast is as low as 2 [92]. Recently, Edagawa *et al.* pointed out that it is possible to reach a complete three-dimensional photonic bandgap in a photonic amorphous diamond structure based on the tight-binding theory [93,94]. Another three-dimensional photonic crystal structure that possesses a complete photonic bandgap is the square spiral photonic crystal, where the square spiral posts wind around the [001] axis of the diamond (or the face centered cubic lattice) [95]. Theoretical calculation implies that a complete photonic bandgap as large as 24% of the gap center frequency can be obtained in a three-dimensional silicon square spiral photonic crystal [96]. Moreover, a photonic crystal heterostructure composed of an inverse square spiral silicon three-dimensional photonic crystal and a two-dimensional silicon photonic crystal defect layer in the center can also provide a complete photonic bandgap [97]. The three-dimensional slanted pore photonic crystal is also very promising to realize the complete photonic bandgap. The slanted pore photonic crystal consists of a set of oriented cylindrical air pores emanating from a two-dimensional mask with a square lattice [98]. A large complete photonic bandgap of up to 25% of the gap center frequency can be expected for a three-dimensional silicon slanted pore photonic crystal [99]. It has been found that three-dimensional photonic crystals with a woodpile structure or a rhombohedral point symmetry structure are also promising candidates for the realization of a complete photonic bandgap [100–102]. Adeva *et al.* also found that a complete three-dimensional photonic bandgap can be obtained in a three-dimensional photonic crystal with a pyrochlore lattice structure, which has a lattice of tetrahedron with atoms situated in the corners [103].

Two-dimensional complete photonic bandgap, which can prohibit the propagation of both TE- and TM-polarized light for any angle of incidence in a plane, can also be reached by specially designing the lattice structure of two-dimensional photonic crystals. Anderson and Giapis found that it is an effective method to achieve a complete two-dimensional photonic bandgap by removing the mode degeneracy of photonic band diagram through breaking the structure symmetry of two-dimensional photonic crystals [104]. For example, two-dimensional circular-air-hole-type photonic crystal slab with a triangular lattice structure can provide a wide fundamental photonic bandgap for the TE-polarized light, while have no fundamental photonic bandgap for the TM-polarized light because of the mode degeneracy between the first and second band at highly symmetric J point [105]. Takayama *et al.* found that the mode degeneracy between the first and second band at highly symmetric J point can be removed by adopting a triangular configuration for air holes, and a complete two-dimensional photonic bandgap can be obtained in the two-dimensional triangular-air-hole photonic crystal slab with a triangular lattice structure [106]. Compared with periodic lattice two-dimensional photonic crystals, the photonic quasicrystals, having no long-range translational symmetry, are more favorable to reach complete two-dimensional photonic bandgap. Zoorob *et al.* and Zhang *et al.* found that a complete two-dimensional photonic bandgap can be achieved in 12-fold symmetric quasicrystals [107,108]. Moreover, two-dimensional photonic quasicrystal heterostructures can also be used to obtain a complete two-dimensional photonic bandgap [109]. Similar to Penrose tilings of quasicrystal structure, the Archimedean tiling structure are regular patterns of polygonal tessellation in plane by using regular polygons, which can fill the whole plane without gaps [110]. Ueda *et al.* found that a complete two-dimensional photonic bandgap can be obtained in an Archimedean tiling structure of $(3^2.4.3.4)$, where two equilateral triangles, a square, an equilateral triangle, and a square gather edge-to-edge around a vertex [110].

2.7.2 The Adopting Novel Material Method

Left-handed materials, metallic materials, semiconductor quantum dots, and dielectric materials with strong anisotropy in dielectricity have novel optical properties. When these novel photonic materials are adopted as component media of a photonic crystal, the dispersion relations of the photonic crystal can be adjusted greatly. This provides an effective approach to reach a complete photonic bandgap.

Metallic microstructures can provide strong plasmonic responses. Christ *et al.* found that a one-dimensional metallic photonic crystal structure, composed of an array of gold nanowires placed on a dielectric optical waveguide, could provide a complete one-dimensional photonic bandgap because of the strong coupling of plasmonic resonance modes and guided photonic modes in the photonic crystal structure [111]. Left-handed materials are artificial microstructure materials having negative dielectric permittivity and negative magnetic permeability simultaneously. Shadrivov *et al.* pointed out that a complete two-dimensional photonic bandgap could be reached in a one-dimensional photonic crystal constructed by alternatively stacking a conventional dielectric layer and a layer made of left-handed material [112]. It has been previously believed that a three-dimensional photonic crystal, which can control the propagation states of photons in all the three spatial dimensions, is needed to achieve a complete three-dimensional photonic bandgap. However, Shadrivov *et al.* found that it is possible to obtain a complete three-dimensional photonic bandgap and photon confinement in a specially designed one-dimensional photonic crystal structure made of a conventional dielectric material, and two different left-handed materials [112].

In nature, there exist a lot of dielectric materials having strong anisotropy in dielectricity, such as various uniaxial crystals, having two different principal refractive indices of the ordinary refractive index n_o and the extraordinary refractive index n_e. It is possible to select the electric-field vector of the TE-polarized modes parallel to the extraordinary axis of the uniaxial crystal material, while that of the TM-polarized modes parallel to the ordinary axis of the uniaxial crystal material [113]. Therefore, the photonic bandgaps for TE- and TM-polarized modes can be adjusted to overlap with each other. As a result, a complete photonic bandgap can be reached. Li *et al.* found that positive uniaxial crystal materials can be used to achieve a complete two-dimensional photonic bandgap in a dielectric-rod-type two-dimensional photonic crystal, while negative uniaxial crystal materials can be used to achieve a complete two-dimensional photonic bandgap in an air-hole-type two-dimensional photonic crystal [113].

Semiconductor quantum dots have strong quantum confinement effect because their size is close to the Bohr radius of excitons [114]. Excitons can be excited when an incident light propagates in semiconductor quantum dot material. The appearance of excitons will change the dielectric constant of semiconductor quantum dots. Zeng *et al.* found that a complete three-dimensional photonic bandgap can be obtained in a three-dimensional semiconductor quantum-dot photonic crystal with a diamond lattice structure [115]. Moreover, the width of the complete photonic bandgap can be modulated by adjusting the filling ratio of the semiconductor quantum dots owing to the coupling of photons and excitons [115].

References

1. J. D. Joannopoulos, R. D. Meade, and J. N. Winn, "*Photonic crystals: Molding the flow of light,*" Princeton University Press, 1995.

2. M. Maksimovic and Z. Jaksic, "Emittance and absorptance tailing by negative refractive index metamaterial-based cantor multilayers," *J. Opt. A: Pure Appl. Opt.* **8**, 355–362 (2006).

3. M. S. Vasconcelos, E. L. Albuquerque, and A. M. Mariz, "Optical localization in quasi-periodic multilayers," *J. Phys.: Condens. Matter.* **10**, 5839–5849 (1998).

4. F. F. de Medeiros, E. L. Albuquerque, and M. S. Vasconcelos, "Optical transmission spectra in quasiperiodic multilayered photonic structure," *J. Phys.: Condens. Matter.* **18**, 8737–8747 (2006).

5. A. Rostami and S. Matloub, "Exactly solvable inhomogeneous Fibonacci-class quasi-periodic structures (optical filtering)," *Opt. Commun.* **247**, 247–256 (2005).

6. J. Li, D. G. Zhao, and Z. Y. Liu, "Zero-n photonic band gap in a quasiperiodic stacking of positive and negative refractive index materials," *Phys. Lett. A* **332**, 461–468 (2004).

7. D. F. Sievenpiper, M. E. Sickmiller, and E. Yablonovitch, "3D wire mesh photonic crystals," *Phys. Rev. Lett.* **76**, 2480 (1996).

8. L. T. Zhang, W. F. Xie, J. Wang, H. Z. Zhang, and Y. S. Zhang, "Optical properties of a periodic one-dimensional metallo-organic photonic crystal," *J. Phys. D: Appl. Phys.* **39**, 2373–2376 (2006).

9. J. M. Pitarke, V. M. Silkin, E. V. Chukov, and P. M. Echenique, "Theory of surface plasmons and surface-plasmon polaritons," *Rep. Prog. Phys.* **70**, 1–87 (2007).

10. J. B. Marion, *Classical Electromagnetic Radiation* (Acadmic, New York, 1965).

11. M. Scalora, M. J. Bloemer, A. S. Pethel, J. P. Dowling, C. M. Bowden, and A. S. Manka, "Transparent metallo-dielectric one-dimensional photonic bandgap structures," *J. Appl. Phys.* **83**, 2377–2383 (1998).

12. A. J. Ward, J. B. Pendry, and W. J. Stewart, "Photonic dispersion surfaces," *J. Phys.: Condens. Matter.* **7**, 2217–2224 (1995).

13. X. C. Xu, Y. G. Xi, D. Z. Han, X. H. Liu, J. Zi, and Z. Q. Zhu, "Effective plasma frequency in one-dimensional metallic-dielectric photonic crystals," *Appl. Phys. Lett.* **86**, 091112 (2005).

14. M. J. Bloemer and M. Scalora, "Transmissive properties of Ag/MgF_2 photonic band gaps," *Appl. Phys. Lett.* **72**, 1676–1678 (1998).

15. Y. H. Ye, G. Bader, and V. V. Truong, "Low-loss one-dimensional metallodielectric photonic crystals fabricated by metallic insertions in a multiplayer dielectric structure," *Appl. Phys. Lett.* **77**, 235–237 (2000).

16. S. Feng, J. M. Elson, and P. L. Overfelt, "Transparent photonic band in metallodielectric nanostructures," *Phys. Rev. B* **72**, 085117 (2005).

17. C. Sibilia, M. Scalora, M. Centini, M. Bertolotti, M. J. Bloemer, and C. M. Bowden, "Electromagnetic properties of periodic and quasi-periodic one-dimensional, metallo-dielectric photonic band gap structures," *J. Opt. A: Pure Appl. Opt.* **1**, 490–494 (1999).

18. A. V. Lavrinenko, S. V. Zhukovsky, K. S. Sandomirski, and S. V. Gaponenko, "Propagation of classical wave in nonperiodic media: Scaling properties of an optical Cantor filter," *Phys. Rev. E* **65**, 036621 (2002).

19. W. Gellermann, M. Kohmoto, B. Sutherland, and P. C. Taylor, "Localization of light waves in Fibonacci dielectric multilayers," *Phys. Rev. Lett.* **72**, 633–636 (1994).

20. J. Kalkman, C. Strohhofer, B. Gralak, and A. Polman, "Surface plasmon polariton modified emission of erbium in a metallodielectric grating," *Appl. Phys. Lett.* **83**, 30–32 (2003).

21. Y. G. Wu, Z. S. Wang, M. Gu, L. Wang, X. Y. Lin, L. Y. Chen, and R. K. Xu, "One-dimensional heterostructural metallodielectric photonic band gap material for th emodification of emission spectrum of BaF$_2$ scintillator," *Appl. Phys. Lett.* **85**, 4337–4339 (2004).

22. H. Shin and S. H. Fan, "All-angle negative refraction and evanescent wave amplification using one-dimensional metallodielectric photonic crystals," *Appl. Phys. Lett.* **89**, 151102 (2006).

23. M. C. Larciprete, C. Sibilia, S. Paoloni, M. Bertolotti, F. Sarto, and M. Scalora, "Accessing the optical limiting properties of metallo-dielectric photonic band gap structures," *Appl. Phys. Lett.* **93**, 5013–5017 (2003).

24. J. M. Bendickson, J. P. Dowling, and M. Scalora, "Analytic expressions for the electromagnetic mode density in finite, one-dimensional, photonic band-gap structures," *Phys. Rev. E* **53**, 4107–4121 (1996).

25. M. Centini, C. Sibilia, M. Scalora, G. D. Aguanno, M. Bertolotti, M. J. Bloemer, C. M. Bowden, and I. Nefedov, "Dispersive properties of finite, one-dimensional photonic band gap structures: applications to nonlinear quadratic interactions," *Phys. Rev. E* **60**, 4891–4898 (1999).

26. M. Scalora, N. Mattiucci, G. D. Aguanno, M. C. Larciprete, and M. J. Bloemer, "Nonlinear pulse propagation in one-dimensional metal-dielectric multiplayer stacks: Ultrawide bandwidth optical limiting," *Phys. Rev. E* **73**, 016603 (2006).

27. J. F. Yu, Y. F. Shen, X. H. Liu, R. T. Fu, J. Zi, and Z. Q. Zhu, "Absorption in one-dimensional metallic-dielectric photonic crystals," *J. Phys.: Condens. Matter.* **16**, L51–L56 (2004).

28. L. M. Li and Z. Q. Zhang, "Multiple-scattering approach to finite-sized photonic band-gap materials," *Phys. Rev. B* **58**, 9587–9590 (1998).

29. G. Tayeb and D. Maystre, "Rigous theoretical study of finite-size two-dimensional photonic crystals doped by microcavity," *J. Opt. Soc. Am. A* **14**, 3323–3332 (1997).

30. R. D. Meade, A. M. Rappe, K. D. Brommer, J. D. Joannopoulos, and O. L. Alerhand, "Accurate theoretical analysis of photonic band-gap materials," *Phys. Rev. B* **48**, 8434–8437 (1993).

31. Y. C. Hsue and T. J. Yang, "Applying a modified plane-wave expansion method to the calculations of transmittivity and reflectivity of a semi-infinite photonic crystal," *Phys. Rev. E* **70**, 016706 (2004).

32. Z. Y. Li, B. Y. Gu, and G. Z. Yang, "Large absolute band gap in 2D anistropic photonic crystals," *Phys. Rev. Lett.* **81**, 2574–2577 (1998).

33. J. B. Pendry and A. Mackinnon, "Calculation of photon dispersion relations," *Phys. Rev. Lett.* **69**, 2772–2775 (1992).

34. M. Sigalas, C. M. Soukoulis, E. N. Economou, C. T. Chen, and K. M. Ho, "Photonic band gaps and defects in two dimensions: studies of the transmission coefficient," *Phys. Rev. B* **48**, 14121–14126 (1993).

35. O. J. F. Martin, C. Girard, and A. Dereux, "Generalized field propagator for electromagnetic scattering and light confinement," *Phys. Rev. Lett.* **74**, 526–529 (1995).

36. O. J. F. Martin and N. B. Piller, "Electromagnetic scattering in polarizable backgrounds," *Phys. Rev. E* **58**, 3909–3915 (1998).

37. S. F. Mingaleev and Y. S. Kivshar, "Self-trapping and stable localized modes in nonlinear photonic crystals," *Phys. Rev. Lett.* **86**, 5474–5477 (2001).

38. L. M. Zhao, X. H. Wang, B. Y. Gu, and G. Z. Yang, "Green's function for photonic crystal slabs," *Phys. Rev. E* **72**, 026614 (2005).

39. A. A. Asatryan, K. Busch, R. C. Mcphedran, L. C. Botten, C. M. D. Sterke, and N. A. Nicorovici, "Two-dimensional Green's function and local density of states in photonic crystals consisting of a finite number of cylinders of infinite length," *Phys. Rev. E* **63**, 046612 (2001).

40. N. Fabre, L. Lalouat, B. Cluzel, X. Melique, D. Lippens, F. D. Fornel, and O. Vanbesien, "Optical near-field microscopy of light focusing through a photonic crystal flat lens," *Phys. Rev. Lett.* **101**, 073901 (2008).

41. H. Kosaka, T. Kawashima, A. Tomita, M. Notomi, T. Tamamura, T. Sato, and S. Kawakami, "Superprism phenomena in photonic crystals," *Phys. Rev. B* **58**, 10096 (1998).

42. M. Notomi, "Theory of light propagation in strongly modulated photonic crystals: Refractionlike behavior in the vicinity of the photonic band gap," *Phys. Rev. B* **62**, 10696 (2000).

43. C. Y. Luo, S. G. Johnson, J. D. Joannopoulos, and J. B. Pendry, "All-angle negative refraction without negative effective index," *Phys. Rev. B* **65**, 201104(R) (2002).

44. B. C. Gupta and Z. Ye, "Disorder effects on the imaging of a negative refractive lens made by arrays of dielectric cylinders," *J. Appl. Phys.* **94**, 2173–2176 (2003).

45. E. Cubukcu, K. Aydin, E. Ozbay, S. Foteinopoulou, and C. M. Soukoulis, "Subwavelength resolution in a two-dimensional photonic-crystal-based superlens," *Phys. Rev. Lett.* **91**, 207401 (2003).

46. Z. Y. Li and L. L. Lin, "Evaluation of lensing in photonic crystal slabs exhibiting negative refraction," *Phys. Rev. B* **68**, 245110 (2003).

47. C. Y. Luo, S. G. Johnson, J. D. Joannopoulos, and J. B. Pendry, "Subwavelength imaging in photonic crystals," *Phys. Rev. B* **68**, 045115 (2003).

48. P. A. Belov, C. R. Simovski, and P. Ikonen, "Canalization of subwavelength images by electromagnetic crystals," *Phys. Rev. B* **71**, 193105 (2005).

49. R. Moussa, S. Foteinopoulou, L. Zhang, G. Tuttle, K. Guven, E. Ozbay, and C. M. Soukoulis, "Negative refraction and superlens behavior in a two-dimensional photonic crystal," *Phys. Rev. B* **71**, 085106 (2005).

50. X. D. Zhang, "Tunable non-near-field focus and imaging of an unpolarized electromagnetic wave," *Phys. Rev. B* **71**, 235103 (2005).

51. X. Wang and K. Kempa, "Effects of disorder on subwavelength lensing in two-dimensional photonic crystal slabs," *Phys. Rev. B* **71**, 085101 (2005).

52. X. D. Zhang and L. M. Li, "Create all-angle negative refraction by using insertion," *J. Appl. Phys.* **86**, 121103 (2005).

53. X. D. Zhang, "Image resolution depending on slab thickness and object distance in a two-dimensional photonic-crystal-based superlens," *Phys. Rev. B* **70**, 195110 (2005).

54. X. D. Zhang, "Tunable non-near-field focus and imaging of an unpolarized electromagnetic wave," *Phys. Rev. B* **71**, 235103 (2005).

55. Y. J. Huang, W. T. Lu, and S. Sridhar, "Alternative approach to all-angle negative refraction in two-dimensional photonic crystals," *Phys. Rev. A* **76**, 013824 (2007).

56. W. T. Lu, Y. J. Huang, P. Vodo, R. K. Banyal, C. H. Perry, and S. Sridhar, "A new mechanism for negative refraction and focusing using selective diffraction from surface corrugation," *Opt. Express* **15**, 9166–9175 (2007).

57. L. Feng, X. P. Liu, Y. F. Tang, Y. F. Chen, J. Zi, S. N. Zhu, and Y. Y. Zhu, "Tunable negative refraction in a two-dimensional active magneto-optical photonic crystal," *Phys. Rev. B* **71**, 195106 (2007).

58. C. R. Pidgeon, *Handbook on Semiconductors*, edited by M. Balkanski, North-Holland, Amsterdam, 1980.

59. C. Kittel, *Introduction to Solid State Physics*, Wiley, New York, 1976.

60. Z. F. Feng, X. G. Wang, Z. Y. Li, and D. Z. Zhang, "Influence of unit cell rotated on the focusing in a two-dimensional photonic-crystal-based flat lens," *J. Appl. Phys.* **101**, 123112 (2007).

61. Z. F. Feng, X. D. Zhang, S. Feng, K. Ren, Z. Y. Li, B. Y. Cheng, and D. Z. Zhang, "Effect of rotational randomness on focusing in a two-dimensional photonic-crystal flat lens," *J. Opt. A: Pure Appl. Opt.* **9**, 101–107 (2007).

62. X. Y. Hu, C. Xin, and Q. H. Gong, "Disorders influences on the focusing effect of all-dielectric photonic crystal slab superlens," *Phys. Lett. A* **373**, 1588–1594 (2009).

63. A. Pimenov and A. Loidl, "Conductivity and permittivity of two-dimensional metallic photonic crystals," *Phys. Rev. Lett.* **96**, 063903 (2006).

64. A. L. Pokrovsky, "Analytical and numerical studies of wire-mesh metallic photonic crystals," *Phys. Rev. B* **69**, 195108 (2004).

65. G. Guida, "Numerical study of band gaps generated by randomly perturbed bidimensional metallic cubic photonic crystals," *Opt. Commun.* **156**, 294–296 (1998).

66. G. Guida, T. Brillat, A. Ammouche, F. Gadot, A. D. Lustrac, A. Priou, "Dissociating the effect of different disturbances on the band gap of a two-dimensional photonic crystal," *J. Appl. Phys.* **88**, 4491–4497 (2000).

67. D. F. Sievenpiper, M. E. Sickmiller, and E. Yablonovitch, "3D wire mesh photonic crystals," *Phys. Rev. Lett.* **76**, 2480–2483 (1996).

68. S. H. Fan, P. R. Villeneuve, J. D. Joannopoulos, "Large omnidirectional band gaps in metallodielectric photonic crystals," *Phys. Rev. B* **54**, 11245–11251 (1996).

69. C. S. Kee, H. Lim, Y. K. Ha, J. E. Kim, and H. Y. Park, "Two-dimensional tunable metallic photonic crystals infiltrated with liquid crystals," *Phys. Rev. B* **64**, 085114 (2001).

70. K. Busch and S. John, "Liquid-crystal photonic-band-gap materials: the tunable electromagnetic vacuum," *Phys. Rev. Lett.* **83**, 967–970 (1999).

71. A. E. Serebryannikov and T. Magath, "Controlling location of opaque ranges in transmission of metallic photonic crystals," *Phys. Rev. A* **76**, 033828 (2007).

72. Y. S. Kim, S. Y. Lin, A. S. P. Chang, J. H. Lee, and K. M. Ho, "Analysis of photon recycling using metallic photonic crystal," *J. Appl. Phys.* **102**, 063107 (2007).

73. J. J. Greffet, R. Carminati, K. Joulain, J. P. Mulet, S. Mainguy, and Y. Chen, "Coherent emission of light by thermal sources," *Nature* **416**, 61–64 (2002).

74. C. M. Cornelius and J. P. Dowling, "Modification of planck blackbody radiation by photonic band-gap structures," *Phys. Rev. A* **59**, 4736–4746 (1999).

75. I. Celanovic, N. Jovanovic, and J. Kassakian, "Two-dimensional tungsten photonic crystals as selective thermal emitters," *Appl. Phys. Lett.* **92**, 193101 (2008).

76. M. U. Pralle, N. Moelders, M. P. Mcneal, I. Puscasu, A. C. Greenwald, J. T. Daly, E. A. Johnson, T. George, D. S. Choi, I. E. Kady, and R. Biswas, "Photonic crystal enhanced narrow-band infrared emitters," *Appl. Phys. Lett.* **81**, 4685–4687 (2002).

77. H. Sai, Y. Kanamori, and H. Yugami, "High-temperature resistive surface grating for spectral control of thermal radiation," *Appl. Phys. Lett.* **82**, 1685–1687 (2003).

78. Y. C. Hsue, A. J. Freeman, and B. Y. Gu, "Extended plane-wave expansion method in three-dimensional anisotropic photonic crystal," *Phys. Rev. B* **72**, 195118 (2005).

79. M. Okano, S. Kako, and S. Noda, "Coupling between a point-defect and a line-defect waveguide in three-dimensional photonic crystal," *Phys. Rev. B* **68**, 235110 (2003).

80. Z. Y. Li and K. M. Ho, "Light propagation in semi-infinite photonic crystals and related waveguide structures," *Phys. Rev. B* **68**, 155101 (2003).

81. Z. Y. Li and L. L. Lin, "Photonic band structures solved by a plane-wave-based transfer-matrix method," *Phys. Rev. E* **67**, 046607 (2003).

82. M. Che, Z. Y. Li, and R. J. Liu, "Tunable optical anisotropy in three-dimensional photonic crystals," *Phys. Rev. A* **76**, 023809 (2007).

83. M. Okano and S. Noda, "Analysis of multimode point-defect cavities in three-dimensional photonic crystals using group theory in frequency and time domain," *Phys. Rev. B* **70**, 125105 (2004).

84. H. B. Chen, Y. Z. Zhu, Y. L. Cao, Y. P. Wang, and Y. B. Chi, "Photonic band gap in the face-centered cubic lattice of spherical shells connected by cylindrical tubes," *Phys. Rev. B* **72**, 113113 (2005).

85. R. Biswas, M. M. Sigalas, G. Subramania, and K. M. Ho, "Photonic band gaps in colloidal systems," *Phys. Rev. B* **57**, 3701–3705 (1998).

86. R. Rengarajan, P. Jiang, V. Colvin, and D. Mittleman, "Optical properties of a photonic crystal of hollow spherical shells," *Appl. Phys. Lett.* **77**, 3517–3519 (2000).

87. H. Miguez, N. Tetreault, S. M. Yang, V. Kitaev, and G. A. Ozin, "A new synthetic approach to silicon colloidal photonic crystals with a novel

topology and an omni-directional photonic bandgap: micromolding in inverse silica opal (MISO)," *Adv. Mater.* **15**, 597–600 (2003).

88. R. Biswas, M. M. Sigalas, K. M. Ho, and S. Y. Lin, "Three-dimensional photonic band gaps in modified simple cubic lattices," *Phys. Rev. B* **65**, 205121 (2002).

89. M. Agio and L. C. Andreani, "Complete photonic band gap in a two-dimensional chessboard lattice," *Phys. Rev. B* **61**, 15519–15522 (2000).

90. M. Qiu and S. L. He, "Large complete band gap in two-dimensional photonic crystals with elliptic air holes," *Phys. Rev. B* **64**, 153108 (2001).

91. Z. Y. Li and Y. N. Xia, "Omnidirectional absolute band gaps in two-dimensional photonic crystals," *Phys. Rev. B* **64**, 153108 (2001).

92. K. M. Ho, C. T. Chan, and C. M. Soukoulis, "Existence of a photonic gap in periodic dielectric structures," *Phys. Rev. Lett.* **65**, 3152–3155 (1990).

93. K. Edagawa, S. Kanoko, and M. Notomi, "Photonic amorphous diamond structure with a 3D photonic band gap," *Phys. Rev. Lett.* **100**, 013901 (2008).

94. S. Imagawa, K. Edagawa, K. Morita, T. Niino, Y. Kagawa, and M. Notomi, "Photonic band-gap formation, light diffusion, and localization in photonic amorphous diamond structures," *Phys. Rev. Lett.* **100**, 013901 (2008).

95. S. R. Kennedy, M. J. Brett, O. Toader, and S. John, "Fabrication of tetragonal square spiral photonic crystals," *Nano Lett.* **2**, 59–62 (2002).

96. O. Toader and S. John, "Square spiral photonic crystals: Robust architecture for microfabrication of materials with large three-dimensional photonic band gaps," *Phys. Rev. E* **66**, 016610 (2002).

97. A. Chutinan, S. John, and O. Toader, "Diffractionless flow of light in all-optical microchips," *Phys. Rev. Lett.* **90**, 123901 (2003).

98. O. Toader, M. Berciu, and S. John, "Photonic bandgaps based on tetragonal lattices of slanted pores," *Phys. Rev. Lett.* **90**, 233901 (2003).

99. O. Toader, and S. John, "slanted-pore photonic-gap materials," *Phys. Rev. E* **71**, 036605 (2005).

100. A. Chutinan and S. John, "Diffractionless flow of light in two- and three-dimensional photonic band gap heterostructurees: theory, design, rules, and simulations," *Phys. Rev. E* **71**, 026605 (2005).

101. E. Ozbay, E. Michel, G. Tuttle, R. Biswas, M. Sigalas, and K. M. Ho, "Micromachined millimeter-wave photonic band-gap crystals," *Appl. Phys. Lett.* **64**, 2059–2061 (1994).

102. D. C. Meisel, M. Wegener, and K. Busch, "Three-dimensional photonic crystals by holographic using the umbrella configuration: symmetries and complete photonic band gaps," *Phys. Rev. B* **70**, 165104 (2004).

103. A. J. G. Adeva, "Band structure of photonic crystals with the symmetry of a pyrochlore lattice," *Phys. Rev. B* **73**, 073107 (2006).

104. C. M. Anderson and K. P. Giapis, "Larger two-dimensional photonic band gaps," *Phys. Rev. Lett.* **77**, 2949–2952 (1996).

105. L. Claudio and D. Gerace, "Photonic-crystal slabs with a triangular lattice of triangular holes investigated using a guided-mode expansion method," *Phys. Rev. B* **73**, 235114 (2006).

106. S. Takayama, H. Kitagawa, and Y. Tanaka, "Experimental semonstration of complete photonic band gap in two-dimensional photonic crystal slabs," *Appl. Phys. Lett.* **87**, 061107 (2006).

107. M. E. Zoorob, M. D. B. Charlton, G. J. Parker, J. J. Baumberg, and M. C. Netti, "Complete photonic bandgaps in 12-fold symmetric quasicrystals," *Nature* **404**, 740–743 (2000).

108. X. D. Zhang, Z. Q. Zhang, and C. T. Chan, "Absolute photonic band gaps in 12-fold symmetric photonic quasicrystals," *Phys. Rev. B* **63**, 081105(R) (2001).

109. M. Florescu, S. Torquato, and P. J. Steinhardt, "Complete band gaps in two-dimensional photonic quasicrystals," *Phys. Rev. B* **80**, 1551121 (2009).

110. K. Ueda, T. Dotera, and T. Gemma, "Photonic band structure calculations of two-dimensional Archimedean tiling patterns," *Phys. Rev. B* **75**, 195122 (2007).

111. A. Christ, S. G. Tikhodeev, N. A. Gippius, J. Kuhl, and H. Giessen, "Waveguide-plasmon polaritons: strong coupling of photonic and electronic resonances in a metallic photonic crystal slab," *Phys. Rev. Lett.* **91**, 183901 (2003).

112. I. V. Shadrivov, A. A. Sukhorukov, and Y. S. Kivshar, "Complete band gaps in one-dimensional left-handed periodic structures," *Phys. Rev. Lett.* **95**, 193903 (2005).

113. Z. Y. Li, B. Y. Gu, and G. Z. Yang, "Large absolute band gap in 2D anisotropic photonic crystals," *Phys. Rev. Lett.* **81**, 2574–2577 (1998).

114. J. Urayama, T. B. Norris, J. Singh, and P. Bhattacharya, "Observation of phonon bottleneck in quantum dot electronic relaxation," *Phys. Rev. Lett.* **86**, 4930–4933 (2001).

115. Y. Zeng, Y. Fu, X. S. Chen, W. Lu, and H. Agren, "Complete band gaps in three-dimensional quantum dot photonic crystals," *Phys. Rev. Lett.* **74**, 115325 (2006).

Questions

2.1 Please derivate the master equation (2.18) of photonic crystals by use of Maxwell equations:

$$\nabla \times \nabla \times \vec{E}(\vec{r}) = (\mu\omega^2)\varepsilon(\vec{r})\vec{E}(\vec{r})$$

2.2 For one-dimensional photonic crystal, the central wavelength of the photonic bandgap is 600 nm. The refractive index of the high and low dielectric layers n_1 and n_2 is 2.3 and 1.6, respectively. Please design the thickness of the high and low dielectric layers d_1 and d_2.

Chapter 3

Fabrication Technique of Photonic Crystals

Photonic crystals can control the propagation states of photons due to their unique photonic bandgap effect. This makes them the promising candidate for the realization of integrated photonic devices and integrated photonic circuits. However, the number and the types of available photonic crystals in the nature are very scarce. Therefore, the fabrication technique of photonic crystals has attracted great attention along with the appearance of the conception of photonic crystals. Various techniques have been developed to fabricate photonic crystals according to different material properties and functional requirements of photonic devices.

3.1 One-Dimensional Photonic Crystals

One-dimensional photonic crystals have a configuration of multi-layers, constructed by alternatively placing high-refractive-index layers and low-refractive-index layers. One-dimensional photonic crystals, also called $(\lambda/4)$ Bragg reflectors, have been used for long time. Usually, the thin-film fabrication technology is used to fabricate one-dimensional photonic crystals, such as the pulsed

Photonic Crystals: Principles and Applications
Qihuang Gong and Xiaoyong Hu
Copyright © 2014 Pan Stanford Publishing Pte. Ltd.
ISBN 978-981-4267-30-4 (Hardcover), 978-981-4364-83-6 (eBook)
www.panstanford.com

laser deposition technology, the molecule beam epitaxy technology, the chemical vapor deposition technology, the radio-frequency sputtering technology, the thermal evaporation technology, the sol-gel method, and so on [1–8]. Recently, the microfabrication etching technology has also been adopted to fabricate one-dimensional photonic crystals, such as the high-density plasma etching technology, the electron beam lithography (EBL) technology, and the focused ion beam (FIB) etching technology, the electrochemical etching, and so on [9–16]. There are also other fabrication methods developed to fabricate one-dimensional photonic crystals.

3.1.1 The Sol-Gel Method

The sol-gel method is a very simple and inexpensive method for the film fabrication on various substrates. In the traditional sol-gel method, the organometallic precursor is hydrolyzed to form a gel solution. Then the gel solution is dipped or spin-coated on the substrates. After the solution is dried and heated, a final film can be obtained. More often than not, a layer of film can be fabricated very conveniently. Chen *et al.* developed the sol-gel method to fabricate one-dimensional SiO_2/TiO_2 photonic crystal in 1999 [17]. Through alternatively spin-coating and heating the precursor gel solution of SiO_2 and TiO_2, a perfect one-dimensional SiO_2/TiO_2 photonic crystal could be obtained. The cross-sectional transmission electron microscopy (TEM) images of the one-dimensional SiO_2/TiO_2 photonic crystal are shown in Fig. 3.1 [17]. The layer thickness can be changed simply by adjusting the spin-coating parameters and the concentration of the precursor gel solution. So, a defect layer can be formed very easily by changing the thickness of the layer in the center of the multilayer structure. Moreover, one-dimensional photonic crystal based on oxide materials, ferroelectrics, and piezoelectrics can be fabricated with the sol-gel method. Therefore, the sol-gel method is a very effective and simple approach to fabricate one-dimensional photonic crystal.

3.1.2 The Spin-Coating Method

As a simple and conventional film fabrication method, the spin-coating method can also be adopted to fabricate one-dimensional photonic crystals, which is composed of alternatively stacked high- and low-dielectric layers. There are two factors that must be taken

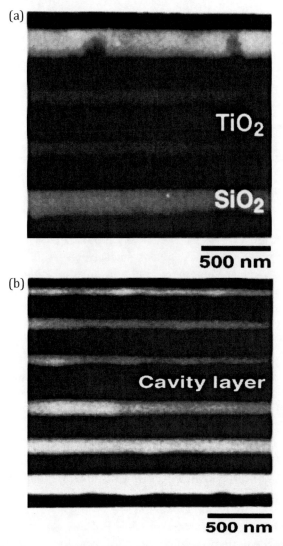

Figure 3.1 Cross-sectional TEM images of the SiO_2/TiO_2 one-dimensional photonic crystal. (a) For photonic crystal without defect. (b) For photonic crystal with defect. Reprinted with permission from *Appl. Phys. Lett.*, **75**(24), K. M. Chen, *et al.*, SiO_2/TiO_2 omnidirectional reflector and microcavity resonator via the sol-gel method. Copyright 1999, American Institute of Physics.

into account when using the spin-coating method. The first factor is that the refractive index contrast between the two component materials must be as high as possible. The high-refractive-index contrast will benefit the formation of high-quality photonic bandgap, including wide bandgap and steep bandedge. Moreover, the photonic bandgap can be opened for a few stacked layers when the refractive index contrast is large. The second factor is that the solvent of one component material cannot dissolve the other component material. The spin-coating method can be adopted to fabricate organic or organic/inorganic hybrid one-dimensional photonic crystal. In 2006, Komikade *et al.* fabricated a one-dimensional polyvinylcarbazole (PVK)/cellulose acetate (CA) photonic crystal by use of the spin-coating method. The photonic crystal was constructed by alternatively stacking polyvinylcarbazole layer and cellulose acetate layer [18]. Chlorobenzene and diacetone alcohol were selected as the solvent for PVK and CA, respectively. Even though the refractive index contrast of PVK and CA is very small, about 0.208, remarkable photonic bandgap centered at 590 nm can still be obtained for 21 stacked layers. Subsequently, Yoon *et al.* fabricated an organic/inorganic hybrid one-dimensional photonic crystal by use of the conventional spin-coating method, with alternative layers of titania nanoparticles and of polymethylmethacrylate (PMMA) [19].

3.1.3 The Electrochemical Micromachining Method

The electrochemical micromachining method was presented by Barillaro *et al.* in 2002 [20–22]. The electrochemical micromachining method adopts the photolithographic and electrochemical etching of an *n*-type silicon wafer to fabricate silicon microstructures. The general fabrication process of the electrochemical micromachining method is as follows: a 100 nm-thick SiO_2 layer is deposited on the surface of an *n*-doped (100) oriented silicon wafer. A standard photolithographic and chemical etching process is used to define the periodic patterns of one-dimensional photonic crystal in SiO_2 layer. Then the electrochemical micromachining process is adopted to obtain the final photonic crystal sample [21,22]. The cross-sectional scanning electron microscopy image of the one-dimensional silicon photonic crystal fabricated by Barillaro *et al.* is shown in Fig. 3.2 [20]. This method allows the fabrication of large-size, high-quality one-dimensional silicon photonic crystal.

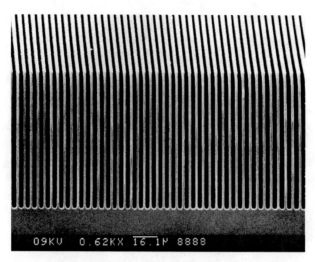

Figure 3.2 Cross-sectional SEM images of the one-dimensional silicon photonic crystal. The width of the silicon and air walls is 1 μm and 2 μm, respectively. Reprinted with permission from *Appl. Phys. Lett.*, **89**(15), G. Barillaro, *et al.*, Silicon micromachined periodic structures for optical applications at λ = 1.55 μm. Copyright 2006, AIP Publishing LLC.

3.1.4   The Electrochemical Anodization Method

The electrochemical anodization method can be used to fabricate porous silicon. Filipek *et al.* proposed that the electrochemical anodization method could also be adopted to fabricate chirped one-dimensional porous silicon photonic crystals with gradually varied lattice constants [23]. The SEM image of the linear chirped porous silicon photonic crystal fabricated by Filipek *et al.* is shown in Fig. 3.3 [23]. The conventional one-dimensional photonic crystal is composed of alternative layers with high refractive index n_h and low refractive index n_l. The thicknesses of the layers, d_h for high-refractive-index and d_l for low-refractive-index layers, and the lattice period, $d_h + d_l$, are constant and do not change. Accordingly, the Bragg wavelength λ_{Bragg} of the photonic crystal is also constant

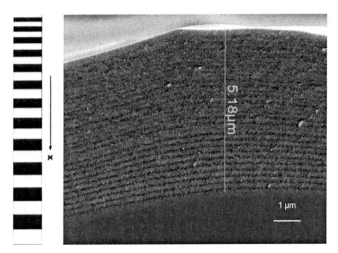

Figure 3.3   SEM image of the linear chirped one-dimensional porous silicon photonic crystal. The refractive index is 2.3 and 3.2 for the layers with large and small voids, respectively. The thickness of porous layers increases with the increase of etching depth of the silicon substrate. Left insert shows the schematic structure of the chirped photonic crystal. Reprinted with permission from *J. Appl. Phys.*, **104**(7), I. J. Kuzma-Filipek, *et al.*, Chirped porous silicon reflectors for thin-film epitaxial silicon solar cells. Copyright 2008, AIP Publishing LLC.

and does not change. While for linear chirped photonic crystals, the lattice constant increases gradually according to a certain scatter parameter s, except that the thickness of the middle double layers is equal to the original lattice constant $d_h + d_l$; s is an integer and its value is not equal to 1 [24]. If the magnitude of the increase of the lattice constant is small, the central wavelength of the photonic bandgap of the linear chirped photonic crystal is almost the same value as that of the unchirped photonic crystal. Contrary to the unchirped photonic crystal, the chirped photonic crystal can promote the constructive interference at the wavelength outside the central wavelength of the photonic bandgap. When increasing the lattice constant linearly, the Bragg wavelength of the photonic crystal will not be a constant. Different Bragg wavelengths will appear, which will lead to an enlargement of the width of the photonic bandgap. In the electrochemical anodization process, the thickness of porous silicon layers can be controlled by changing the current density and the etching time. When the etching time of each porous silicon layer is linearly increased, a chirped photonic crystal can be obtained [23]. Chirped one-dimensional photonic crystal may find important potential applications in the fields of broadband dispersion compensators in femtosecond lasers [25].

3.1.5 The Phase-Controlled Holographic Lithography Method

The multiple laser-beam holographic lithography method has been widely used to fabricate periodic dielectric microstructures [26,27]. In 2008, Li *et al.* developed a one-step phase-controlled holographic lithography method to fabricate one-dimensional photonic crystals [28]. In the one-step phase-controlled holographic lithography method, each of the interfering beams is large enough in the transverse dimension so as to neglect the propagation diffraction. A variable phase retardation mask is used to control the laser beams so as to obtain the required interference pattern. This is the space domain analogy of time domain pulse shaping [29,30]. When a tunable liquid-crystal spatial light modulator is used to control the phase properties of multiple laser beams, the inter-ference pattern can be dynamically tunable [31]. The intensity distribution of the interference pattern can be written as [32].

$$I(\vec{r}, t) = <\sum_{i}^{n} \vec{E}_i^2 > + \sum_{i<j}^{n} 2\vec{E}_i \cdot \vec{E}_j \cos((\vec{k}_i - \vec{k}_j) \cdot \vec{r} + (\varphi_i - \varphi_j)) \qquad (3.1)$$

where i and j represent different laser beams, $\vec{E}_i$ and φ_i are the electric-field and the initial phase of the laser beams, respectively. $\vec{k}_i$ is the wave vector. The interference pattern can be controlled by changing the polarization and the amplitude of the electric field, the wave vector, and the phase of the incident laser beams. This makes it possible to fabricate various photonic crystal structures. When the photoresist is exposed by the interference pattern, a photonic crystal sample can be obtained after development.

3.2   Two-Dimensional Photonic Crystals

Two-dimensional photonic crystals have the periodic distribution of dielectric constant in two normal directions. Two-dimensional photonic crystals, promising to be integrated into the traditional silicon integrated circuits, have attracted great attention. More often than not, the microfabrication technology, including the focused ion-beam (FIB) etching technology [33] and the electron-beam lithography (EBL) technology [34–36], are widely used to fabricate two-dimensional semiconductor photonic crystals. But the microfabrication technology is very expensive and time consuming. Therefore, people have also developed other methods to fabricate two-dimensional photonic crystals, such as the nanoimprint lithography method [37], the holographic lithography method [38–41], and so on.

3.2.1   The Electron-Beam Lithography Method

The electron-beam lithography method is a very important and mature microfabrication technology, which is widely used in the opt-electronic industry. The electron-beam lithography method can be used to fabricate two-dimensional photonic crystals with a relatively large size and high resolution. The primary fabrication process is as follows: First, a layer of electron-beam resist is coated on the surface of the needed films. Then the periodic patterns are formed in the resist layer in the subsequent electron-beam lithography process. Finally, the reactive ion beam etching process is adopted to transfer the periodic patterns from the resist layer to the needed films. Then a photonic crystal sample can be obtained when the resist layer is removed. Two-dimensional photonic crystals based on semiconductor, ferroelectric, and even organic materials can be fabricated by use of the electron-beam lithography method. Moreover, both the air-hole type and the dielectric-rod type two-dimensional photonic crystal can be fabricated based on the electron-beam lithography method [42].

Fu *et al.* reported a two-dimensional silicon photonic crystal fabricated by the electron-beam lithography method, which consists of hexagonal arrays of silicon circular columns connected to its nearest neighbors by slender rectangular rods [43]. The patterned area was about 100 μm × 100 μm. The thickness and the lattice constant of the photonic crystal were 1 µm and 2.5 µm, respectively. The fabrication process was simply described as follows: First, a periodical hexagonal array of circular columns connected to its nearest neighbors by slender rectangular rods of openings was prepared in the ZEP-520 layer, a positive electron-beam resist, on silicon substrates with the electron-beam lithography method. Second, a silicon dioxide layer was deposited on the ZEP-520 layer. Then the resist layer was removed and the hexagonal lattice of silicon dioxide remained on the silicon substrates. Finally, the two-dimensional photonic crystal sample was obtained by use of inductively coupled plasma reactive ion etching technique [43]. Figure 3.4 shows a two-dimensional poly(methyl methacrylate) (PMMA) photonic crystal fabricated with the electron-beam lithography method [44]. A palladium hard mask was deposited on the PMMA film to obtain the needed periodic patterns. Figure 3.4a is a square lattice two-dimensional PMMA photonic crystal. The lattice constant and the diameter of air holes were 150 nm and 75 nm, respectively. Figure 3.4b was a hexagonal lattice two-dimensional PMMA photonic crystal. The lattice constant and the diameter of air holes were 150 nm and 85 nm, respectively [44].

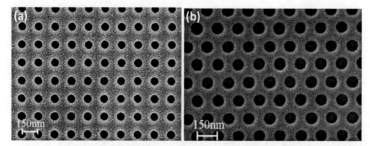

Figure 3.4 SEM images of two-dimensional PMMA photonic crystal fabricated by the electron-beam lithography method. (a) For square lattice. (b) For hexagonal lattice. Reprinted with permission from *Nano Lett.*, **7**, V. A. Parekh, *et al.*, Close-packed noncircular nanodevice pattern generation by self-limiting ion-mill process, 3246–3248. Copyright 2007, American Chemical Society.

3.2.2 The Focused Ion-Beam Etching Method

The focused ion-beam etching method is another very important microfabrication technique widely used in the opt-electronic industry. Various materials can be adopted to fabricate two-dimensional photonic crystals by use of the focused ion-beam etching method, such as semiconductor, polymer, and ferroelectric materials, and so on [45–48]. The fabrication process is very simple. The sample is placed in an ultrahigh vacuum chamber, where a Ga⁺ beam is used to etch the sample directly. The sample can be precisely translated and rotated, so that the periodic pattern can be obtained. The focused ion-beam etching method can also be adopted to fabricate air-hole type or dielectric-rod type two-dimensional photonic crystals. However, there are two limiting factors for the applications of the focused ion-beam etching method. One limitation is that the sputtered material will remain in the bottom of deep holes during the etching process, which will reduce the depth of drilled holes. The other limitation is the reflection of ion beam at the walls of the drilled holes, also called the self-focusing effect [49]. Figure 3.5 shows the SEM image of a two-dimensional photonic crystal composed of periodic arrays of diamond nanocones. The crystal is fabricated by means of focused ion-beam etching and hot-filament chemical vapor deposition [50,51]. The lattice constant is 20 μm. The height, the apex radius, and the base radius are 9 μm, 100 nm, and 1 μm, respectively [51].

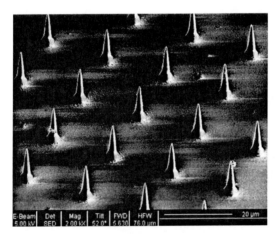

Figure 3.5 SEM images of two-dimensional diamond nanocone photonic crystal fabricated by the focused ion-beam etching and chemical vapor deposition method. Reprinted from *Microelectron. Eng.*, **78–79**, Z. Wang, *et al.*, A novel method for making high aspect ratio solid diamond tips, 353–358, Copyright (2005), with permission from Elsevier.

3.2.3 The Nanoimprint Lithography Method

Nanoimprint lithography is a simple and effective method to achieve two-dimensional periodic patterns with a nanometer-scale resolution over a large area, providing an intriguing approach for the fabrication of functional microstructures [52]. Owing to its outstanding advantages of high-resolution, high-throughput patterning capability, and extremely low cost, nanoimprint lithography has been considered to be one of the most promising techniques for nanometer-scale microstructure fabrication [37]. The primary fabrication process of the nanoimprint lithography method can be described as follows: First, the imprint stamp, which contains a reversal pattern of the surface protrusion structure of the two-dimensional photonic crystal, is fabricated. Second, the imprint resin is coated on the substrate. Third, the imprinting process is performed at a temperature of about 120°C [37]. Then the photonic crystal sample can be obtained. Figure 3.6 shows the SEM image of a two-dimensional GaN photonic crystal fabricated with nanoimprint lithography and inductively coupled plasma etching [37]. The photonic crystal consists of a rectangle array of GaN pillars with a diameter of 130 nm. The lattice constant of the photonic crystal and the height of the pillars are 520 nm and 120 nm, respectively. Figure 3.7 shows the SEM image of a two-dimensional InP photonic crystal composed of hexagonal arraysof InP nanowires fabricated with the nanoimprint lithography

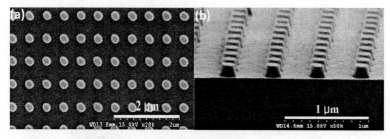

Figure 3.6   SEM images of two-dimensional *p*-GaN photonic crystal fabricated by nanoimprint lithography and inductively coupled plasma etching. (a) For top views. (b) For tilted views. Reprinted with permission from *Appl. Phys. Lett.*, **91**(9), K.-J. Byeon, *et al.*, Fabrication of two-dimensional photonic crystal patterns on GaN-based light-emitting diodes using thermally curable monomer-based nanoimprint lithography, 091106. Copyright 2007, AIP Publishing LLC.

and self-assembly method [53]. The diameter and length of InP nanowires are 290 nm and 1.5 μm, respectively. The nearest neighbor distance in the hexagonal array is 1 μm.

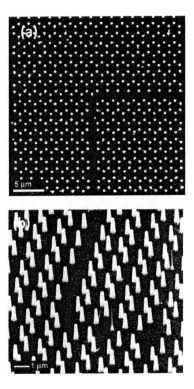

Figure 3.7   SEM images of two-dimensional InP photonic crystal fabricated by nanoimprint lithography and self-assembly. (a) For top views. (b) For tilted views. Reprinted with permission from *Nano Lett.*, **4**, T. Martensson, *et al.*, Nanowire arrays defined by nanoimprint lithography, 699–702. Copyright 2004, American Chemical Society.

3.2.4 The Electrochemical Etching Method

The electrochemical etching in hydrofluoric acid has been widely used to fabricate porous silicon with a regular pattern of uniform pores, with a diameter in the micrometer range and a depth of several hundreds of micrometers [54,55]. Gruning *et al.* adopted the electrochemical etching method to fabricate two-dimensional photonic crystal [54]. When a polished silicon substrate was anodized in a hydrofluoric acid solution with the backside of the wafer illuminated, the pores would be formed in a random pattern. The pits generated by the standard lithography and the alkaline etching were adopted to make the pores formed periodically. Then a two-dimensional photonic crystal made of periodic arrays of micropores could be fabricated. The SEM image of the two-dimensional photonic crystal fabricated by Gruning *et al.* is shown in Fig. 3.8 [54]. The lattice constant of the photonic crystal was 2.3 μm. The pore diameter and the thinnest parts of the pore walls were 2.13 μm and 170 nm, respectively [54].

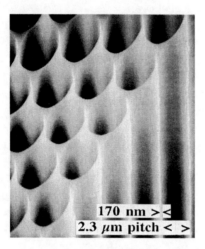

Figure 3.8   SEM image of two-dimensional microporous silicon photonic crystal fabricated by electrochemical etching. Reprinted with permission from *Appl. Phys. Lett.*, **68**, U. Gruning, *et al.*, Macroporous silicon with a complete two-dimensional photonic band gap centered at 5 μm, 747–749. Copyright 1996, AIP Publishing LLC.

3.2.5   The Holographic Lithography Method

The holographic lithography method is a very economical and powerful tool to fabricate large-area periodic lattice two-dimensional photonic crystals and two-dimensional photonic quasicrystals [56–59]. Generally, three laser beams can be used to generate a two-dimensional interference pattern with the holographic lithography method. Wu *et al.* proposed the concept of creating heterogeneity into the interference pattern by introducing auxiliary beams [60]. They fabricated two-dimensional hetero-binary and honeycomb photonic crystals by introducing one or three auxiliary beams into the three basic beams to form a regular hexagonal structure. The size contrast between the center rod and its six neighbors in a hetero-binary structure can be tuned by adjusting the intensity contrast between the auxiliary beam and basic beams. Lai *et al.* proposed an approach to fabricate periodic nanovein structures by mixing a negative and a positive photoresist in the holographic lithography method, which could be used as templates for fabricating photonic crystals possessing a large complete photonic bandgap [61]. The structures were obtained by using either a pure negative photoresist doped with very low concentration of photoinitiator or a mixture of a negative and a positive photoresist. The fabricated periodic structures were constructed by square or triangular arrays of rods connected with narrow nanoveins, which was not a simple duplication of the interference pattern. In 1995, Mao *et al.* found that two-dimensional compound photonic lattices could also be fabricated by use of four noncoplanar beams with the holographic lithography method [62]. Subsequently, they reported an approach of single-exposure holographic lithography for controllable fabrication of large-scale two-dimensional square and hexagonal compound polymer photonic crystals [63]. Compound structures with required micropores on a 100 nm scale could be generated controllably by adjusting the elliptical polarizations.

3.3 Three-Dimensional Photonic Crystals

Traditionally, there are two main kinds of approaches to fabricate three-dimensional photonic crystals: the bottom-up self-assembly method for the fabrication of colloidal photonic crystals, and the top-down lithography method for the fabrication of non-colloidal photonic crystals [64]. The fabrication process of the bottom-top self-assembly method is very simple and easy. Very large-size three-dimensional colloidal photonic crystals in the centimeter order can be obtained by use of this method. However, the relatively short range of ordering and the incapacity to form controllable defect structures greatly limit the practical applications of the bottom-top self-assembly method [65,66]. The top-down lithography method adopts the complicated lithography techniques, direct laser writing technique, and so on [64].

3.3.1 The Microfabrication Method

The microfabrication method, including the focused ion-beam etching method and the electron-beam lithography method, can be used to fabricate three-dimensional photonic crystals when combined with other techniques. Usually, the combination of plasma or chemical etching with focused ion-beam etching has been considered to be a promising method to fabricate various three-dimensional photonic crystals [67]. For example, the Yablonovite-like three-dimensional silicon photonic crystals can be fabricated by combining focused ion-beam etching method with the porous silicon fabrication technique [67]. The lattice period of about 0.75 μm and the central wavelength of the photonic bandgap close to 3 μm could be obtained [67]. Point-defect microcavities with the resonant wavelength in the optical communication range can also be introduced in a three-dimensional silicon photonic crystal by using the fabrication process combining the electron-beam lithography, the reactive-ion etching, and the planarization [68].

Lin *et al.* reported the construction of a three-dimensional infrared photonic crystal on a silicon wafer with a layer-by-layer periodic structure by use of relatively standard microelectronics fabrication technology [69]. The layer-by-layer structure consists of layers of one-dimensional rods in a stacking sequence that repeats itself every four layers with a repeat distance c. Within each layer, the axes of the rods are parallel to each other with a pitch d. The orientations of the axes are rotated by 90° between adjacent layers. Between every other layer, the rods are shifted relative to each other by $0.5d$. The resulting structure has a face-centered-tetragonal lattice symmetry [69]. For the special case of $(c/d)^2 = 2$, the lattice can be derived from a face-centered-cubic unit cell with a basis of two rods. The resulting structure can also be derived by replacing the <110> chains of atoms in the diamond structure by the rods [70–71]. A comprehensive five-level stacking process was developed to fabricate the three-dimensional silicon photonic crystal. Within each layer, silicon dioxide was first deposited, patterned, and then etched to the desired depth. The resulting trenches were then filled with polycrystalline silicon. Following this, the surfaces of the wafers were made flat using chemical mechanical polishing, and the process was then repeated. After the five-level process was completed, the wafer was immersed in an HF/water solution for the final SiO_2 removal [69]. Based on the fabrication of three-dimensional layer-by-layer silicon photonic crystals, Fleming *et al.* fabricated all-metallic three-dimensional tungsten photonic crystals with a large infrared bandgap [72,73]. The tungsten photonic crystal was fabricated by selectively removing silicon from the already formed polysilicon/ SiO_2 structures, and by back-filling the resulting mould with chemical vapor deposited tungsten [73]. The width of tungsten rods, the rod-to-rod spacing, and the filling fraction of tungsten rods were 1.2 μm, 4.2 μm, and 28%, respectively [73]. This method could also be extended to create almost any three-dimensional single-crystal metallic photonic crystal in the near-infrared range [73].

Recently, Wu *et al.* proposed a method to fabricate three-dimensional polymer photonic crystal by use of single-step electron-beam lithography [74]. The material system used in their experiment was a polymethylmethacrylate/polydimethylglu-tarimide-based polymer (PMMA/LOR) system. LOR has much higher charge sensitivity than that of PMMA, which leads to a very large contrast in the exposure sensitivity between PMMA and LOR. The high-energy electron-beam and the large contrast in the exposure sensitivity between PMMA and LOR provides relatively large penetration depth in resists, which makes it possible to fabricate a three-dimensional photonic crystal with a multiplayer structure by use of a single-step electron-beam lithography [74]. SEM images of the three-dimensional polymer photonic crystal are shown in Fig. 3.9 [74]. The square PMMA roofs were 770 nm on each side and 200 nm in thickness. The LOR pillars were 550 nm on each side and 500 nm in height. The inter-roof distance was 230 nm and the lattice constant was 1 μm [74].

Figure 3.9   SEM images of the three-dimensional polymer photonic crystals fabricated by the electron-beam lithography method. (a) For the cross-sectional view. (b) For the top view. Reprinted with permission from *Adv. Mater.*, **19**, C. S. Wu, *et al.*, Polymer-based photonic crystals fabricated with single-step electron-beam lithography, 3052–3056. Copyright © 2007 WILEY-VCH Verlag GmbH & Co. KGaA, Weinheim.

3.3.2   The Direct Laser Writing Method

3.3.2.1   Based on two- and multiple-photon polymerization

When the laser intensity is far less than the damage threshold of the dielectric material, the direct laser writing method adopts a tightly focused femtosecond laser beam to scan in liquid resins or photoresists to fabricate complicated three-dimensional micro-structures based on the two-photon or multiple-photon poly-merization [75,76]. The rate of the two-photon absorption is proportional to the square of the local light intensity. Owing to its threshold response properties, the two-photon absorption only occurs in a small volume within the depth of focus, which leads to a sub-diffraction-limit resolution [77]. For example, a lateral spatial resolution down to 120 nm could be achieved conveniently by use of objective lens with a high numerical aperture for an 800 nm femtosecond laser source [78–80]. Ideally, any complicated three-dimensional periodic macrostructures can be fabricated by use of the computer controlled direct laser writing method in all kinds of materials [81–83]. Therefore, the direct laser writing method has been considered as one of the most promising candidates for the fabrication of various high-quality three-dimensional photonic crystals [84,85].

3.3.2.1.1   *Based on two-photon polymerization*

Using inorganic-organic hybrid materials: ORMOCER-1 is a high-quality inorganic-organic hybrid optical material. ORMOCER-1 contains Irgacure 369 initiator and is very sensitive to the ultraviolet radiation. ORMOCER-1 is transparent in the near-infrared range, and allows a 780 nm femtosecond laser beam to be focused into the volume of the liquid resin. Two-photon absorption is caused by the high photon density in the focus position, which leads to the polymerization of the resin [86]. When the laser focus is moved through the resin in three dimensions, polymerization will occur along the trace of the focus, which makes it possible to obtain a three-dimensional photonic crystal structure [87]. The photonic crystal is composed of dielectric rods stacked in cubic lattice. The diameter of the dielectric rods and the lattice constant are 200 nm and 450 nm, respectively [86].

Using organic materials: Many complicated three-dimensional photonic crystal structures can be obtained in organic materials by use of the direct laser writing method. Figure 3.10 is the SEM image of a three-dimensional horizontal circular spiral SU-8 photonic crystal fabricated by use of the direct laser writing method based on two-photon absorption [88]. The width, depth, and height of the circular spiral photonic crystal are 40, 31, and 20 μm, respectively [88]. The photonic bandgap is around 1 μm. The SEM image of a three-dimensional bi-chiral SU-8 photonic crystal fabricated by use of direct laser writing based on two-photon absorption is shown in Fig. 3.11 [89]. The bi-chirality refers to structures consisting of left- or right-handed circular dielectric spirals that are arranged along the three orthogonal spatial axes of a cubic lattice with left- or right-handed corners [90]. The cubic lattice constant and the diameter of each spiral are 4 and 3.6 μm, respectively [89].

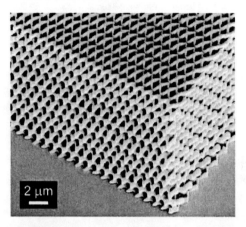

Figure 3.10　SEM image of the circular spiral photonic crystal fabricated by the direct laser writing method. Reprinted with permission from *Appl. Phys. Lett.*, **88**, K. K. Seet, *et al.*, Three-dimensional horizontal circular spiral photonic crystals with stop gaps below 1 μm, 221101. Copyright 2006, AIP Publishing LLC.

Figure 3.11　SEM image of the bi-chiral photonic crystal with right-handed corner and left-handed spirals. Reprinted with permission from *Adv. Mater.*, **21**, M. Thiel, *et al.*, Three-dimensional bi-chiral photonic crystals, 4680–4682. Copyright © 2009 WILEY-VCH Verlag GmbH & Co. KGaA, Weinheim.

The photoresist etched by the direct laser writing can also be adopted as a template for the fabrication of silicon inverse photonic crystal. Figure 3.12 is the SEM image of a three-dimensional silicon inverse photonic quasicrystal fabricated by Ledermann *et al.* by use of the direct laser writing method combined with a silicon inversion procedure [91,92]. They adopted SU-8 as the photoresist. The fabrication process was as follows: First, the templates of the three-dimensional SU-8 photonic quasicrystal with a local five-fold real-space symmetry axis were fabricated by use of the direct laser writing method. The interatomic spacing, i.e., the length of the connections, was 2 µm. Second, the polymeric templates were coated with a thin film of silicon dioxide and then infiltrated with silicon by the chemical vapor deposition technique. Finally, the silicon dioxide and SU-8 were removed by hydrofluoric acid and calcinations, respectively [93]. As a result, only the silicon constructing the inverse of the template was obtained. Shir *et al.* reported the fabrication of three-dimensional silicon photonic crystals using polymer templates defined by two-photon exposure through a layer of photopolymer with relief molded on its surface [94]. In the fabrication process, the molded photopolymer layer served as both the optical phase mask and the recording media for three-dimensional patterning. The exposure of this material followed by the removal of the unexposed regions in a development process produced high-quality three-dimensional photonic crystal over large areas via two-photon absorption [94]. High-contrast exposure could be achieved due to the quadratic intensity dependence of two-photon absorption, and the efficient optical coupling of the exposure light into the photopolymer could be reached due to the usage of molded relief [94].

The two-photon absorption initiated isomerization effect can also be adopted to fabricate three-dimensional polymer photonic crystals via the direct laser writing method [95]. For example, *cis*-form rich poly[2,5-dihexyloxy-1,4-phenylene vinylene-alt-2,5-diphenyl-1,4-phenylene vinylene] (DPO-PPV) is a precursory material, which will undergo molecular conformational change upon the near-infrared femtosecond laser irradiation. The photoisomerization of DPO-PPV can be described as follows: Under excitation of a focused 800 nm femtosecond laser beam, each molecule of the *cis*-olefine simultaneously absorbs two infrared photons, which leads to the molecular configuration changes from *cis*-form to *trans*-form with rotational motion of the double bonds of the molecule. The *cis*-form is soluble in common organic solvents because of the twisting configuration of the *cis*-vinylene segments, while the *trans*-form is insoluble due to the stronger intermolecular attractions arising from the more-coplanar molecule conformation [95]. This method may create a new way to fabricate complicated three-dimensional polymer photonic crystal structures.

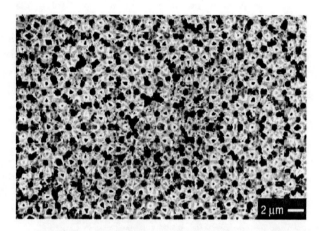

Figure 3.12   SEM image of the three-dimensional silicon inverse photonic quasicrystal fabricated by the direct laser writing method. Reprinted by permission from Macmillan Publishers Ltd: *Nat. Mater.*, **5**(12), A. Ledermann, *et al.*, Three-dimensional silicon inverse photonic quasicrystals for infrared wavelengths, 942–945, copyright (2006).

3.3.2.1.2   *Based on multiple-photon polymerization*

In the multiple-photon polymerization, a photoresist is illuminated by laser light at a frequency below the single-photon polymerization threshold. More often than not, the laser beam is tightly focused inside the photoresist, which makes the laser intensity in a small focus volume exceed the threshold of multiple-photon polymerization. The size and the shape of the formed voxels depend on the iso-intensity surfaces of the microscope objective, and the exposure threshold of the multiple-photon polymerization of the photoresist [96]. When photoresists are used as the matrix of the photonic crystal, the combination of the direct laser writing method with the holographic lithography method makes it possible to realize large-scale and complicated three-dimensional photonic crystal structures.

Dubel *et al.* reported a large-scale and high-quality three-dimensional SU-8 photonic crystal with a face-centered-cubic layer-by-layer structure, fabricated by use of the direct laser writing method combined with the holographic lithography method [97]. The photonic crystal was composed of stacks of parallel polymer rods with a period of 650 nm and a lattice constant of 920 nm [97]. The size of the photonic crystal sample was as large as 100 μm × 100 μm. Deubel *et al.* also reported the fabrication and characterization of the recently proposed slanted pore SU-8 photonic crystals via direct laser writing by multiple-photon polymerization [98]. They found that this slanted pore geometry allows for the controlling of the surface termination of the photonic crystal. For a slanted pore (SP_n) photonic crystal, there exist n pores in each unit cell. SP_n photonic crystals can be defined using a two-dimensional mask lattice and n pore axes associated with each unit cell of the mask lattice. In each of the SP_n photonic crystals, the mask lattice has a two-point basis. A SP_2 photonic crystal can be constructed from a square lattice mask with primitive vectors $\vec{a}_1$ and $\vec{a}_2$ with a length a. $\vec{c}$ is an independent vector with a length c perpendicular to the mask plane; $\vec{a}_1$, $\vec{a}_2$, and $\vec{c}$ form the unit cell of the tetragonal three-dimensional slanted pore photonic crystal [99]. An 800 nm femtosecond laser beam with a pulse width of 120 fs was tightly focused inside the SU-8 film. The SEM image of the slanted pore photonic crystal is shown in Fig. 3.13. The values of a and c of the unit cell were 1.0 and 1.4 μm, respectively. The rod diameter was 360 nm [98].

Sun *et al.* reported a method to fabricate photonic crystals by use of the voxels of an intensely modified refractive index, generated by multiple-photon absorption at the focus of femto-second laser pulses in Ge-doped silicon dioxide, to construct a three-dimensional photonic crystal [100]. Materials at the small volume of focus were excited to the free-carrier plasma state primarily by multiple-photon absorption, and partially by impact ionization during the pulse duration in their experiment. The highly excited and tightly localized plasma expands in an explosive way before it can transfer the energy to a lattice, generating a hollow core surrounded by a denser material [100]. If the laser energy and the focusing condition are properly chosen, the voxels of subwavelength or near-wavelength size will take the form of a well-defined shape similar to that of a sphere [100]. A 400 nm laser beam with a 150-fs pulse duration was tightly focused inside the Ge-doped silica slab in their experiment. The voxels had an outer diameter of 1.0 µm and a core diameter of 300 nm. They fabricated a three-dimensional photonic crystal with a face-centered cubic lattice, with the total size of 40 µm long, 40 µm wide, and 20 µm thick.

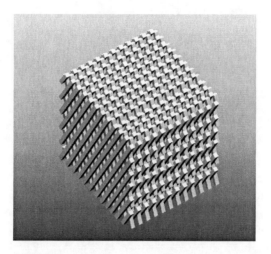

Figure 3.13 Schematic structure of the slanted pore photonic crystal fabricated by the direct laser writing method.

3.3.2.2 Based on localized laser-induced microexplosion

When an intense laser beam is tightly focused in a dielectric material, microexplosion will occur in the focus position inside the bulk dielectric material due to the strong localized laser field in the focus. The mechanism of microexplosion could be understood as follows: the laser beam is focused into a volume about λ^3 by a high numerical aperture objective, where λ is the wavelength of the laser pulse [101]. Usually, the energy of the laser pulse is absorbed in a volume of 0.1 µm^3 order, which creates a high pressure and a high temperature [102]. The material within this small volume is rapidly atomized, ionized, and subsequently converted into a tiny super-hot dense cloud of expanding plasma. The plasma generated in the focal region increases the absorption coefficient and produces a fast energy release in a very small volume. A strong shock wave is generated in the interaction region and propagated into the surrounding cold material. The propagation of the shock wave is companied by compression of the solid material at the wave front and by decompression behind it, leading to the formation of a void inside the material [101–103]. The voxel-based direct laser writing method is a flexible and cheap fabrication technology of three-dimensional photonic crystals.

Using inorganic materials

Ferroelectric materials, such as lithium niobate ($LiNbO_3$), have a relatively high refractive index in the visible and near-infrared range. Therefore, the spherical aberration induced by the refractive index mismatch is strong and affects the fabrication process of the direct laser writing method very seriously. Zhou and Gu reported the fabrication of three-dimensional photonic crystals in a high refractive index $LiNbO_3$ crystal based on the femtosecond laser-induced microexplosion mechanism [104]. When a laser beam is tightly focused into a $LiNbO_3$ crystal, the diffraction pattern at the focal region will be distorted compared with the diffraction-limited pattern in a refractive index matched medium. The focus region becomes broader slightly in the transverse direction. Moreover, there exist a focus shift and several sidelobes along the axial direction. The lower-order sidelobes may produce unwanted effects if the laser intensity is so high that the intensity of the sidelobes gets higher than the ionized threshold [104]. They used a threshold fabrication technique to reduce the influences of the refractive index mismatch-induced spherical aberration. The laser power was selected so that only the main focus was above the ionized threshold, and all the sidelobes were below the ionized threshold. Then the focal distortion could be reduced greatly by use of this technique. A 750 nm femtosecond laser beam (with a pulse duration of 100 fs and a repetition rate of 82 MHz) was tightly focused by an Olympus 60× oil immersion objective with a numerical aperture of 1.4 inside the $LiNbO_3$ slab in their experiment. Three-dimensional photonic crystals with a face-centered-cubic periodic structure were fabricated by stacking quasi-spherical voids layer by layer. The lattice constant was 4.5 μm. Photonic bandgaps with a suppression rate of up to 30% in the transmittance spectra were achieved [104].

Laser crystal and ceramic materials can also be used to fabricate three-dimensional photonic crystal by use of the microexplosion based direct laser writing method. Rodenas *et al.* reported a three-dimensional photonic crystal in a neodymium doped yttrium aluminum garnet laser ceramic [105]. An 800 nm femtosecond laser beam was focused inside the ceramic material by a 60× oilimmersion objective with a numerical aperture (NA) of 1.4 in the experiment. Voxels with a size of micrometer or submicron scale could be obtained by adjusting the focused depth and the laser energy. For a focused depth of 6 μm, voxels with a shape factor of about 1.1, with the lateral and the axial size of 495 nm and 532 nm, respectively, could be achieved [105]. The three-dimensional photonic crystal was composed of a 35-layer face-centered-cubic lattice periodic voxels with a lattice constant of 1 μm in the background material of neodymium doped yttrium aluminum garnet laser ceramic [105]. The photonic bandgap was centered at 2.4 μm.

Using organic materials

Compared with other dielectric materials, the damaging threshold of organic polymer materials induced by a tightly focused femtosecond laser is much lower. Therefore, it is easier to achieve voids in organic polymer materials generated by laser-induced microexplosion [106,107]. Ventura *et al.* reported the fabrication of a void channel infrared photonic crystal in resin solid by use of laser-induced microexplosion [108]. Norland NOA63 resin, a polyurethane oligomer having C=C unsaturated bond, and cross-linked by a mercapto-ester oligomer, was adopted as the organic polymer material in their experiment. A 540 nm femtosecond laser beam (with a repetition rate of 76 MHz and a pulse width of 200 fs) was tightly focused by a 40× oil immersion objective lens. The elliptical void channels with a cross section of 0.7–1.3 μm in the lateral diameter and an elongation in the focusing direction of approximately 50% could be achieved [108]. A 20-layer woodpile-type photonic crystal with a 1.7 μm layer spacing and a 1.8 μm in-plane channel spacing was obtained. The central wavelength of the photonic bandgap was 4.8 μm. A suppression of infrared transmission in the stacking direction by 85% was achieved [108].

More often than not, photonic crystals with higher-order photonic bandgaps are difficult to fabricate due to the requirement of highly correlated structural arrangements with a minimum of defects. Straub *et al.* reported the fabrication of highly correlated void-channel polymer photonic crystals by femtosecond laser induced microexplosion in solid polymer blocks [109]. The high-quality void channels, and various unit cell geometries make this method suitable for the generation of higher-order photonic bandgaps in the near-infrared range. The polymer shrinkage does not play a significant role in this method [109]. The number of higher-order photonic bandgaps could also be selected by changing the ratio between the layer spacing and the in-plane channel spacing of a woodpile structure. Moreover, the ratio of gap center to midgap width could be tuned by changing the filling ratio of the photonic crystals. They adopted Norland NOA63 to fabricate a three-dimensional photonic crystal, which consists of 20 layers of void channels stacked in woodpile structures, with layer spacing varying from 1.6 to 2.7 μm and in-plane channel spacing ranging from 1.1 to 2.6 μm [109].

However, even though the woodpile structures based on void channels in polymer can be used to achieve higher-order photonic bandgaps in the stack direction, it is difficult to obtain different types of lattice structure [109]. While it is possible to arrange void dots to fabricate arbitrary lattice structures. The face-centered cubic lattice structure has a square tightly packed arrangement in the [100] direction and a hexagonal tightly packed arrangement in the [111] direction. Zhou *et al.* reported the fabrication of three-dimensional photonic crystals with a face-centered cubic lattice in a solid polymer material by use of the ultrafast laser induced microexplosion method [110]. A solidified optical adhesive NOA63 was used as the polymer material in their experiment. A 740 nm femtosecond laser was focused by a 60× Olympus oil-immersion objective with a numerical aperture of 1.45 inside the polymer material. The power of the laser beam and the exposure time were set at 40 mW and 10 ms, respectively. Spherical void dots with a diameter of 1.2–1.8 μm were obtained, which is different from the elliptical cross section of the rods fabricated in the two-photon polymerization method [81]. The reason may be that in the case of microexplosion high pressure gas exists in a void and a spherical shape is a stable state [110]. Totally 22 rows of void dots were fabricated at different depths with a dot spacing and a row spacing of 6 and 3 μm, respectively [110]. They fabricated a three-dimensional NOA63 photonic crystal with a face-centered cubic lattice structure by stacking void dots layer by layer. Both the fundamental photonic bandgap as well as the second-order bandgap were observed.

3.3.3 The Holographic Lithography Method

For the fabrication of three-dimensional photonic crystals by use of the electron-beam lithography method, the required patterns are obtained on a photoresist by selective illumination, and subsequent transfer of the image to the photonic crystal film [70,111]. The holographic lithography method combines the holography technique and the photoinduced polymerization technique, which is considered to be a promising approach for the fabrication of three-dimensional microstructures [112]. The photonic crystal structures are defined by the isointensity surfaces of interference of several coherent laser beams. The mask is not indispensable for the fabrication of three-dimensional photonic crystals by using the holographic lithography method. Compared with other fabrication methods, the holographic lithography method possesses many remarkable advantages, such as high resolution, low cost, and easy-to-construct various lattice structures [113]. Moreover, it has been pointed out that photonic crystals with all 14 Bravais lattices, including the monoclinic, the triclinic, the orthorhombic, the tetragonal, the cubic, the trigonal, and the hexagonal lattices, can be generated by the holographic lithography method [114,115]. Furthermore, a specific type of three-dimensional photonic crystals, such as a Kagome-like lattice with a *p6mm* symmetry on the top surface and a chain-like lattice on a cleavage plane, can be fabricated by using a special single-exposure, two-set laser holographic lithography configuration [116]. The holographic lithography method may provide an approach to realize cheap and mass-production of three-dimensional photonic crystals. However, there may also exist a remarkable disadvantage: the pre-designed structural defects are difficult to introduce in three-dimensional photonic crystals fabricated by the holographic lithography method.

3.3.3.1 Without phase mask

The holographic lithography method can be adopted to fabricate three-dimensional photonic crystals without the help of a phase mask to adjust the phase properties of the incident laser beams. Two basic configurations of the holographic lithography method are the single exposure combined with multiple-beam interference, and the multiple-exposure combined with two-beam interference. The electric-field distribution of the interference pattern of N laser beams can be written as [117]

$$\vec{E}(\vec{r}) = \sum_{n=0}^{N-1} \vec{A}_n e^{i\vec{k}_n \cdot \vec{r}} \tag{3.2}$$

where $\vec{A}_n$ and $\vec{k}_n$ are the amplitude of the electric field and the wave vector of the incident laser beam. The intensity can be written as [117]

$$I(\vec{r}) \propto \sum_{n,m=0}^{N-1} (\vec{A}_n \cdot \vec{A}_m^*) e^{i(\vec{k}_n - \vec{k}_m)\vec{r}} \tag{3.3}$$

3.3.3.1.1 *Single exposure*

Theoretically, it has been pointed out that the interference patterns of four laser beams are possible to form any three-dimensional Bravais lattice [117]. Therefore, the minimum number of laser beams required to generate a three-dimensional photonic crystal is four for the holographic lithography method.

Four-beam interference

To form a three-dimensional diamond structure, the polarization and the incident angle of the four laser beams must be carefully designed to make the interference pattern take on the form [117]

$$I(\vec{r}) \propto 2\sqrt{3} - \cos\left(\frac{2\pi}{a}[111]\cdot\vec{r}\right) + \cos\left(\frac{2\pi}{a}[\bar{1}11]\cdot\vec{r}\right) + \cos\left(\frac{2\pi}{a}[1\bar{1}1]\cdot\vec{r}\right)$$

$$+ \cos\left(\frac{2\pi}{a}[11\bar{1}]\cdot\vec{r}\right) \tag{3.4}$$

where [111], [$\bar{1}$11], [1$\bar{1}$1], [11$\bar{1}$] are reciprocal lattice vectors. The constant-intensity surfaces of the interference pattern will form a three-dimensional photonic crystal with an $Fd\bar{3}m$ diamond symmetry structure. For this high-symmetry diamond-like three-dimensional photonic crystal, there exists a complete photonic bandgap between the second and third bands when the photonic crystal is formed by a dielectric material with a refractive index of 2.5.

To form a three-dimensional face-centered cubic lattice structure, the wave vector of the four laser beams can have the form [118]

$$\vec{k}_1 = \frac{\pi}{a}[2, 0, 1] \tag{3.5}$$

$$\vec{k}_2 = \frac{\pi}{a}[-2, 0, 1] \tag{3.6}$$

$$\vec{k}_3 = \frac{\pi}{a}[0, -2, -1] \tag{3.7}$$

$$\vec{k}_4 = \frac{\pi}{a}[0, 2, -1] \tag{3.8}$$

Escuti *et al.* reported the fabrication of a tunable face-centered-cubic lattice photonic crystal formed in holographic polymer dispersed liquid crystal materials by use of the holographic lithography method [118]. The holographic polymer dispersed liquid crystal materials are often composed of a photosensitive monomer, a photoinitiator, and liquid crystal [118]. Escuti *et al.* adopted the holographic polymer dispersed liquid crystal material composed of a blend of urethane acrylate trifunctional and hexafunctional oligomers mixed with a nematic liquid crystal BL038, and photoinitiators at a weight ratio of 50:36:14 [118]. The holographic polymer dispersed liquid crystal material was mixed into a homogeneous solution, sandwiched between planar substrates with transparent electrodes, and exposed to the face-centered-cubic lattice interference patterns formed by four coherent laser beams. The structured light leads to an intensity driven mass-transfer process that results in a composite of spherical liquid crystal domains with a diameter of about 50 nm arrayed at lattice nodes in a polymer matrix [118]. An applied electric field can realign the liquid crystal droplets and exert a modulation in the effective refractive index of the photonic crystal. Therefore, it is possible to realize an electrically tunable three-dimensional photonic crystal.

Five-beam interference

Three-dimensional photonic crystals with a tetragonal or cubic symmetry can be fabricated by use of the holographic lithography design of five-beam symmetric umbrella configuration [119]. The configuration has a central beam and four ambient beams symmetrically placed around the central one with the same apex angle. If we set that the central beam is along the z axis, the wave vectors of the five laser beams can be written as [120]

$$\vec{k}_0 = k(0, 0, 1) \tag{3.9}$$

$$\vec{k}_1 = k(-\sin\varphi, 0, \cos\varphi) \tag{3.10}$$

$$\vec{k}_2 = k(0, -\sin\varphi, \cos\varphi) \tag{3.11}$$

$$\vec{k}_3 = k(\sin\varphi, 0, \cos\varphi) \tag{3.12}$$

$$\vec{k}_4 = k(0, \sin\varphi, \cos\varphi) \tag{3.13}$$

where $k = 2\pi/\lambda$, λ is the wavelength of the laser source, φ is the angle between the side beams $\vec{k}_i$ and the central beam, $\vec{k}_0$ is the wave vector of the central beam, and $\vec{k}_i$ is the wave vector of side beams, $i = 1, 2, 3, 4$. The intensity distribution of the interference pattern can be expressed as [120]

$$I(\vec{r}) = \sum_{l,m} \vec{E}_l \cdot \vec{E}_m^* e^{-i(\vec{k}_l - \vec{k}_m)\cdot\vec{r} - i(\delta_l - \delta_m)} \tag{3.14}$$

where δ is the phase of the laser beams. When the apex angle $\varphi = 41.8°$, a three-dimensional photonic crystal with a woodpile structure can be fabricated. When the apex angle $\varphi = 70.53°$, a three-dimensional face-centered cubic photonic crystal can be obtained. When the apex angle $\varphi = 90°$, a three-dimensional bulk-centered cubic photonic crystal can be obtained.

Seven- and ten-beam interference

Alike quasicrystals of metal alloys, photonic quasicrystals have rotational symmetries and no translational symmetries [121]. It has been found that since photonic quasicrystals have higher rotational symmetries than ordinary periodic crystals, they have the optimal conditions for constructing a complete isotropic photonic bandgap [122]. Xu *et al.* reported the fabrication and optical characterization of icosahedral quasicrystals using a holographic lithography method for the visible range [123]. Quasicrystals can be classified as physical three-dimensional projects of higher-dimensional periodic structures or in terms of wave vectors in the reciprocal space corresponding to the diffraction patterns of quasicrystals [124]. The reciprocal space approach provides the basis for fabricating quasicrystals based on the laser interference holography method [124]. They adopted a seven-beam laser interference holography technique to fabricate the icosahedral patterns in SU-8 photoresist and high-resolution dichromate gelatin holographic plates. The seven-beam laser interference configuration consists of five evenly spaced side beams surrounding two opposite central beams with incidence angle φ [125,126]. Wang *et al.* reported the fabrication of three-dimensional photonic quasicrystal exhibiting the quasiperiodicity in the *x–y* plane and periodic along the *z* axis by use of 10-beam visible light holographic lithography [127]. The 10 beams can be represented by [126]

$$\vec{K}_n = k\left(\cos\frac{2n\pi}{5}\sin\varphi, \sin\frac{2n\pi}{5}\sin\varphi, \cos\varphi\right) \qquad (3.15)$$

$$\vec{K}'_n = k\left(\cos\frac{2n\pi}{5}\sin\varphi, \sin\frac{2n\pi}{5}\sin\varphi, -\cos\varphi\right) \qquad (3.16)$$

where $n = 0, 1, 2, 3, 4$, $k = (2\pi/\lambda)$, λ is the wavelength of the laser beam inside the photoresists, φ is the angle between each beam with the vertical *z* axis direction, and $\vec{K}'_n$ denotes the reflection images of the $\vec{K}_n$ from the *x–y* plane. The intensity distribution of the interference pattern can be written as [127]

$$I(\vec{r}) = \sum_{n,m}(\vec{E}_n e^{-i\vec{k}_n\cdot\vec{r}} + \vec{E}_n e^{-i\vec{k}'_n\cdot\vec{r}}) \cdot (\vec{E}^*_m e^{i\vec{k}_m\cdot\vec{r}} + \vec{E}^*_m e^{i\vec{k}'_m\cdot\vec{r}}) \qquad (3.17)$$

where n, m = 0, 1, 2, 3, 4. The interference pattern can be characterized by quasiperiodicity in the x–y plane and periodicity along the z axis direction with a period of $(\lambda/2)\cos\varphi$. When the five beams are placed symmetrically at 72° from neighboring beams around the z axis, three-dimensional photonic quasicrystals have a 10-fold-symmetry Penrose pattern in the x–y plane and a periodicity in the z axis direction with a period of $\cos\varphi(\lambda/2)$ [127].

Special design for wide photonic gap

To realize wide, or even complete photonic bandgap is a long-term dream of the study of photonic crystals. Ho *et al.* pointed out that photonic crystals exhibit complete photonic bandgaps in the woodpile and diamond structures with high dielectric contrast [128]. Photonic quasicrystals possessing higher order symmetry than the periodic structures are another possible way to achieve complete photonic bandgaps [129]. Huang *et al.* reported the fabrication of concentric spherical layer structures, a special kind of three-dimensional photonic crystals, exhibiting complete photonic bandgaps in the visible range in dichromate gelatin emulsions by use of the holographic lithography method [130]. They found that the complete bandgap does not originate from the high dielectric contrast but from the fact that the concentric spherical layer structure has the highest symmetry of all structures and is isotropic with equal spacing in all accessible directions. The concentric spherical pattern can be obtained by the interference between the light from a point source placed at the center of a spherical mirror and the reflected light from the spherical mirror. The arrangement of their experiment is as follows: a 488 nm laser beam was focused by a microscopic objective and then reflected by a spherical mirror placed at a distance equal to the radius of the curvature of the mirror from the focal point of the objective [130]. The divergent beam and the reflected beam would interfere to form a concentric spherical pattern that could be recorded as spherical layers with spacing given by $d = \lambda/2n$ in a plate of dichromate gelatin emulsions, placed in the position between the objective and the spherical mirror, where n = 1.57 is the refractive index of the dichromate gelatin emulsions, λ is the laser wavelength [130]. The curvature radius of the spherical layers varied by sandwiching the plate of dichromate gelatin emulsions placed between two glass spacers in contact with the objective and the planoconvex lens [130]. A complete photonic bandgap centered in the range of 500–600 nm can be obtained in this three-dimensional photonic crystal with a spherical layer structure and low refractive contrast. The dichromate gelatin has unique swelling properties, which makes dichromate gelatin very sensitive to the environmental humidity and temperature [131]. It is possible to fabricate planar structures with wide photonic bandgaps by controlling the swelling of dichromate gelatin [132,133].

It was also pointed out that three-dimensional spiral photonic crystals, breaking the chiral symmetry and resembling the classical diamond photonic crystal structure, possess a large complete photonic bandgap [134]. Pang *et al.* reported the fabrication of three-dimensional spiral photonic crystals with the holographic lithography method based on the interference of six equally spaced circumpolar linear polarized side beams and a circular polarized central beam [135]. The pitch and the separation of the spiral structure could vary by changing the angle between the side and the central beams. The wave vectors of six side beams can be written as [135].

$$\vec{k}_n = k\left(\cos\frac{2(n-1)\pi}{6}\sin\varphi, \sin\frac{2(n-1)\pi}{6}\sin\varphi, \cos\varphi\right) \qquad (3.18)$$

where $k = 2\pi/\lambda$ is the wave vector, n = 1, 2, 3, 4, 5, 6 and φ is the incident angle between the side beams with the vertical axis (z direction). The wave vector of the central beam is [135]

$$\vec{k}_0 = k(0, 0, 1) \qquad (3.19)$$

The intensity profile of the interference pattern can be written as [135]

$$I(\vec{r}) = \sum_{n,m} \vec{E}_n e^{-i\vec{k}_n \cdot \vec{r} - i\delta_n} \cdot \vec{E}_m^* e i\vec{k}_m \cdot \vec{r} + i\delta_m \qquad (3.20)$$

where n, m = 0–6, $\vec{E}_n$ and δ_n are the electric field and the phase of each beam, respectively. The angle between the electric field $\vec{E}_n$ and the incident plane is defined as ω_n. The chiral properties originate from the central beam with the electric field [135]

$$\vec{E}_0 = \frac{E_0}{\sqrt{2}}(1,+i,0) \quad \text{for the right-handed chirality} \qquad (3.21)$$

$$\vec{E}_0 = \frac{E_0}{\sqrt{2}}(1,-i,0) \quad \text{for the left-handed chirality} \qquad (3.22)$$

They adopted a 325 nm He-Cd laser, which was split into seven beams by a grating. The six side beams were linearly polarized light with $\omega_n = 90°$, placed symmetrically at 60° from neighboring beams and making an angle of 37.7° with the central axis [135]. The central beam was circularly polarized. SU8 was used as the photoresist.

3.3.3.1.2 *Multiple exposures*

The holographic lithography method based on the multiple-exposure of two-beam interference technique can also be adopted to fabricate three-dimensional photonic crystals. Different types of periodic lattice structures can be created depending on the number of exposure and the rotation angle of the sample for each exposure. The electric field of two laser beams with the same polarization and intensity can be written as [136].

$$E_{1\alpha\beta,2\alpha\beta} = E_0 \cos[kz\cos(\theta \mp \beta) \pm k\sin(\theta \mp \beta)(x\cos\alpha + y\sin\alpha) - \omega t]$$
$$(3.23)$$

where E_0 is the amplitude of the electric field of laser beams, k is the wave vector, θ is the semi-angle between two laser beams, and α and β are the rotation angles around the z-axis and y-axis directions, respectively. ω is the angular frequency of the laser beams. The intensity distribution of the interference pattern oriented at angle α and β can be written as [136]

$$I_{\alpha\beta} = 2E_0^2 \cos^2[k\sin\theta(z\sin\theta + (x\cos\alpha + y\sin\alpha)\cos\beta)] \qquad (3.24)$$

The period Λ of the formed structures is determined by [136]

$$\Lambda = \frac{\lambda}{2\sin\theta} \qquad (3.25)$$

where λ is the wavelength of the laser beams. For the multiple-exposure, the total intensity I_{total} of the interference pattern is the sum of each exposure at different angles α and β. Three-dimensional photonic crystals with different periodic structures can be formed by three-exposure two-beam interference. A three-dimensional periodic rectangular-square structure can be obtained by setting (α, β) = (90°, 0°), (0°, 45°), and (180°, 45°) for three exposures. A three-dimensional periodic rectangular-hexagonal structure can be obtained by setting (α, β) = (90°, 0°), (0°, 30°), and (180°, 30°) for three exposures. A three-dimensional periodic hexagonal-hexagonal structure can be obtained by setting (α, β) = (60°, 0°), (0°, 30°), and (180°, 30°) for three exposures [136].

More often than not, the alignment for each step is of great importance in the multistep lithographic process. Any new feature on the sample should be precisely located relative to the exciting ones [137]. Feigel *et al.* proposed a method of aligning a two-beam interference patterns to a previously fabricated reference grating to fabricate three-dimensional photonic crystals [114]. The proposed alignment method is based on the interference of scattered light by use of the moire effect [138]. The reference grating acts as an ordinary diffraction grating, leading to several diffraction orders from each beam. The alignment can be completed in two steps by observation of only two specific diffraction orders that correspond to different incident beams: the first step is to adjust the period and the direction of the fringes, and the second step is to fit the relative shift [137,138]. This simple moire-like alignment technique has a translation resolution of better than 20 nm and a rotation resolution of 45 μrad [138]. They fabricated a three-dimensional chalcogenide glass photonic crystal by use of this unique holographic lithography method.

3.3.3.2 With phase mask

The phase mask can be designed to realize various multiple-beam interference patterns. Three-dimensional photonic crystals having a body-centered tetragonal, a face-centered tetragonal, a body-centered cubic, a face-centered cubic, or a face-centered-orthorhombic structure can be fabricated by use of the properly designed phase mask [139,140].

3.3.3.2.1 *Using one-dimensional phase mask*

More often than not, three-dimensional photonic crystals can be fabricated by use of the holographic lithography method with a one-dimensional phase mask combined with two separate exposures [141,142]. For one-dimensional phase mask composed of surface relief grating in fused silica, three diffractive beams with zero order, +1 order, and –1 order can be generated. The interference of diffractive beams will generate a pattern with array of rods parallel to the grating direction, with the lattice constant a in the direction perpendicular to the grating direction equal to the grating period of the phase mask, and the lattice constant c in the direction of perpendicular to the grating plane depending on the ratio of grating period and the laser wavelength, respectively [139]. To fabricate a three-dimensional woodpile-type photonic crystal, the photoresist is first exposed to the interference patterns generated by the phase mask. Second, the phase mask is rotated by 90° and the photoresist is shifted by a distance of $c/2$ away from the phase mask. Then the photoresist is exposed by the interference pattern once again. A woodpile-type photonic crystal will be generated after being developed. When the ratio c/a equates 1, a body-centered cubic structure can be formed. When the ratio of c/a is close to 1, a body-centered tetragonal structure can be generated. When the ratio of c/a equates $\sqrt{2}$, a face-centered cubic structure can be formed. When the ratio of c/a is close to $\sqrt{2}$, a face-centered tetragonal structure can be generated [139].

3.3.3.2.2 *Using two-dimensional phase mask*

Three-dimensional photonic crystals can also be fabricated by the holographic lithography method with a two-dimensional phase mask combined with a single exposure [143]. Lin *et al.* proposed a design of two-dimensional phase mask for a single exposure fabrication of microsphere-type photonic crystals by use of the holographic lithography method [139]. The two-dimensional phase mask is composed of square lattice of arrays of grooves in fused silica, which can diffract the incident laser beam into one zeroth-order beam and four first-order beams [144]. The isointensity surfaces of the interference patterns of the five beams have the shape of microsphere. Chan *et al.* proposed an approach for fabricating three-dimensional photonic crystals with large photonic bandgaps using a single-exposure and single beam holographic lithography based on diffraction of light through a phase mask [145]. The phase mask consists of two orthogonally oriented binary gratings separated by a layer of homogeneous material. Illuminating the phase mask with a normally incident beam produces a five-beam diffraction pattern. A diamond-like (face-centered cubic) structure can be formed in a single exposure and single beam holographic lithography using this phase mask. Recently, Lin *et al.* reported the fabrication of two-layer integrated phase mask for single-beam and single-exposure fabrication of three-dimensional photonic crystal templates [146]. The photoresist used in their experiment were composed of dipentaerythritol penta/hexaacrylate (DPHPA) monomer, BL111 liquid crystal, photoinitiator rose Bengal, co-initiator *N*-phenyl glycine, and chain extender *N*-vinyl pyrrolidinone. The phase mask was fabricated by two exposures of 532 nm laser. Two interfering laser beams coming from the glass substrate side for the first exposure, and two interfering laser beams coming from the sample side, rotated by 90°, for the second exposure [146]. The phase mask consists of two-layers of orthogonally oriented gratings with high diffraction efficiency on a single substrate. The vertical spatial separation between two layers produces a phase shift among diffractive laser beams. The grating period was 1.06 μm. When a laser beam propagates through this phase mask, five low-order diffractive beams will be generated [145].

$$\vec{E}_1(\vec{r},t) = \vec{E}_1 \cos(\vec{k}_1 \cdot \vec{r} - \omega t + \delta_1) \qquad (3.26)$$

$$\vec{E}_2(\vec{r},t) = \vec{E}_2 \cos(\vec{k}_2 \cdot \vec{r} - \omega t + \delta_1) \qquad (3.27)$$

$$\vec{E}_3(\vec{r},t) = \vec{E}_3 \cos(\vec{k}_3 \cdot \vec{r} - \omega t + \delta_1) \qquad (3.28)$$

$$\vec{E}_4(\vec{r},t) = \vec{E}_4 \cos(\vec{k}_4 \cdot \vec{r} - \omega t + \delta_2) \qquad (3.29)$$

$$\vec{E}_5(\vec{r},t) = \vec{E}_5 \cos(\vec{k}_5 \cdot \vec{r} - \omega t + \delta_2) \qquad (3.30)$$

where $\vec{k}$, ω, and δ are the wave vector, the angular frequency, and the initial phase of the laser beam, respectively. E is the electric field intensity. If the five beams have the same initial phase, three-dimensional photonic crystals having a face-center-cubic or face-center-tetragonal structure can be fabricated [146]. They fabricated a three-dimensional face-center-tetragonal photonic crystal in SU-8 photoresist by use of a single 532 nm laser and a single exposure with two-layer phase mask. Chanda *et al.* reported the fabrication of diamond-like three-dimensional photonic crystal by use of the holographic lithography method based on the single exposure and through a tunable two-dimensional phase mask [147]. The two-dimensional phase mask was constructed by orthogonal combination of two gratings etched in two silica substrates, the structural parameters were identical for both gratings. By adjusting the gap between two gratings, a variable phase shift could be achieved to control the interlacing position of two orthogonally rotated periodic structures. The transition of formation of diamond-like woodpile structures having tetragonal symmetry to structures having body-centered-tetragonal symmetry and variations in between could be reached [147].

3.3.3.2.3 *Using multidimensional phase mask*

The multidimensional phase mask can also be adopted to fabricate three-dimensional photonic crystals with the holographic lithography method. Lin *et al.* reported the fabrication of multi-dimensional phase mask for the holographic fabrication of three-dimensional photonic crystals [148]. The photoresist used for the fabrication of the multidimensional phase mask consists of a mixture of dipentaerythritol penta/hexaacrylate (DPHPA) monomer, BL111 liquid crystal, photoinitiator rose bengal, coinitiator *N*-phenyl glycine, and chain extender *N*-vinyl pyrrolidinone with a weight percentage of 65%, 25%, 0.3%, 0.5%, and 9.2%, respectively [148]. Three micrometer-thick photoresist mixture films were spin-coated on transparent glass substrates. Then the photoresist mixture film was exposed to the interference pattern of two 532 nm laser beams. A periodic distribution of liquid-crystal-rich domain, corresponding to the dark regions of the interference pattern, and the polymer-rich grating structure could be formed based on the polymerization induced phase separation effect [149]. For the first exposure, two laser beams came from the substrate side, and only the photoresist near the substrate surface was polymerized because of the short exposure time of 0.5 s. While for the second exposure, the sample was rotated by 90° and two laser beams came from the upside. The phase mask consisting of two layers of orthogonally oriented gratings with a period of 1.06 μm could be formed [148]. The vertical spatial separation between two layers of gratings produces a phase difference among diffracted laser beams. Three-dimensional photonic crystals can be fabricated by exposing photoresist to the five beam interference patterns generated through the multidimensional phase mask.

Berger *et al.* reported the fabrication of another type of multi-dimensional phase mask for the holographic fabrication of three-dimensional photonic crystals [150]. The phase mask consists of three diffraction gratings with a grating period of 4 μm, etched in a Cr film deposited on quartz substrates. The direction between neighboring gratings was 120°. An additional triangular Cr layer was deposited in the middle of the three gratings in order to block the directly transmitting light. The intensity distribution of the interference pattern through the phase mask can be written as [150]

$$I(x,y) \propto 1 + 4\left[\cos\left(\pi\sqrt{3}\frac{x}{d}\right) + \cos\left(3\pi\frac{y}{d}\right)\right]\cos\left(\pi\sqrt{3}\frac{x}{d}\right) \qquad (3.31)$$

where d is grating period. It is very clear that the interference pattern was mainly determined by the structural parameters of the grating, and the usage of a monochromatic laser was not necessary [150]. They found that this phase mask could be used to fabricate three-dimensional photonic crystals. Divliansky *et al.* also reported a multi-dimensional phase mask for the fabrication of three-dimensional photonic crystals based on single-exposure holographic lithography method [151]. The mask had a central opening surrounded by three diffraction gratings oriented 120° relative to one another, so that the three first order diffracted beams and the nondiffracted laser beam give a three-dimensional spatial light intensity pattern [151]. The intensity distribution of the interference pattern can be written as [151]

$$\begin{aligned}
I = \frac{1}{2}E_0^2 + \frac{1}{2}E_1^2 + \frac{1}{2}E_2^2 + \frac{1}{2}E_3^2 + [E_0 \cdot E_1 \cos(k_0 r - k_1 r) \\
+ E_0 \cdot E_2 \cos(k_0 r - k_2 r) + E_0 \cdot E_3 \cos(k_0 r - k_3 r) \\
+ E_1 \cdot E_2 \cos(k_1 r - k_2 r) + E_1 \cdot E_3 \cos(k_1 r - k_3 r) \\
+ E_2 \cdot E_3 \cos(k_2 r - k_3 r)]
\end{aligned} \qquad (3.32)$$

where E_i and k_i are the amplitude of the electric field and the wave vector of i-th laser beam. The phase mask was fabricated in a thin polymer film on silicon dioxide substrate. The diameter of the central opening was 7.0 mm and the grating period was 1.0 μm [151].

3.3.3.2.4 *Using special phase mask*

Shao and Chen reported the surface-plasmon-assisted three-dimensional nanolithography (3D-SPAN) method for the fabrication of three-dimensional photonic crystals [152]. Owing to the momentum mismatch, surface plasmons could not be generated directly with light in a smooth surface of metal film. The periodic arrays of holes or apertures etched in the metal film could provide momentum compensation and allow the light to surface plasmon conversion [153]. They found that when apertures are positioned at resonant periodicities, the incident light could generate a strong near-field interference pattern on the opposite side of the apertures [152]. For a 365 nm incident light, the optical field intensity distribution after a 3D-SPAN mask with one-dimensional aluminum grating would show a structure with a triangular lattice and very strong intensity contrast when the aperture width and grating periodicity is 150 and 300 nm, respectively [152]. The interference pattern results from the phase modulation by the aluminum mask through the coupling of surface plasmons and light, which is different from the interference pattern created by the traditional holographic lithography method through maintaining phase differences among several laser beams by optical path differences [152]. Only the coherent polarized laser source could be used in the traditional holographic lithography method because of the need of controlling the optical path difference. An ultraviolet lamp could be used in the 3D-SPAN method because surface plasmons can be excited by a non-coherent light source. They used a 365 nm ultraviolet lamp as the light source. A 50 nm thick aluminum layer was deposited on quartz substrate with the electron-beam evaporation method. The focused ion beam etching method was used to pattern the one-dimensional aluminum mask with the aperture width of 150 nm and the grating periodicity of 300 nm. Then a 3 μm-thick SU-8 photoresist was spin-coated on the aluminum mask

[152]. The 3D-SPAN method ensures high intensity contrast for the interference patterns with the help of surface plasmons. The post-exposure bake process can be omitted [152]. The SEM image of three-dimensional SU-8 photonic crystal fabricated by the 3D-SPAN method is shown in Fig. 3.14 [152]. Moreover, line defects could be introduced in the three-dimensional photonic crystals through properly designing the mask, which overcomes the limitation of the traditional holographic lithography method that only defect-free photonic crystals can be fabricated. This method may find great potential applications in the fabrication of photonic crystal devices [154].

Figure 3.14   SEM image of the SU-8 three-dimensional photonic crystal fabricated by the 3D-SPAN method. The scale bar is 2.0 μm. Reprinted with permission from *Nano Lett.*, **6**, D. B. Shao, S. C. Chen, Direct patterning of three-dimensional periodic nanostructures by surface-plasmon-assisted nanolithography, 2279–2283. Copyright 2006, American Chemical Society.

3.3.4   The Self-Assembly Method

In 1983, Pieranski reported that the monodispersed colloidal nanospheres could spontaneously organize themselves into regular, periodic, and long-range-ordered crystalline structure, which is also called colloidal crystals or artificial opals, because of the static electric interactions between colloidal nanospheres [155]. This is the significant basis of the self-assembly method. Three-dimensional colloidal photonic crystals with a face-centered cubic or a cubic-close-packed structure over a large area of several cm^2 order could be achieved conveniently by use of the self-assembly method [156]. Compared with other methods, the self-assembly method may be the simplest and cheapest approach for the fabrication of three-dimensional photonic crystals. Various approaches have been developed based on self-assembling of colloidal nanospheres, such as the gravity sedimentation method [157–159], the electrophoretic deposition method [160–162], the electrostatic interaction method [163,164], and so on. However, the self-assembly method has remarkable disadvantages. First, intrinsically large concentration of lattice defects and stacking faults limits the practical applications of self-assembly method. Irregular and polycrystalline photonic crystal structures are often obtained when using the self-assembly method [165]. Second, polystyrene or silicon dioxide nanospheres are generally used in the self-assembly method. Owing to the relatively small refractive-index contrast, the photonic bandgap of a colloidal crystal is relatively narrow. Moreover, the lattice defects and stacking faults can easily close the gap by filling it with localized photonic states [166,167]. Therefore, it is still a great challenge to fabricate large-scale, high-quality, single-crystal three-dimensional colloidal photonic crystals.

3.3.4.1 Xia's method

Xia's method is a simple and efficient approach, allowing the rapid formation of colloidal photonic crystal structures over a large area of several cm^2 order [168]. The sample cell is constructed by a square frame of photoresist sandwiched between two glass substrates. The photoresist frame is etched with small channels in only one side, so as to retain the colloidal nanospheres while allowing the solvent to flow out. A small hole with a diameter of 3 mm was etched in the top glass substrate, and a glass tube with a diameter of 3 mm was attached to the hole by using an epoxy adhesive. The schematic structure of the sample cell is shown in Fig. 3.15. When the monodispersed aqueous solution of polystyrene nanospheres is injected into the sample cell, the external positive pressure of N$_2$ gas and the additional ultrasonic vibration help

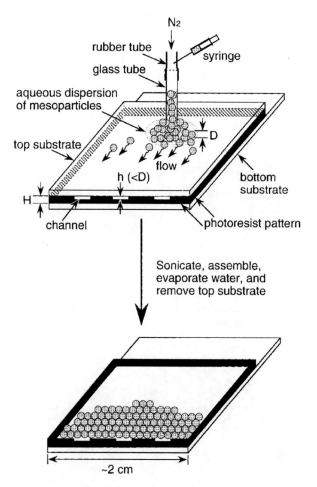

Figure 3.15 Schematic structure of the sample cell and the growth process of Xia's method. Reprinted with permission from *Langmuir*, **15**, S. H. Park, Y. N. Xia, Assembly of mesoscale particles over large areas and its application in fabricating tunable optical filters, 266–273. Copyright 1999, American Chemical Society.

to accelerate the formation of cubic-close-packed colloidal photonic crystal. Polystyrene colloidal photonic crystals with an area of over 1 cm^2 can be obtained in about 48 hours, with the (111) face parallel to the surface of the glass substrates. The SEM image of a fabricated polystyrene photonic crystal by use of Xia's method is shown in Fig. 3.16 [168]. The diameter of polystyrene nanospheres was 480 nm. Moreover, the thickness of the photonic crystal can be well controlled by adjusting the distance between two glass substrates [169]. Usually, the lattice structure of three-dimensional photonic crystal fabricated by Xia's method is a face-center cubic structure with (111) plane perpendicular to the growth direction. To avoid the expensive microfabrication requirement of the photoresist frame of Xia's method, Ni *et al.* developed an improved sample cell to fabricate three-dimensional colloidal photonic crystals [170]. The sample cell was made up of two glass substrates with a ring pad between them. The distance between two substrates was determined by the thickness of the ring pad. In order to let solvent flow out of the sample cell, some small ducts were cut into the ring. The suspension of colloidal nanospheres was injected through a small glass tube connected to the sample cell. It took 3~4 weeks to grow a three-dimensional photonic crystal with a size of 1 cm^2 and a thickness of 100 μm [170].

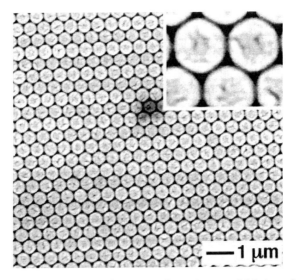

Figure 3.16 SEM image of polystyrene photonic crystal fabricated by use of Xia's method. The diameter of polystyrene was 480 nm. Reprinted with permission from *Adv. Mater.,* **10**, S. H. Park, D. Qin, Y. N. Xia, Crystallization of mesoscale particles over large areas, 1028–1032. Copyright © 1998 WILEY-VCH Verlag GmbH, Weinheim.

One limitation of the three-dimensional photonic crystals fabricated by use of Xia's sample cell with flat substrates is that almost all optical measurements are limited to the [111] direction of the face-center cubic lattice, also called the *L* point of its reciprocal lattice [171]. Three-dimensional photonic crystals with a crystallographic direction other than [111] also have very important applications in practice. Various approaches have been proposed to control the growth direction of colloidal crystals, such as by using an external electrical, magnetic, or optical field [172–174]. Adopting substrates with appropriate arrays of relief structures on surfaces as templates is also a promising approach to control the crystallographic orientation of the photonic crystal [175,176]. The mechanism of this template-directed growth process is analogous to the mesoscopic epitaxial growth at an atomic scale. Ozin and Xia also presented a kind of templates composed of arrays of square pyramidal pits or V-shaped grooves etched in substrates, which could fabricate colloidal photonic crystals with their (100) planes oriented parallel to the substrates [177,178]. The pyramidal pits and V-shaped grooves can be prepared in the surfaces of the substrates by use of conventional photolithography, followed by a wet etching process [179]. The growth mechanism is as follows: Due to the concentration gradient across the substrate surface originating from the solvent evaporation of colloidal solution, colloidal nanospheres were confined within the template pits and were assembled into pyramid-shaped cubic-close-packed lattices with their (100) planes parallel to the surface of substrates. After the pits were completely filled, the upper layer could form a long-range-ordered (100) plane if the period of templates was a multiple of the diameter of colloidal nanospheres. The width of the ridges on top of the template was much smaller than the diameter of colloidal nanospheres, so as to eliminate the nucleation and growth of (111) planes on the ridge surfaces. Then the colloidal crystal could epitaxially grow without changing its crystallographic orientation [179]. Figure 3.17 shows the SEM image of a (100)-oriented colloidal crystal fabricated by template-directed growth of polystyrene nanospheres with a diameter of 1.0 μm by use of substrates etched with two-dimensional arrays of square pyramidal pits with a width of 4.0 μm [179].

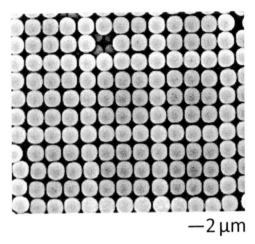

$-2\,\mu m$

Figure 3.17 SEM image of a (100)-oriented colloidal crystal fabricated by templating polystyrene microspheres against two-dimensional arrays of square pyramidal pits with a width of 4.0 µm. Reprinted with permission from *Langmuir*, **19**, Y. D. Yin, *et al.*, Template-directed growth of (100)-oriented colloidal crystals, 622–631. Copyright 2003, American Chemical Society.

3.3.4.2 The vertical deposition method

3.3.4.2.1 *Mechanisms of the vertical deposition method*

In 1992, Denko *et al.* reported that two-dimensional patterns of periodic arrays of polystyrene nanospheres could be formed in a horizontal glass substrate immersed in the colloidal suspension of polystyrene nanospheres [180]. He found that the major factors governing the ordering process were the attractive capillary forces originating from the menisci formed around the polystyrene nanospheres, and the convective flux transporting polystyrene nanospheres from the colloidal suspension toward the ordered region arising from the solvent evaporation. The growth process of the two-dimensional colloidal photonic crystals could be controlled to some extent [181]. Based on these foundations, Dimitrov and Nagayama developed a vertically self-assembling growth method for the fabrication of three-dimensional colloidal photonic crystals, the method also being referred to as the vertical deposition method [182]. When a clean and hydrophilic substrate is vertically dipped into a colloidal suspension containing monodispersed polystyrene nanospheres and water, a meniscus region is formed between the substrate and the horizontal surface of the suspension due to the wetting action of water. Colloidal nanospheres in the meniscus region are forced to construct an ordered face-centered cubic structure on the surface of the substrate by the lateral capillary forces and the hydrodynamic pressure of the water influx as the meniscus is slowly swept downwards across the substrate with the evaporation of solvent [183,184]. The schematic sample cell and growth mechanism is shown in Fig. 3.18. The vertical deposition method possesses some remarkable advantages: First, the growth process of the photonic crystal can be controlled in part by adjusting the external parameters, such as the surrounding temperature and the relative humidity. So, there are fewer lattice defects in photonic crystals fabricated by the vertical deposition method compared with the gravity sedimentation method. Under appropriate growth conditions, high-quality single-crystal three-dimensional photonic crystals could be fabricated by use of the vertical deposition method. Second, tailoring of the lattice structure, orientation, and size of the photonic crystal can also be achieved by sedimentation of the colloidal nanospheres onto patterned substrates [185]. Moreover, the thickness of the photonic crystal can be controlled by changing the suspension concentration or lifting the substrate [186,187].

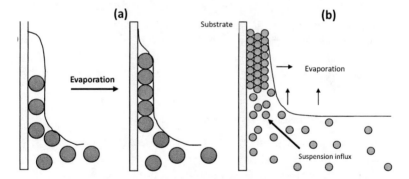

Figure 3.18 Schematic structure of the growth mechanism of vertical deposition method. (a) Formation of the leading edge of the lattice. (b) Growth of the lattice.

More often than not, the growth process of the vertical deposition method can be divided into two stages [182]: First, colloidal nanospheres are transported from the bulk suspension to the meniscus site by the convective flux caused by the solvent evaporation of the colloidal suspension. Second, the periodic lattice is formed in the meniscus and continuously grows along with the meniscus sweeping downward the substrate due to the solvent evaporation. Therefore, the meniscus is also called the leading edge of the lattice. The mechanism of the lattice growth dynamics can be analyzed as follows: owing to the solvent evaporation of the colloidal suspension, a pressure gradient is formed from the colloidal suspension toward the meniscus site, which leads to an influx of colloidal microspheres from the bulk suspension to the meniscus. This is the origination of the forces transporting the colloidal nanospheres for the meniscus. The origination of the formation and growth of the periodic lattice of colloidal nanospheres lies in the interactions of the pressure gradient caused by the solvent evaporation of the meniscus, the lateral capillary, and the electrostatic forces of colloidal microspheres [182].

3.3.4.2.2 *How to achieve high-quality colloidal photonic crystal?*

To achieve a high-quality colloidal photonic crystal, the growth parameters, including the environmental temperature, the diameter of the colloidal microspheres, and the surrounding humidity, must be carefully controlled. The surrounding humidity influences the evaporation rate of the colloidal suspension remarkably. A proper value of the surrounding humidity is of great importance for the growth process of colloidal photonic crystal. Usually, the surrounding humidity is set in a range from 30% to 50% in the experiment.

Proper environmental temperature

The environmental temperature affects the evaporation rate of the colloidal suspension very seriously. Theoretical calculations have indicated that the thermodynamically stable crystalline phase is face-centered cubic stacking [183]. However, the difference of the Gibbs free energy between the hexagonal close-stacked phase and the face-centered cubic phase is very small, only $0.05RT$ per mol, where R is the universal gas constant and T is temperature [183]. Various structure defects are very easy to introduce during the growth process of the colloidal photonic crystal. Therefore, large-area, single-crystal colloidal photonic crystals are difficult to obtain. For the fabrication of a high-quality single-crystal colloidal photonic crystal, the growth rate of the periodic lattice should be slow and strictly controlled, because the surrounding temperature influences the solvent evaporation rate, and subsequently the growth rate of the periodic lattice. Ye *et al.* studied the effect of the evaporation temperature on the crystalline quality of colloidal photonic crystals [188]. They found that the increase of the evaporation temperature from 45°C to 55°C helped achieve a balance between the lattice growth rate and the microsphere transportation rate. The fabricated colloidal photonic crystals possess few structure defects. When the evaporation temperature was higher than 55°C, a large number of structure defects appeared in the photonic crystal. High evaporation temperature results in a fast lattice growth rate and nanosphere transportation rate, so that the nanospheres in the meniscus do not have enough time to shift to the favorite lattice sites [188].

Proper colloidal nanosphere diameter

The primary requirement of the vertical deposition technique is that the velocity of the solvent evaporation must match the sedimentation velocity of colloidal nanospheres. So it has been believed that this technique is limited to nanospheres with a diameter of about 400 nm. Many improved methods have been proposed to fabricate the three-dimensional colloidal photonic crystal with a nanosphere diameter of much less or larger than 400 nm.

For colloidal microspheres with a small diameter of less than 300 nm, many irregularly placed stacks of nanospheres are easily formed on the surface of the substrate due to the faster transportation rate of the colloidal nanospheres than the growth rate of the crystal lattice. This difficulty can be overcome partly by lifting the substrate out of the colloidal suspension at a speed equal to the growth rate of the crystal lattice [187]. Zhou *et al.* proposed a flow-controlled vertical deposition method based on the Langmuir–Blodgett technique [189]. The sample cell was connected with a microtube, so that the suspension could be withdrawn from the sample cell by using a variable-flow peristaltic pump. The dropping velocity of the suspension surface can be controlled by adjusting the flow rate of the pump [189]. Colloidal nanospheres with a very large range of diameter, changing from 0.5 to 1.5 μm, can be adopted to fabricate high-quality three-dimensional colloidal photonic crystal with a controllable and uniform thickness by using this method. Meng *et al.* also proposed a simple and effective approach to improve the traditional vertical deposition method for the fabrication of high-quality three-dimensional colloidal photonic crystal made of polystyrene nanospheres with a diameter of less than 300 nm [190]. The self-assembling growth process is as follows: a clean microslide was dipped vertically into a vial containing a colloidal suspension made from monodispersed polystyrene nanospheres and water. Then, the vial was covered with a plastic film punched through with a few fine holes to keep the airflow out. The vial was kept in a constant-temperature and -humidity oven. The temperature and the humidity were set to be 50°C and 30%, respectively. The concentration of the suspension was 0.036%. After about one week, high-quality polystyrene colloidal photonic crystals could be obtained.

For the colloidal photonic crystals applied in the optical communication range, i.e., the wavelength of 1300 or 1550 nm, the colloidal nanospheres with a diameter of about 800 nm is needed. For colloidal nanospheres with such a large diameter, the velocity of the sedimentation caused by the gravity may be faster than the growth velocity of the periodic lattice. Therefore, a temperature gradient must be added to induce a convective flow, which can resist the sedimentation of the nanospheres and provide a continuous flow of nanospheres to the leading edge of the lattice to maintain the continuous growth of the crystal lattice [191]. Vlasov *et al.* reported the fabrication of large-scale, and single-crystal SiO_2 colloidal photonic crystals by use of the vertical deposition method combined with a temperature gradient [192]. They used a silicon wafer as the substrate, which was vertically placed in a vial containing an ethanolic suspension of silicon dioxide nanospheres with a diameter of 855 nm and a concentration of 1%. The convective flow was achieved by placing a temperature gradient across the vial, from 80°C at the bottom to 65°C near the top [192]. The SEM image obtained by the three-dimensional SiO_2 colloidal photonic crystal is shown in Fig. 3.19 [192]. The colloidal photonic crystal has a defect density of 1% stacking faults and 10^{-3} order point defects per unit cell, much lower than that of the photonic crystal obtained by use of the gravity sedimentation method, which usually being that of 20% stacking faults and 10^{-2} order point defects per unit cell [156]. Moreover, the colloidal photonic crystal was up to 20-layer thick and had a large size of several centimeter order, with the single crystalline areas of 1 mm~1 cm order. The obtained single crystalline areas were 10~100 times larger than that in the best photonic crystal obtained by use of the gravity sedimentation method [192]. The colloidal photonic crystal had a unity reflectance around a wavelength of 1300 nm.

Figure 3.19 SEM image of the three-dimensional SiO_2 colloidal photonic crystal fabricated by the vertical deposition technique. Reprinted by permission from Macmillan Publishers Ltd: *Nature*, **414**(6861), Y. A. Vlasov, *et al.*, On-chip natural assembly of silicon photonic bandgap crystals, 289–293, copyright (2001).

3.3.4.3 Fabrication of inverse opal

The refractive index of polystyrene and silicon dioxide is 1.59 and 1.45, respectively. The small refractive index contrast will make the photonic bandgap very narrow. Titania (TiO_2) and silicon possess relatively high refractive index. Titania or silicon can be infiltrated in the air voids of the colloidal photonic crystal template. Accordingly, a high-refractive-index contrast inverse opal can be obtained when the template is removed. This provides a promising approach to fabricate three-dimensional photonic crystals with complete photonic bandgap. The traditional procedure for the fabrication of an inverse opal is as follows: First, fabricating a high-quality three-dimensional colloidal photonic crystal template. Then, infiltrating silicon or titania in the air voids of the photonic crystal template. Finally, removing the photonic crystal template and an inverse opal can be obtained. However, high-quality three-dimensional photonic crystal templates are not easy to fabricate on the basis of the self-assembly method. Complete infilling of the air voids with titania or silicon is also not easy to achieve due to small size of air voids, which is often in several-nanometer scale. Moreover, lattice shrinkage and crack may occur after removal of the photonic crystal template. Therefore, it is still a great challenge to fabricate high-quality and large-size inverse opal.

Wijnhoven *et al.* reported the fabrication of three-dimensional photonic crystal composed of air spheres in titania with radii between 120 and 1000 nm, achieved by filling titania in the voids of a polystyrene opal by use of the liquid-phase chemical reaction method [193]. The three-dimensional photonic crystal template were fabricated by the self-assembly method based on the gravity sedimentation of colloidal nanospheres. The colloidal suspensions, composed of monodisperse polystyrene nanospheres with a radius between 180 and 1460 nm in water, were loaded in long and flat glass capillaries. The thickness and the width of glass capillaries were 0.3 and 3 mm, respectively. Dense and face-centered cubic colloidal crystals were grown by sedimentation of the polystyrene nanospheres at an acceleration of about 400 g. The opal was obtained from the colloidal crystals by evaporating the water from the capillary. After the opal was obtained, a liquid-phase chemical reaction method was used to fill the voids of the opal with titania. The precursor, a solution of tetrapropoxy-titane in ethanol, penetrated the voids in the opal by capillary forces. Then the tetrapropoxy-titane reacted with water to form titania. Finally, the sample was slowly heated to 450°C to remove polystyrene and the inverted opal was obtained [193].

Meng *et al.* proposed a simple idea of infilling ultrafine nanoparticles into the voids of photonic crystal templates by means of capillary forces [194]. The colloidal suspension was composed of polystyrene microspheres, titania and silica ultrafine particles. The average diameters of the silica and titania ultrafine nanoparticles were 7 and 13 nm, respectively. A clean substrate was vertically dipped into the glass vial containing the colloidal suspension. Then the glass vial was covered with a plastic film punched through with several holes to keep out any external airflow. The glass vial was placed in a constant-temperature and -humidity chamber, the temperature and humidity of which were set at 50°C and 30%, respectively. After 24 hours, a three-dimensional photonic crystal with a size of 0.5 cm² and a thickness of about 6 μm could be obtained. The removal of the photonic crystal template was realized by calcinations. The sample was slowly heated to 450°C for 7 hours and high-quality inverse opal was obtained.

Ni *et al.* also proposed a simple approach to fabricate titania inverse opal [170]. The polystyrene template was prepared in the sample cell made up of two optical plates with a ring pad between them. A mixture of $Ti(OC_2H_5)_4$ and ethanol with a volume ratio of 1:100 was used as the infiltration liquid. The liquid penetrated the air voids due to the strong capillarity. The sample was heated to 50°C to accelerate the chemical reaction. After the polystyrene template was removed by heating the sample, the high-quality titania inverse opal was obtained. The diameter of air spheres was 170 nm. Blanco *et al.* also reported a silicon inverse opal with a complete three-dimensional bandgap near 1.5 micrometers [195]. Silicon was infiltrated in the air voids of silica opal template by means of chemical vapor deposition method. The diameter of the silica nanospheres was 500 nm. The infiltration degree up to 100% was achieved. After infiltration, the sample was heated to 600°C to improve the silicon crystallization. The silica template was removed by use of a fluoride-based etching method.

References

1. Z. M. Jiang, B. Shi, D. T. Zhao, J. Liu, and X. Wang, "Silicon-based photonic crystal heterostructure," *Appl. Phys. Lett.* **79**, 3395–3397 (2001).

2. H. Y. Lee, H. Makino, T. Yao, and A. Tanaka, "Si-based omnidirectional reflector and transmission filter optimized at a wavelength of 1.55 μm," *Appl. Phys. Lett.* **81**, 4502–4504 (2002).

3. Y. Park, Y. G. Roh, C. O. Cho, H. Jeon, M. G. Sung, and J. C. Woo, "GaAs-based near-infrared omnidirectional reflector," *Appl. Phys. Lett.* **82**, 2770–2772 (2003).

4. T. Baba, H. Makino, T. Mori, T. Hanada, T. Yao, and H. Y. Lee, "Experimental demonstration of Fano-type resonance in photoluminescence of ZnS: Mn/SiO$_2$ one-dimensional photonic crystals," *Appl. Phys. Lett.* **87**, 171106 (2005).

5. C. Becker, M. Wegener, S. Wong, and G. V. Freymann, "Phase-matched nondegenerate four-wave mixing in one-dimensional photonic crystals," *Appl. Phys. Lett.* **89**, 131122 (2006).

6. X. C. Sun, J. J. Hu, C. Y. Hong, J. F. Viens, X. M. Duan, R. Das, A. M. Agarwal, and L. C. Kimerling, "Multispectral pixel performance using a one-dimensional photonic crystal design," *Appl. Phys. Lett.* **89**, 223522 (2006).

7. M. Patrini, M. Galli, M. Belotti, L. C. Andreani, G. Guizzetti, G. Pucker, A. Lui, and P. Bellutti, and L. Pavesi, "Optical response of one-dimensional $(Si/SiO_2)_m$ photonic crystal," *J. Appl. Phys.* **92**, 1816–1820 (2002).

8. F. Osullivan, I. Celanovic, N. Jovanovic, J. Kassakian, S. Akiyama, and K. Wada, "Optical characteristics of one-dimensional Si/SiO_2 photonic crystals for thermophotovoltaic applications," *J. Appl. Phys.* **97**, 033529 (2005).

9. L. Chen, Y. Suzuki, and G. E. Kohnke, "Integrated platform for silicon photonic crystal devices at near-infrared wavelengths," *Appl. Phys. Lett.* **80**, 1514–1516 (2002).

10. V. Agarwal and A. D. Rio, "Tailoring the photonic band gap of a porous silicon dielectric mirror," *Appl. Phys. Lett.* **82**, 1512–1514 (2003).

11. A. Bruyant, G. Lerondel, P. J. Reece, and M. Gal, "All-silicon omnidirectional mirrors based on one-dimensional photonic crystals," *Appl. Phys. Lett.* **82**, 3227–3229 (2003).

12. I. V. Soboleva, E. M. Murchikova, A. A. Fedyanin, and O. A. Aktsipetrov, "Second- and third- harmonic generation in birefringent photonic crystals and microcavities based on anisotropic porous silicon," *Appl. Phys. Lett.* **87**, 241110 (2005).

13. P. Murzyn, A. Z. G. Deniz, D. O. Kundys, A. M. Fox, J. P. R. Wells, D. M. Whittaker, M. S. Skolnick, T. F. Krauss, and J. S. Roberts, "Control of the nonlinear carrier response time of AlGaAs photonic crystal waveguides by sample design," *Appl. Phys. Lett.* **88**, 141104 (2006).

14. J. Semmel, L. Nahle, S. Hofling, and A. Forchel, "Edge emitting quantum cascade microlasers on InP with deeply etched one-dimensional photonic crystals," *Appl. Phys. Lett.* **91**, 071104 (2007).

15. E. Descrovi, F. Frascella, B. Sciacca, F. Geobaldo, L. Dominici, and F. Michelotti, "Coupling of surface waves in highly defined one-dimensional porous silicon photonic crystals for gas sensing applications," *Appl. Phys. Lett.* **91**, 241109 (2007).

16. H. Park, J. H. Dickerson, and S. M. Weiss, "Spatially localized one dimensional porous silicon photonic crystals," *Appl. Phys. Lett.* **92**, 011113 (2008).

17. K. M. Chen, A. W. Sparks, H. C. Luan, D. R. Lim, K. Wada, and L. C. Kimerling, "SiO$_2$/TiO$_2$ omnidirectional reflector and microcavity resonator via the sol-gel method," *Appl. Phys. Lett.* **75**, 3805–3807 (1999).

18. T. Komikado, S. Yoshida, and S. Umegaki, "Surface-emitting distributed-feedback dye laser of a polymeric multiplayer fabricated by spin coating," *Appl. Phys. Lett.* **89**, 061123 (2006).

19. J. Yoon, W. Lee, J. M. Caruge, M. Bawendi, E. L. Thomas, S. Kooi, and P. N. Prasad, "Defect-mode mirrorless lasing in dye-doped organic/inorganic hybrid one-dimensional photonic crystal," *Appl. Phys. Lett.* **88**, 091102 (2006).

20. G. Barillaro, A. Diligenti, M. Benedetti, and S. Merlo, "Silicon micro-machined periodic structures for optical applications at λ = 1.55 µm," *Appl. Phys. Lett.* **89**, 151110 (2006).

21. G. Barillaro, A. Nannini, and M. Piotto, "Electrochemical etching in HF solution for silicon micromachining," *Sens. Actuators A* **102**, 195–201 (2002).

22. G. Barillaro, A. Diligenti, A. Nannini, and G. Pennelli, "A thick silicon dioxide fabrication process based on electrochemical trenching of silicon," *Sens. Actuators A* **107**, 279–284 (2003).

23. I. J. K. Filipek, F. Duerinckx, E. V. Kerschaver, K. V. Nieuwenhuysen, G. Beaucarne, and J. Poortmans, "Chirped porous silicon reflectors for thin-film epitaxial silicon solar cells," *J. Appl. Phys.* **104**, 073529 (2008).

24. R. Szipocs, K. Ferencz, C. Spielmann, and F, Krausz, "Chirped multiplayer coatings for broadband dispersion control in femtosecond lasers", *Opt. Lett.* **19**, 201–203 (1994).

25. E. Yablonovitch, "Engineered omnidirectional external-reflectivity spectra from one-dimensional layered interference filters," *Opt. Lett.* **19**, 1648–1649 (1994).

26. M. Campbell, D. N. Sharp, M. T. Harrison, R. G. Denning, and A. J. Turberfield, "Fabrication of photonic crystals for the visible spectrum by holographic lithography," *Nature* **404**, 53–56 (2000).

27. L. Wu, Y. Zhong, C. T. Chen, K. S. Wong, and G. P. Wang, "Fabrication of large area two- and three-dimensional polymer photonic crystals using single refracting prism holographic lithography," *Appl. Phys. Lett.* **86**, 241102 (2005).

28. J. T. Li, Y. K. Liu, X. S. Xie, P. Q. Zhang, B. Liang, L. Yan, J. Y. Zhou, G. Kurizki, D. Jacobs, K. S. Wong, and Y. C. Zhong, "Fabrication of photonic crystals with functional defects by one-step holographic lithography," *Opt. Express* **16**, 12899–12904 (2008).

29. T. Kondo, S. Juodkazis, V. Mizeikis, S. Matsuo, and H. Misawa, "Fabrication of three-dimensional periodic microstructures in photoresist SU-8 by phase-controlled holographic lithography," *New J. Phys.* **8**, 250 (2006).

30. X. S. Xie, M. Li, J. Guo, B. Liang, Z. X. Wang, A. Sinistkii, Y. Xiang, and J. Y. Zhou, "Phase manipulated multi-beam holographic lithography for tunable optical lattices," *Opt. Express* **15**, 7032–7037 (2007).

31. G. Zito, B. Piccirillo, E. Santamato, A. Marino, V. Tkachenko, and G. Abbate, "Two-dimensional photonic quasicrystals by single beam computer-generated holography," *Opt. Express* **16**, 5164–5170 (2008).

32. L. Z. Cai, X. L. Yang, and Y. R. Wang, "Formation of a microfiber bundle by interference of three noncoplanar beams," *Opt. Lett.* **26**, 1858–1860 (2001).

33. J. Tian, M. Yan, M. Qiu, C. G. Ribbing, Y. Z. Liu, D. Z. Zhang, and Z. Y. Li, "Direct characterization of focusing light by negative refraction in a photonic crystal flat lens," *Appl. Phys. Lett.* **93**, 191114 (2008).

34. G. Vecchi, F. Raineri, I. Sagnes, K. H. Lee, S. Guilet, L. L. Gratiet, A. Talneau, A. Levenson, and R. Raj, "High contrast reflection modulation near 1.55 μm in InP 2D photonic crystals on silicon wafer," *Opt. Express* **15**, 1254–1260 (2007).

35. A. R. Alija, L. J. Martinez, P. A. Postigo, J. S. Dehesa, M. Galli, A. Politi, M. Patrini, L. C. Andreani, C. Seassal, and P. Viktorovitch, "Theoretical and experimental study of the Suzuki-phase photonic crystal lattice by angle-resolved photoluminescence spectroscopy," *Opt. Express* **15**, 704–713 (2007).

36. A. Lavrinenko, P. I. Borel, L. H. Frandsen, M. Thorhauge, A. Harpoth, M. Kristensen, T. Niemi, and H. M. H. Chong, "Comprehensive FDTD modeling of photonic crystal waveguide components," *Opt. Express* **12**, 234–248 (2004).

37. K. J. Byeon, S. Y. Hwang, and H. Lee, "Fabrication of two-dimensional photonic crystal patterns on GaN-based light-emitting diodes using thermally curable monomer-based nanoimprint lithography," *Appl. Phys. Lett.* **91**, 091106 (2007).

38. G. Zito, B. Piccirillo, E. Santamato, A. Marino, V. Tkachenko, and G. Abbate, "Two-dimensional photonic quasicrystals by single beam computer-generated holography," *Opt. Express* **16**, 5164–5170 (2008).

39. L. J. Wu, Y. C. Zhong, K. S. Wong, G. P. Wang, and L. Yuan, "Fabrication of hetero-binary and honeycomb photonic crystals by one-step holographic lithography," *Appl. Phys. Lett.* **88**, 091115 (2006).

40. C. Lu, X. K. Hu, I. V. Mitchell, and R. H. Lipson, "Diffraction element assisted lithography: Pattern control for photonic crystal fabrication," *Appl. Phys. Lett.* **86**, 193110 (2005).

41. Y. Yang and G. P. Wang, "Realization of periodic and quasiperiodic microstructures with sub-diffraction-limit feature sizes by far-field holographic lithography," *Appl. Phys. Lett.* **89**, 111104 (2006).

42. S. Assefa, P. T. Rakich, P. Bienstman, S. G. Johnson, G. S. Petrich, J. D. Joannopoulos, L. A. Kolodziejski, E. P. Ippen, and H. I. Smith, "Guiding 1.5 μm light in photonic crystals based on dielectric rods," *Appl. Phys. Lett.* **85**, 6110–6112 (2004).

43. H. K. Fu, Y. F. Chen, R. L. Chern, and C. C. Chang, "Connected hexagonal photonic crystals with largest full band gap," *Opt. Express* **13**, 7854–7860 (2005).

44. V. A. Parekh, A. Ruiz, P. Ruchhoeft, S. Brankovic, and D. Litvinov, "Closed-packed noncircular nanodevice pattern generation by self-limiting ion-mill process," *Nano Lett.* **7**, 3246–3248 (2007).

45. M. Sun, J. Tian, S. Z. Han, Z. Y. Li, B. Y. Cheng, D. Z. Zhang, A. Z. Jin, and H. F. Yang, "Effect of the subwavelength hole symmetry on the enhanced optical transmission through metallic films," *J. Appl. Phys.* **100**, 024320 (2006).

46. N. W. Liu, A. Datta, C. Y. Liu, and Y. L. Wang, "High-speed focused-ion-beam patterning for guiding the growth of anodic alumina nanochannel arrays," *Appl. Phys. Lett.* **82**, 1281–1283 (2003).

47. R. Li and M. Levy, "Bragg grating magnetic photonic crystal waveguides," *Appl. Phys. Lett.* **86**, 251102 (2005).

48. N. A. Paraire, P. G. Filloux, and K. Wang, "Patterning and characterization of 2D photonic crystals fabricated by focused ion beam etching of multiplayer membranes," *Nanotechnology* **15**, 341–346 (2004).

49. A. Chelnokov, K. Wang, S. Roweon, P. Garoche, and J. M. Lourtioz, "Near-infrared Yablonovite-like photonic crystals by focused-ion-beam etching of macroporous silicon," *Appl. Phys. Lett.* **77**, 2943–2945 (2000).

50. Z. L. Wang, C. Z. Gu, J. J. Li, and Z. Cui, "Anovel method for making high aspect ratio solid diamond tips," *Microelectron. Eng.* **78–79**, 353–358 (2005).

51. Z. L. Wang, Q. Wang, H. J. Li, J. J. Li, P. Xu, Q. Luo, A. Z. Jin, H. F. Yang, and C. Z. Gu, "The field emission properties of high aspect ratio diamond nanocone arrays fabricated by focused ion beam milling," *Sci. Technol. Adv. Mater.* **6**, 799–803 (2005).

52. J. S. Gwag, M. Ohe, M. Yoneya, H. Yokoyama, H. Satou, and S. Itami, "Advanced nanoimprint lithography using a graded functional imprinting material tailored for liquid crystal alignment," *J. Appl. Phys.* **102**, 063501 (2007).

53. T. Martensson, P. Carlberg, M. Borgstrom, L. Montelius, W. Seifert, and L. Samuelson, "Nanowire arrays defined by nanoimprint lithography," *Nano Lett.* **4**, 699–702 (2004).

54. U. Gruning, V. Lehmann, S. Ottow, and K. Busch, "Macroporous silicon with a complete two-dimensional photonic band gap centered at 5 μm," *Appl. Phys. Lett.* **68**, 747–749 (1996).

55. V. Lehmann, "The physics of macropore formation in low doped n-type silicon," *J. Electrochem. Soc.* **140**, 2836–2843 (1993).

56. G. Zito, B. Piccirillo, E. Santamato, A. Marino, V. Tkachenko, and G. Abbate, "Two-dimensional photonic quasicrystals by single beam computer-generated holography," *Opt. Express* **16**, 5164–5170 (2008).

57. Y. Yang, Q. Li, and G. P. Wang, "Fabrication of periodic complex photonic crystals constructed with a portion of photonic quasicrystals by interference lithography," *Appl. Phys. Lett.* **93**, 061112 (2008).

58. C. Lu, X. K. Hu, I. V. Mitchell, and R. H. Lipson, "Diffraction element assisted lithography: Pattern control for photonic crystal fabrication," *Appl. Phys. Lett.* **86**, 193110 (2005).

59. Y. Yang and G. P. Wang, "Realization of periodic and quasiperiodic microstructures with sub-diffraction-limit feature sizes by far-field holographic lithography," *Appl. Phys. Lett.* **89**, 111104 (2006).

60. L. J. Wu, Y. C. Zhong, K. S. Wong, G. P. Wang, and L. Yuan, "Fabrication of hetero-binary and honeycomb photonic crystals by one-step holographic lithography," *Appl. Phys. Lett.* **88**, 091115 (2006).

61. N. D. Lai, Y. D. Huang, J. H. Lin, D. B. Do, and C. C. Hsu, "Fabrication of periodic nanovein structures by holigraphy lithography technique," *Opt. Express* **17**, 3362–3369 (2009).

62. W. D. Mao, J. W. Dong, Y. C. Zhong, G. Q. Liang, and H. Z. Wang, "Formation principles of two-dimensional compound photonic lattices by one-step holographic lithography," *Opt. Express* **13**, 2994–2999 (2005).

63. W. D. Mao, G. Q. Liang, H. Zou, and H. Z. Wang, "Controllable fabrication of two-dimensional compound photonic crystals by single-exposure holographic lithography," *Opt. Lett.* **31**, 1708–1710 (2006).

64. D. Y. Xia, J. Y. Zhang, X. He, and S. R. J. Brueck, "Fabrication of three-dimensional photonic crystal structures by interferometric lithography and nanoparticle self-assembly," *Appl. Phys. Lett.* **93**, 071105 (2008).

65. X. Y. Hu, Y. H. Liu, B. Y. Cheng, D. Z. Zhang, and Q. B. Meng, "Fabrication of high quality three-dimensional photonic crystals," *Chin. Phys. Lett.* **21**, 1289–1291 (2004).

66. C. Lopez, "Materials aspects of photonic crystals," *Adv. Mater.* **15**, 1679–1704 (2003).

67. A. Chelnokov, K. Wang, S. Rowson, P. Garoche, and J. M. Lourtioz, "Near-infrared Yablonovite-like photonic crystals by focused-ion-beam etching of macroporous silicon," *Appl. Phys. Lett.* **77**, 2943–2945 (2000).

68. M. H. Qi, E. Lidorikis, P. T. Rakich, S. G. Johnson, J. D. Joannopoulos, E. P. Ippen, and H. I. Smith, "A three-dimensional optical photonic crystal with designed point defects," *Nature* **429**, 538–542 (2004).

69. S. Y. Lin, J. G. Fleming, D. L. Hetherington, B. K. Smith, R. Biswas, K. M. Ho, M. M. Sigalas, W. Zubrzycki, S. R. Kurtz, and J. Bur, "A three-dimensional photonic crystal operating at infrared wavelengths," *Nature* **394**, 251–253 (1998).

70. S. H. Fan, P. R. Villeneuve, R. D. Meade, and J. D. Joannopoulos, "Design of three-dimensional photonic crystals at sub micron lengthscales," *Appl. Phys. Lett.* **65**, 1466–1468 (1994).

71. E. Ozbay, A. Abeyta, G. Tuttle, M. Tringides, R. Biswas, C. T. Chan, C. M. Soukoulis, and K. M. Ho, "Measurement of a three-dimensional photonic band gap in a crystal structure made of dielectric rods", *Phys. Rev. B* **50**, 1945–1948 (1994).

72. K. M. Ho, C. T. Chan, C. M. Soukoulis, R. Biswas, and M. Sigalas, "Photonic band gaps in three dimensions: New layer-by-layer periodic structures," *Solid State Commun.* **89**, 413–416 (1994).

73. J. G. Fleming, S. Y. Lin, I. E. Kady, R. Biswas, and K. M. Ho, "All-metallic three-dimensional photonic crystals with a large infrared bandgap," *Nature* **417**, 52–55 (2002).

74. C. S. Wu, C. F. Lin, H. Y. Lin, C. L. Lee, and C. D. Chen, "Polymer-based photonic crystals fabricated with single-step electron-beam lithography," *Adv. Mater.* **19**, 3052–3056 (2007).

75. V. D. Blanco, J. Siegel, A. Ferrer, A. R. D. L. Cruz, and J. Solis, "Deep subsurface waveguides with circular cross section produced by femtosecond laser writing," *Appl. Phys. Lett.* **91**, 051104 (2007).

76. B. H. Cumpston, S. P. Ananthavel, S. Barlow, D. L. Dyer, J. E. Ehrllch, L. L. Ersklne, A. A. Helkal, S. M. Kuebler, I. Y. S. Lee, D. M. Maughon, J. Qin, H. Rockel, M. Ruml, X. L. Wu, S. R. Marder, and J. W. Perry, "Two-photon polymerization initiators for three-dimensional optical data storage and microfabrication," *Nature* **398**, 51–54 (1999).

77. S. Maruo, O. Nakamura, and S. Kawata, "Three-dimensional microfabrication with two-photon-absorbed photopolymerization," *Opt. Lett.* **22**, 132–134 (1997).

78. T. Tanaka, H. B. Sun, and S. Kawata, "Rapid sub-diffraction-limit laser micro/nanoprocessing in a threshold material system," *Appl. Phys. Lett.* **80**, 312–314 (2002).

79. K. Takada, K. Kaneko, Y. D. Li, S. Kawata, Q. D. Chen, and H. B. Sun, "Temperature effects on pinpoint photopolymerization and polymerized micronanostructures," *Appl. Phys. Lett.* **92**, 041902 (2008).

80. M. A. White, *Properties of Materials*, Oxford University Press, New York, 1999.

81. B. H. Jia, H. Kang, J. F. Li, and M. Gu, "Use of radially polarized beams in three-dimensional photonic crystal fabrication with the two-photon polymerization method," *Opt. Lett.* **34**, 1918–1920 (2009).

82. J. Serbin and M. Gu, "Experimental evidence for superism effects in three-dimensional polymer photonic crystals," *Adv. Mater.* **18**, 221–224 (2006).

83. M. Straub and M. Gu, "Near-infrared photonic crystals with higher-order bandgaps generated by two-photon photopolymerization", *Opt. Lett.* **27**, 1824–1826 (2002).

84. S. Juodkazis, V. Mizeikis, and H. Misawa, "Three-dimensional microfabrication of materials by femtosecond lasers for photonics applications," *J. Appl. Phys.* **106**, 051101 (2009).

85. H. B. Sun, S. Matsuo, and H. Misawa, "Three-dimensional photonic crystal structures achieved with two-photon-absorption photopolymerization of resin," *Appl. Phys. Lett.* **74**, 786–788 (1999).

86. J. Serbin, A. Egbert, A. Ostendorf, B. N. Chichkov, R. Houbertz, G. Domann, J. Schulz, C. Cronauer, L. Frohlich, and M. Popall, "Femtosecond laser-induced two-photon polymerization of inorganic-organic hybrid materials for applications in photonics," *Opt. Lett.* **28**, 301–303 (2003).

87. J. Serbin, A. Ovsianikov, and B. Chichkov, "Fabrication of woodpile structures by two-photon polymerization and investigation of their optical properties," *Opt. Express* **12**, 5221–5228 (2004).

88. K. K. Seet, V. Mizeikis, S. Juodkazis, and H. Misawa, "Three-dimensional horizontal circular spiral photonic crystals with stop gaps below 1 μm," *Appl. Phys. Lett.* **88**, 221101 (2006).

89. M. Thiel, M. S. Rill, G. V. Freymann, and M. Wegener, "Three-dimensional bi-chiral photonic crystals," *Adv. Mater.* **21**, 4680–4682 (2009).

90. S. Kawata, H. B. Sun, T. Tanaka, and K. Takada, "Finer features for functional microdevices," *Nature* **412**, 697–698 (2001).

91. A. Ledermann, L. Cademartiri, M. Hermatschweiler, C. Toninelli, G. A. Ozin, D. S. Wiersma, M. Wegener, and G. V. Freymann, "Three-dimensional silicon inverse photonic quasicrystals for infrared wavelengths," *Nat. Mater.* **5**, 942–945 (2006).

92. N. Tetreault, G. V. Freymann, M. Deubel, M. Hermatschweiler, F. P. Willard, S. John, M. Wegener, and G. A. Ozin, "New route to three-dimensional photonic bandgap materials: silicon double inversion of polymer templates," *Adv. Mater.* **18**, 457–460 (2006).

93. C. T. Chan, "Quasicrystals enter third dimension," *Nat. Photon.* **1**, 91–92 (2007).

94. D. Shir, E. C. Nelson, Y. C. Chen, A. Brzezinski, H. Liao, P. V. Braun, P. Wiltzius, K. H. A. Bogart, and J. A. Rogers, "Three dimensional silicon photonic crystals fabricated by two photon phase mask lithography," *Appl. Phys. Lett.* **94**, 011101 (2009).

95. H. Xia, W. Y. Zhang, F. F. Wang, D. Wu, X. W. Liu, L. Chen, Q. D. Chen, Y. G. Ma, and H. B. Sun, "Three-dimensional micronanofabrication via two-photon-excited photoisomerization," *Appl. Phys. Lett.* **95**, 083118 (2009).

96. S. Kawata, H. B. Sun, T. Tanaka, and K. Takada, "Finer features for functional microdevices," *Nature* **412**, 697–698 (2001).

97. M. Deubel, G. Freymann, M. Wegener, S. Pereira, K. Busch, and C. M. Soukoulis, "Direct laser writing of three-dimensional photonic-crystal templates for telecommunications," *Nature Mater.* **3**, 444–447 (2004).

98. M. Deubel, M. Wegener, A. Kaso, and S. John, "Direct laser writing and characterization of "slanted pore" photonic crystals," *Appl. Phys. Lett.* **85**, 1895–1897 (2004).

99. O. Toader, M. Berciu, and S. John, "Photonic band gaps based on tetragonal lattices of slanted pores," *Phys. Rev. Lett.* **90**, 233901 (2003).

100. H. B. Sun, Y. Xu, S. Juodkazis, K., Sun, M. Watanabe, S. Matsuo, H. Misawa, and J. Nishii, "Arbitrary-lattice photonic crystals created by multiphoton microfabrication," *Opt. Lett.* **26**, 325–327 (2001).

101. E. G. Gamaly, S. Juodkazis, K. Nishimura, H. Misawa, B. L. Davis, L. Hallo, P. Nicolai, and V. T. Tikhonchuk, "Laser-matter interaction in the bulk of a transparent solid: Confined microexplosion and void formation," *Phys. Rev. B* **73**, 214101 (2006).

102. A. Rodenas, J. A. Sanz, D. Jaque, G. A. Torchia, and L. Roso, and P. Moreno, "Femtosecond laser induced micromodifications in Nd: SBN crystals: amporphization and luminescence inhibition," *J. Appl. Phys.* **100**, 113517 (2006).

103. E. N. Glezer and E. Mazur, "Ultrafast-laser driven micro-explosions in transparent materials," *Appl. Phys. Lett.* **71**, 882–884 (1997).

104. G. Y. Zhou and M. Gu, "Direct optical fabrication of three-dimensional photonic crystals in a high refractive index $LiNbO_3$ crystal," *Opt. Lett.* **31**, 2783–2783 (2006).

105. A. Rodenas, G. Zhou, D. Jaque, and M. Gu, "Direct laser writing of three-dimensional photonic structures in Nd:yttrium aluminum garnet laser ceramics," *Appl. Phys. Lett.* **93**, 151104 (2008).

106. D. Day and M. Gu, "Formation of voids in a doped methacrylate polymer," *Appl. Phys. Lett.* **80**, 2404–2406 (2002).

107. G. Y. Zhou, M. J. Ventura, M. Straub, M. Gu, A. Ono, S. Kawata, X. H. Wang, and Y. Kivshar, "In-plane and out-of-plane band-gap properties of a two-dimensional triangular polymer-based void channel photonic crystal," *Appl. Phys. Lett.* **84**, 4415–4417 (2004).

108. M. J. Ventura, M. Straub, and M. Gu, "Void channel microstructures in resin solids as an efficient way to infrared photonic crystals," *Appl. Phys. Lett.* **82**, 1649–1651 (2003).

109. M. Straub, M. Ventura, and M. Gu, "Multiple higher-order stop gaps in infrared polymer photonic crystals," *Phys. Rev. Lett.* **91**, 043901 (2003).

310. G. Y. Zhou, M. J. Ventura, M. R. Vanner, and M. Gu, "Fabrication and characterization of face-centered-cubic void dots photonic crystals in a solid polymer material," *Appl. Phys. Lett.* **86**, 011108 (2005).

111. S. Noda, K. Tomoda, N. Yamamoto, and A. Chutinan, "Full three-dimensional photonic bandgap crystals at near-infrared wavelengths," *Science* **289**, 604–606 (2000).

References | **139**

112. A. Feigel, Z. Kotler, B. Sfez, A. Arsh, M. Klebanov, and V. Lyubin, "Chalcogenide glass-based three-dimensional photonic crystals", *Appl. Phys. Lett.* **77**, 3221–3223 (2000).

113. O. Toader, T. Y. M. Chan, and S. John, "Photonic band gap architectures for holographic lithographiy," *Phys. Rev. Lett.* **92**, 043905 (2004).

114. L. Z. Cai, X. L. Yang, and Y. R. Wang, "All fourteen Bravais lattices can be formed by interference of four noncoplanar beams," *Opt. Lett.* **27**, 900–902 (2002).

115. S. Shoji, R. P. Zaccaria, H. B. Sun, and S. Kawata, "Multi-step multi-beam laser interference patterning of three-dimensional photonic lattices," *Opt. Express* **14**, 2309–2316 (2006).

116. W. D. Mao, G. Q. Liang, Y. Y. Pu, H. Z. Wang, and Z. H. Zeng, "Completed three-dimensional photonic crystals fabricated by holographic lithography," *Appl. Phys. Lett.* **91**, 261911 (2007).

117. D. N. Sharp, A. J. Turberfield, and R. G. Denning, "Holographic photonic crystals with diamond symmetry," *Phys. Rev. B* **68**, 205102 (2003).

118. M. J. Escuti, J. Qi, and G. P. Crawford, "Tunable face-centered-cubic photonic crystal formed in holographic polymer dispersed liquid crystals," *Opt. Lett.* **28**, 522–524 (2003).

119. G. Y. Dong, L. Z. Cai, X. L. Yang, X. X. Shen, X. F. Meng, X. F. Xu, and Y. R. Wang, "Holographic design and band gap evolution of photonic crystals formed with five-beam symmetric umbrella configuration," *Opt. Express* **14**, 8096–8102 (2006).

120. Y. K. Pang, J. C. W. Lee, C. T. Ho, and W. Y. Tam, "Realization of woodpile structure using optical interference holography," *Opt. Express* **14**, 9113–1991 (2006).

121. W. D. Mao, G. Q. Liang, H. Zou, R. Zhang, H. Z. Wang, and Z. H. Zeng, "Design and fabrication of two-dimensional holographic photonic quasi crystals with high-order symmetries," *J. Opt. Soc. Am. B* **23**, 2046–2050 (2006).

122. W. Man, M. Megens, P. J. Steinhardt, and P. M. Chaikin, "Experimental measurement of the photonic properties of icosahedral quasicrystals," *Nature* **436**, 993–996 (2005).

123. J. Xu, R. Ma, X. Wang, and W. Y. Tam, "Icosahedral quasicrystals for visible wavelengths by optical interference holography," *Opt. Express* **15**, 4287–4295 (2007).

124. P. J. Steinhardt and S. Ostlund, "The Physics of Quasicrystals", Singapore: World Scientific Press, 1987.

125. D. S. Rokhsar, D. C. Wright, and N. D. Mermin, "Scale equivalence of quasicrystallographic space groups," *Phys. Rev. B* **37**, 8145–8149 (1988).

126. W. Y. Tam, "Icosahedral quasicrystals by optical interference holography," *Appl. Phys. Lett.* **89**, 251111 (2006).

127. X. Wang, J. Xu, J. C. W. Lee, Y. K. Pang, W. Y. Tam, C. T. Chan, and P. Shen, "Realization of optical periodic quasicrystals using holographic lithography," *Appl. Phys. Lett.* **88**, 051901 (2006).

128. K. M. Ho, C. T. Chan, and C. M. Soukoulis, "Existence of a photonic gap in periodic dielectric structures," *Phys. Rev. Lett.* **65**, 3152 (1990).

129. X. D. Zhang, Z. Q. Zhang, and C. T. Chan, "Absolute photonic band gaps in 12-fold symmetric photonic quasicrystals," *Phys. Rev. B* **63**, 081105 (2001).

130. J. Huang, M. H. Kok, and W. Y. Tam, "Complete photonic bandgaps in the visible range from spherical layer structures in dichromate gelatin emulsions," *Appl. Phys. Lett.* **94**, 014102 (2009).

131. G. M. Naik, A. Mathur, and S. V. Pappu, "Dichromated gelatin holograms: an investigation of their environmental stability," *Appl. Opt.* **29**, 5292–5297 (1990).

132. T. Kubota, "Control of the reconstruction wavelength of Lippmann holograms recorded in dichromated gelatin," *Appl. Opt.* **28**, 1845–1849 (1989).

133. R. Ma, J. Xu, and W. Y. Tam, "Wide band gap photonic structures in dichromate gelatin emulsions," *Appl. Phys. Lett.* **89**, 081116 (2006).

134. A. Chutinan and S. Noda, "Spiral three-dimensional photonic-band-gap structure," *Phys. Rev. B* **57**, R2006–R2008 (1998).

135. Y. K. Pang, J. C. W. Lee, H. F. Lee, W. Y. Tam, C. T. Chan, and P. Sheng, "Chiral microstructures (spirals) fabrication by holographic lithography," *Opt. Express* **13**, 7615–7620 (2005).

136. N. D. Lai, W. P. Liang, J. H. Lin, C. C. Hsu, and C. H. Lin, "Fabrication of two- and three-dimensional periodic structures by multi-exposure of two-beam interference technique," *Opt. Express* **13**, 9605–9611 (2005).

137. A. Feigel, Z. Kotler, and B. Sfez, "Scalable interference lithography alignment for fabrication of three-dimensional photonic crystals", *Opt. Lett.* **27**, 746–748 (2002).

138. M. C. King and D. H. Berry, "Photolithographic mask alignment using moiré technique," *Appl. Opt.* **11**, 2455–2459 (1972).

139. Y. Lin, P. R. Herman, and K. Darmawikarta, "Design and holographic fabrication of tetragonal and cubic photonic crystals with phase mask: toward the mass-production of three-dimensional photonic crystals," *Appl. Phys. Lett.* **86**, 071117 (2005).

140. Y. K. Lin, D. Rivera, and K. P. Chen, "Woodpile-type photonic crystals with orthorhombic or tetragonal symmetry formed through phase mask techniques," *Opt. Express* **14**, 887–892 (2006).

141. D. Chanda, L. Abolghasemi, and P. R. Herman, "One-dimensional diffractive optical element based fabrication and spectral characterization of three-dimensional photonic crystal templates," *Opt. Express* **14**, 8568–8577 (2006).

142. Z. Poole, D. Xu, K. P. Chen, I. Olvera, K. Ohlinger, and Y. K. Lin, "Holographic fabrication of three-dimensional orthorhombic and tetragonal photonic crystal templates using a diffractive optical element", *Appl. Phys. Lett.* **91**, 251101 (2007).

143. G. Y. Zhou and F. S. Chau, "Three-dimensional photonic crystal by holographic contact lithography using a single diffraction mask", *Appl. Phys. Lett.* **90**, 181106 (2007).

144. Y. Lin, P. R. Herman, and E. L. Abolghasemi, "Proposed single-exposure holographic fabrication of microsphere-type photonic crystal through phase mask techniques," *J. Appl. Phys.* **97**, 096102 (2005).

145. T. Y. M. Chan, O. Toader, and S. John, "Photonic band-gap formation by optical-phase-mask lithography," *Phys. Rev. E* **73**, 046610 (2006).

146. Y. K. Lin, A. Harb, D. Rodriguez, K. Lozano, D. Xu, and K. P. Chen, "Fabrication of two-layer integrated phase mask for single-beam and single-exposure fabrication of three-dimensional photonic crystal," *Opt. Express* **16**, 9165–9172 (2008).

147. D. Chanda and P. R. Herman, "Phase tunable multilevel diffractive optical element based single laser exposure fabrication of three-dimensional photonic crystal templates," *Appl. Phys. Lett.* **91**, 061122 (2007).

148. Y. K. Lin, A. Harb, D. Rodriguez, K. Lozano, D. Xu, and K. P. Chen, "Holographic fabrication of photonic crystals using multidimensional phase masks," *J. Appl. Phys.* **104**, 113111 (2008).

149. R. Jakubiak, L. V. Natarajan, V. Tondiglia, G. S. He, and P. N. Prasad, "Electrically switchable lasing from pyrromethene 597 embedded holographic-polymer dispersed liquid crystals," *Appl. Phys. Lett.* **85**, 6095–6097 (2004).

150. V. Berger, O. G. Lafaye, and E. Costard, "Photonic band gaps and holography," *J. Appl. Phys.* **82**, 60–64 (1997).

151. I. Divliansky, T. S. Mayer, K. S. Holliday, and V. H. Crespi, "Fabrication of three-dimensional polymer photonic crystal structures using single diffraction element interference lithography," *Appl. Phys. Lett.* **82**, 1667–1669 (2003).

152. D. B. Shao and S. C. Chen, "Direct patterning of three-dimensional periodic nanostructures by surface-plasmon-assisted nanolithography," *Nano Lett.* **6**, 2279–2283 (2006).

153. W. L. Barnes, A. Dereux, and T. W. Ebbesen, "Surface plasmon subwavelength optics," *Nature* **424**, 824–830 (2003).

154. C. K. Ullal, M. Maldovan, M. Wohlgemuth, E. L. Thomas, C. A. White, and S. Yang, "Triply periodic bicontinuous structures through interference lithography: a level-set approach," *J. Opt. Soc. Am. A* **20**, 948–954 (2003).

155. P. Pieranski, "Colloidal crystals," *Contemp. Phys.* **24**, 25–73 (1983).

156. Y. A. Vlasov, V. N. Astratov, A. V. Baryshev, A. A. Kaplyanskii, O. Z. Karimov, and M. F. Limonov, "Manifestation of intrinsic defects in optical properties of self-organized opal photonic crystals," *Phys. Rev. E* **61**, 5784–5793 (2000).

157. R. Mayoral, J. Requena, J. S. Moya, C. Lopez, A. Cintas, H. Miguez, F. Meseguer, L. Vazquez, M. Holgado, and A. Blanco, "3D long-range ordering in ein SiO_2 submicrometer-sphere sintered superstructure," *Adv. Mater.* **9**, 257–260 (1997).

158. L. N. Donselaar, A. P. Philipse, and J. Suurmond, "Concentration-dependent sedimentation of dilute magnetic fluids and magnetic silica dispersions," *Langmuir* **13**, 6018–6025 (1997).

159. H. Miguez, F. Meseguer, C. Lopez, A. Mifsud, J. S. Moya, and L. Vazquez, "Evidence of FCC crystallization of SiO_2 nanospheres," *Langmuir* **13**, 6009–6011 (1997).

160. M. Giersig and P. Mulvaney, "Preparation of ordered colloid monolayers by electrophoretic deposition," *Langmuir* **9**, 3408–3413 (1993).

161. M. Trau, D. A. Saville, and I. A. Aksay, "Field-induced layering of colloidal crystals," *Science* **272**, 706–709 (1996).

162. S. R. Yeh, M. Seul, and B. I. Shraiman, "Assembly of ordered colloidal aggregates by electric-field-induced fluid flow," *Nature* **386**, 57–59 (1997).

163. H. B. Sunkara, J. M. Jethmalani, and W. T. Ford, "Composite of colloidal crystals of silica in poly(methyl methacrylate)," *Chem. Mater.* **6**, 362–364 (1994).

164. A. E. Larsen and D. G. Grier, "Like-charge attractions in metastable colloidal crystallites," *Nature* **385**, 230–233 (1997).

165. Z. Y. Li and Z. Q. Zhang, "Fragility of photonic band gaps in inverse-opal photonic crystals," *Phys. Rev. B* **62**, 1516–1519 (2000).

166. H. S. Sozuer, J. W. Haus, and R. Inguva, "Photonic bands: Convergence problems with the plane-wave method," *Phys. Rev. B* **45**, 13962–13972 (1992).

167. K. Busch and S. John, "Photonic band gap formation in certain self-organizing system," *Phys. Rev. E* **58**, 3896–3908 (1998).

168. S. H. Park and Y. N. Xia, "Assembly of mesoscale particles over large areas and its application in fabricating tunable optical filters", *Langmuir* **15**, 266–273 (1999).

169. S. H. Park, D. Qin, and Y. N. Xia, "Crystallization of mesoscale particles over large areas," *Adv. Mater.* **10**, 1028–1032 (1998).

170. P. G. Ni, B. Y. Cheng, and D. Z. Zhang, "Inverse opal with an ultraviolet photonic gap," *Appl. Phys. Lett.* **80**, 1879–1881 (2002).

171 I. I. Tarhan and G. H. Watson, "Photonic band structure of fcc colloidal crystals," *Phys. Rev. Lett.* **76**, 315–318 (1996).

172. R. C. Hayward, D. A. Saville, and I. A. Aksay, "Electrophoretic assembly of colloidal crystals with optically tunable micropatterns", *Nature* **404**, 56–59 (2000).

173. B. A. Grzybowski, H. A. Stone, and G. M. Whitesides, "Dynamic self-assembly of magnetized, millimeter-sized objects rotating at a liquid-air interface," *Nature* **405**, 1033–1036 (2000).

174. M. Hildebrand, A. S. Mikhailov, and G. Erti, "Traveling nanoscale structures in reactive adsorbates with attractive lateral interactions," *Phys. Rev. Lett.* **81**, 2602–2605 (1998).

175. K. H. Lin, J. C. Crocker, V. Prasad, A. Schofield, D. A. Weitz, T. C. Lubensky, and A. G. Yodh, "Entropically driven colloidal crystallization on patterned surfaces," *Phys. Rev. Lett.* **85**, 1770–1773 (2000).

176. A. V. Blaaderen and P. Wiltzius, "Growing large, well-oriented colloidal crystals," *Adv. Mater.* **9**, 833–835 (1997).

177. G. A. Ozin and M. Y. Yang, "The race for the photonic chip: colloidal crystal assembly in silicon wafers," *Adv. Funct. Mater.* **11**, 95–104 (2001).

178. Y. Yin and Y. Xia, "Growth of large colloidal crystals with their (100) planes oriented parallel to the surfaces of supporting substrates," *Adv. Mater.* **14**, 605–608 (2002).

179. Y. D. Yin, Z. Y. Li, and Y. Xia, "Template-directed growth of (100)-oriented colloidal crystals," *Langmuir* **19**, 622–631 (2003).

180. N. D. Denkov, O. D. Velev, P. A. Kralchevsky, I. B. Ivanov, H. Yoshimura, and K. Nagayama, "Mechanism of formation of two-dimensional crystals from latex particles on substrates," *Langmuir* **8**, 3183–3190 (1992).

181. A. S. Dimitrov, C. D. Dushkin, H. Yoshimure, and K. Nagayama, "Observations of latex particle two-dimensional-crystal nucleation in wetting films on mercury, glass, and mica," *Langmuir* **10**, 432–440 (1994).

182. A. S. Dimitrov and K. Nagayama, "Continuous convective assembling of fine particles into two-dimensional arrays on solid surfaces," *Langmuir* **12**, 1303–1311 (1996).

183. L. Woodcock, "Entropy difference between the face-centred cubic and hexagonal close-packed crystal structures," *Nature* **385**, 141–143 (1997).

184. A. Bruce, N. Wilding, and G. Ackland, "Free energy of crystalline solids: a lattice-switch monte carlo method," *Phys. Rev. Lett.* **79**, 3002–3005 (1997).

185. A. V. Blaaderen, R. Ruel, and P. Wwiltzius, "Template-directed colloidal crystallization," *Nature* **385**, 321–324 (1997).

186. P. Jiang, J. F. Bertone, K. S. Hwang, and V. L. Colvin, "Single-crystal colloidal multilayers of controlled thickness," *Chem. Mater.* **11**, 2132–2140 (1999).

187. Z. Z. Gu, A. Fujishima, and O. Sato, "Fabrication of high-quality opal films with controllable thickness," *Chem. Mater.* **14**, 760–765 (2002).

188. Y. H. Ye, F. Leblanc, A. Hache, and V. V. Truong, "Self-assembling three-dimensional colloidal photonic crystal structure with high crystalline quality," *Appl. Phys. Lett.* **78**, 52–54 (2001).

189. Z. C. Zhou and X. S. Zhao, "Flow-controlled vertical deposition method for the fabrication of photonic crystals," *Langmuir* **20**, 1524–1526 (2004).

190. X. Y. Hu, Y. H. Liu, B. Y. Cheng, D. Z. Zhang, and Q. B. Meng, "Fabrication of high quality three-dimensional photonic crystals," *Chin. Phys. Lett.* **21**, 1289–1291 (2004).

191. J. D. Joannopoulos, "Self-assembly lights up," *Nature* **414**, 257–258 (2001).

192. Y. A. Vlasov, X. Z. Bo, J. C. Sturm, and D. J. Norris, "On-chip natural assembly of silicon photonic bandgap crystals," *Nature* **414**, 289–293 (2001).

193. J. E. G. J. Wijnhoven and W. L. Vos, "Preparation of photonic crystals made of air spheres in titania," *Science* **281**, 802–804 (1998).

194. Q. B. Meng, Z. Z. Gu, O. Sato, and A. Fujishima, "Fabrication of highly ordered porous structure," *Appl. Phys. Lett.* **77**, 4313–4315 (2000).

195. A. Blanco, E. Chomsld, S. Grabtchak, M. Lblsate, S. John, S. W. Leonard, C. Lopez, F. Meseguer, H. Mlguez, J. P. Mondia, C. A. Ozin, O. Toader, and H. M. V. Driel, "Large-scale synthesis of a silicon photonic crystal with a complete three-dimensional bandgap near 1.5 micrometer," *Nature* **405**, 437–440 (2000).

Questions

3.1 What method can be adopted to fabricate a two-dimensional silicon photonic crystal slab with the photonic bandgap in the optical communication range?

3.2 For a polystyrene opal with a photonic bandgap centered at 800 nm, please calculate the diameter of polystyrene sphere according to the Bragg equation

$$\lambda = 2\sqrt{f_1\varepsilon_1 + f_2\varepsilon_2} \cdot d$$

where λ is the center wavelength of the photonic bandgap, f_1 and f_2 are the filling rates of air and polystyrene, respectively, ε_1 and ε_2 are the dielectric constants of air and polystyrene, respectively, and d is the spacing of (111) crystal planes.

Chapter 4

Photonic Crystal All-Optical Switching

Owing to their unique properties of controlling the propagation states of photons, photonic crystals have been considered to be promising candidates for the study of future integrated photonic devices and integrated photonic chips. Along with the extensive research on the fabrication of high-quality one-, two-, and three-dimensional photonic crystals, various kinds of integrated photonic devices based on photonic crystals have attracted great attention, and have been studied widely and intensively. Nonlinear photonic crystals constructed by third-order nonlinear optical materials are the very important basis for the realization of photonic devices.

4.1 Mechanism of Photonic Crystal All-Optical Switching

Analogous to its electric counterpart, all-optical switching can permit (or prohibit) the propagation of a signal light under the trigger of a control light. Photonic crystal all-optical switching can modulate the propagation states of the signal light based on the interactions of light and matter. As an essential integrated photonic device, photonic crystal all-optical switching plays a very important role in the fields of integrated photonic circuits, optical

Photonic Crystals: Principles and Applications
Qihuang Gong and Xiaoyong Hu
Copyright © 2014 Pan Stanford Publishing Pte. Ltd.
ISBN 978-981-4267-30-4 (Hardcover), 978-981-4364-83-6 (eBook)
www.panstanford.com

interconnection networks, and optical computing systems. In 1994, Scalora *et al.* presented the concept of photonic crystal all-optical switching [1]. It can be understood simply as following: at the beginning, a signal light is reflected completely by a photonic crystal and cannot propagate through it. The optical switching is in the "OFF" state. Under the excitation of a control light, the signal light can propagate through the photonic crystal. Then the optical switching is in the "ON" state. The propagation states of the signal light are dominated fully by the appearance of the control light. Up to now, various mechanisms have been presented to demonstrate photonic crystal all-optical switching, such as the photonic bandgap shift mechanism, defect mode shift mechanism, and so on. Third-order nonlinear photonic crystals are the essential basis for the realization of all-optical switching devices.

4.1.1   Photonic Bandgap Shift Method

The photonic bandgap shift method was proposed by Scalora *et al.* in 1994 [1]. When the nonlinear medium constructing the photonic crystal has a positive third-order nonlinear susceptibility, the frequency of the probe light can be set at the high-frequency edge of the photonic bandgap. At first, the probe light is reflected completely by the photonic crystal and the optical switching is in the "OFF" state. When the pump light is switched on, the refractive index of the nonlinear medium in the photonic crystal increases due to the nonlinear optical Kerr effect, which leads to the increase of the effective refractive index of the photonic crystal. As a result, the photonic bandgap shifts in the low-frequency direction. Then the frequency of the probe light drops into the pass band and can propagate through the photonic crystal. Then the optical switching is in the "ON" state.

When the nonlinear medium constructing the photonic crystal has a negative third-order nonlinear susceptibility, the frequency of the probe light should be set at the low-frequency edge of the photonic bandgap. At first, the probe light is reflected completely by the photonic crystal and the optical switching is in the "OFF" state. When the pump light is switched on, the refractive index of the nonlinear medium in the photonic crystal decreases due to the nonlinear optical Kerr effect, which leads to the decrease of the effective refractive index of the photonic crystal. As a result, the photonic bandgap shifts in the high-frequency direction. Then the frequency of the probe light drops into the pass band and can propagate through the photonic crystal. Now the optical switching is in the "ON" state. The schematic structure of the photonic bandgap shift mechanism is shown in Fig. 4.1.

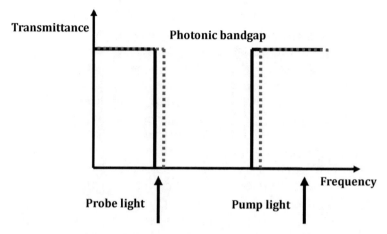

Figure 4.1 Schematic structure of the photonic bandgap shift method.

The ultrashort signal pulses with a pulse width of few cycles order may find great potential applications in the future integrated photonic technology. However, ultrashort signal light with wide frequency spectrum will be severely distorted at the photonic bandedges where the group velocity dispersion is generally very large. Lan *et al.* proposed that this limitation could be overcome by using the impurity bandgap shift [2]. When a structural defect is introduced in a perfect photonic crystal, the spatial periodicity of the distribution of the dielectric materials in the photonic crystal is destroyed. According to the photon localization theory, an electromagnetic mode with a certain resonant frequency will be confined in the defect site, which leads to the appearance of the defect modes in the photonic bandgap [3,4]. An impurity band with very high transmittance can be formed in the photonic bandgap of the photonic crystal based on the coupling of a series of periodically placed identical defect modes [2]. A slight periodic modulation of the dielectric properties of these defects can open up a very deep bandgap in the impurity band. The reason lies in that the strong concentration of electromagnetic fields in the defect regions makes the electromagnetic waves extremely sensitive to the small changes of the defects. Therefore, under the excitation of a pump light, the dielectric properties of the defects change, which makes the photonic bandgap of the impurity band shift. So, an all-optical switching can be achieved based on the impurity bandgap shift. Moreover, if the bandwidth of the impurity band is much larger than the line width of the ultrashort signal light, the frequency of the signal light can be set at the center of the impurity band, so that the problem of the pulse distortion of ultrashort signal pulse near the bandgap can be overcome [2].

The photonic bandgap shift method is very simple, efficient, and feasible in practice. This mechanism is widely used by researchers in their experiments to demonstrate photonic crystal all-optical switching. The photonic bandgap shift mechanism has a stringent requirement of the photonic bandedge: the photonic bandedge should be very steep. When the photonic bandedge is steep, it is possible to reach a high transmittance contrast of the "ON" and "OFF" state for a small frequency shift of the photonic bandgap. Accordingly, a high-quality photonic crystal sample is indispensable.

4.1.2 Defect Mode Shift Method

The defect mode shift method was proposed by Tran in 1997 [5]. The frequency of the probe light can be set at the photonic bandgap, but close to the center of the defect mode. When the nonlinear medium constructing the photonic crystal has a negative third-order nonlinear susceptibility, the frequency of the probe light should be larger than that of the defect mode. At the beginning, the probe light cannot propagate through the photonic crystal due to the photonic bandgap effect. The photonic crystal all-optical switching is in the "OFF" state. When the pump light is switched on, the refractive index of the nonlinear medium in the photonic crystal decreases, which makes the effective refractive index of the photonic crystal decrease and the photonic crystal bandgap shift in the high-frequency direction. Accordingly, the defect modes shift in the same direction as that of the photonic bandgap. As a result, the frequency of the probe light is in the center of the defect mode, and the probe light can propagate through the photonic crystal. Then the photonic crystal all-optical switching is in the "ON" state.

When the nonlinear medium constructing the photonic crystal has a positive third-order nonlinear susceptibility, the frequency of the probe light should be less than that of the defect mode. At the beginning, the probe light cannot propagate through the photonic crystal due to the photonic bandgap effect. The photonic crystal all-optical switching is in the "OFF" state. When the pump light is switched on, the refractive index of the nonlinear medium in the photonic crystal increases, which makes the effective refractive index of the photonic crystal increase and the photonic crystal bandgap shift in the low-frequency direction. Accordingly, the defect modes shift in the same direction as that of the photonic bandgap. As a result, the frequency of the probe light is in the center of the defect mode, and the probe light can propagate through the photonic crystal. Then the photonic crystal all-optical switching is in the "ON" state. The schematic structure of the defect mode shift mechanism is shown in Fig. 4.2.

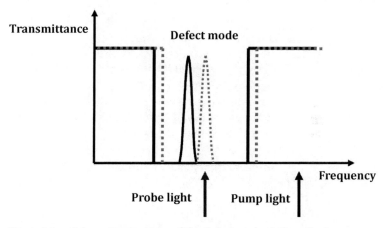

Figure 4.2 Schematic structure of the defect mode shift method.

Villeneuve *et al.* proposed an improved defect mode shift mechanism to realize the all-optical switching effect [6]. This approach adopts two photonic crystal microcavities in series, one for triggering the "ON" state and the other for triggering the "OFF" state. The resonant frequency is $\omega_0 + \Delta\omega$ and ω_0 for the first and second microcavity, respectively. When the pump light excites the first microcavity, the resonant frequency of the first microcavity will shift from $\omega_0 + \Delta\omega$ to ω_0. Then the signal light with the frequency of ω_0 is located in the center of the microcavity mode and can propagate through the photonic crystal. The optical switching is in the "ON" state. When the pump light excites the second microcavity, the resonant frequency of the second microcavity will shift from ω_0 to $\omega_0 - \Delta\omega$. As a result, the signal light with the frequency of ω_0 drops into the stop band and cannot propagate through the photonic crystal. The optical switching is in the "OFF" state.

The defect mode shift method is also very simple and feasible in practice. This mechanism is also widely used by researchers in their study of photonic crystal all-optical switching. Just like the photonic bandgap shift mechanism, the defect mode shift mechanism also has a rigid requirement to the photonic crystal. When the line width of the defect is very narrow, it is possible to reach a high transmittance contrast of the "ON" and "OFF" state for a small frequency shift of the defect mode. Therefore, defect modes with a high quality factor are needed.

4.1.3   Optical Bistable Switching

All-optical switching can also be realized based on the optical bistable effect of the nonlinear photonic crystal microcavity [7,8]. The photonic crystal microcavity can be formed by introducing a lattice defect, including point defect, line defect, and so on, in a photonic crystal; the defect mode will appear in the photonic bandgap. The frequency of the signal light is set in the photonic bandgap, but near the resonant frequency of the defect mode. There is a small frequency detuning between the defect mode and the signal light. At the beginning, the signal light cannot propagate through the photonic crystal because of the strong photonic bandgap effect. The optical switching is in the "OFF" state. The transmittance of the signal light increases with the increment of incident power of the signal light. The strong photon localization effect of the photonic crystal microcavity and the third-order nonlinearity can offer nonlinear feedback for the incident light. As a result, the transmittance spectrum of the signal light takes on the properties of optical bistability. When the incident power exceeds a threshold value, the transmittance of the signal light becomes very large abruptly. Then the all-optical switching is in the "ON" state. On the contrary, the transmittance of the signal light reduces with the decrease of the incident power. When the incident power is less than a threshold value, the transmittance of the signal light becomes very small quickly. Then the all-optical switching is in the "OFF" state again. A schematic structure of a photonic crystal bistable switching is shown in Fig. 4.3, which is composed of a microcavity structure inserted in a photonic crystal waveguide [9]. The electric field distribution shows that the optical switching is in the "ON" state.

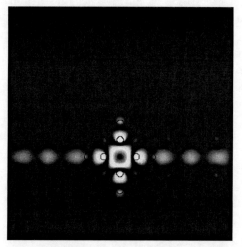

Figure 4.3   Schematic structure of the photonic crystal bistable switching. The electric field distribution shows that the optical switching is in the "ON" state.

The physical mechanism of the optical bistable switching can be explained as follows [9]: With the increase of the incident power, the refractive index of the nonlinear medium in the photonic crystal changes, which makes the effective refractive index of the photonic crystal microcavity change. Accordingly, the resonant frequency of the microcavity mode ω_{res} shifts toward the frequency of the signal light ω_0. As a result, the transmittance of the signal light increases. Owing to the strong photon localization effect of the microcavity structure, the coupling between the signal light and the microcavity mode is enhanced greatly when the frequency of the signal light is in resonance with that of the microcavity mode. This provides a very strong nonlinear positive feedback, which makes the transmittance of the signal light increase sharply. On the other hand, when the incident power decreases, the refractive index of the nonlinear medium varies, which makes the effective refractive index of the photonic crystal microcavity change. Accordingly, the resonant frequency of the microcavity mode ω_{res} shifts away from the frequency of signal light ω_0. The nonlinear coupling between the signal light and the microcavity mode is reduced greatly when the frequency of the defect mode is off-resonant with that of the incident light. This provides a very strong nonlinear negative feedback and the transmittance of the signal light decreases remarkably. Therefore, three parameters are needed to characterize an optical bistable switching: the frequency detuning between the resonant frequency of the microcavity modes and the signal light $\Delta\omega$, the quality factor Q of the photonic crystal microcavity, and the nonlinear feedback parameter κ. A large frequency detuning $\Delta\omega$ requires a relatively high incident power and large third-order nonlinear optical coefficient to obtain the large frequency shift of the microcavity mode. While a large transmittance contrast between the "ON" and "OFF" state is difficult to achieve for a very small frequency detuning $\Delta\omega$. More often than not, high quality factor Q is required when the optical bistability effect is used. High quality factor implies a strong photon localization effect of photonic crystal microcavity, which will result in an intense interaction of light and matter in the microcavity. The nonlinear feedback parameter κ is mainly determined by the coupling of the signal light and the microcavity mode, and the interactions of signal light and the nonlinear medium in the photonic crystal microcavity.

The physical mechanism of the bistable optical switching can also be simply explained in terms of the effective refractive index changes of a one-dimensional photonic crystal microcavity containing a nonlinear defect layer [10]. According to the third-order nonlinear optical Kerr effect, the refractive index of the defect layer is related to the incident power of the signal light. On the one hand, the refractive index of the defect layer increases under the excitation of the signal light if the defect layer has a positive nonlinear refractive index. On the other hand, the increase of the refractive index of the defect layer will change the electric field distribution of the signal light in the defect structure. This will influence the increase of the refractive index of the defect layer conversely. In the case of a low incident power, the refractive index of the defect layer changes slightly and the photonic crystal microcavity operates in the low-refractive-index state. When the incident power exceeds a high threshold value, the photonic crystal microcavity system is forced to jump to a higher refractive index state due to strong nonlinear interactions. The high-refractive-index state is kept until the incident power is reduced to a low threshold value [10]. This phenomenon forms a hysteresis cycle for the refractive index of the nonlinear defect layer, and accordingly for the transmission spectrum of the photonic crystal microcavity. An all-optical switching can be realized when the signal light transits between the two threshold intensities.

Chen *et al.* also presented the concept of bistable optical switching through photonic bandedge mode shifting [11]. Owing to the strong photon localization effect, large enhancement of optical nonlinearity, and low group velocity of the photonic bandedge mode, it is possible to realize the bistable optical switching by use of gap-edge mode shifting. The nonlinear feedback is provided by the strong photon localization of the photonic bandedge mode and by the intense nonlinear interactions of light and matter in the defect structure. A bistable loop in the transmittance spectrum of the photonic bandedge mode as a function of the incident power can be obtained. Accordingly, a photonic crystal bistable switching can be realized.

The optical bistable switching method is also widely adopted in experiments. However, high transmittance contrast between the "ON" and "OFF" states is not easy to achieve by use of this mechanism. The most serious limitation is that a large shift magnitude is very difficult to achieve for a defect mode without the help of an intense pump light. Moreover, the fabrication of a defect mode with an ultrahigh quality factor is very difficult. These limitations restrict the practical applications of the optical bistable switching.

4.1.4 Waveguide–Microcavity Coupling Method

The waveguide–microcavity coupling method was proposed by Yanik *et al.* in 2003 [12]. The photonic crystal has a configuration of a wide-band line waveguide side coupled with a single-mode microcavity. The photonic crystal microcavity is composed of third-order nonlinear optical material. The resonant frequency of the photonic crystal microcavity is ω_0. The transmission spectrum of the total photonic crystal system takes on a Lorentzian line shape with the resonant frequency of ω_0. There is a frequency detuning between the probe light and microcavity mode. At the beginning, the probe light can propagate through the photonic crystal waveguide. Then the optical switching is in the "ON" state. Under the excitation of a pump light, the effective refractive index of the photonic crystal microcavity changes, which leads to a shift of the resonant frequency of the microcavity mode. When the frequency of the microcavity mode is in resonance with that of the probe light, the energy of the probe light will be stored in the microcavity completely. Then the probe light cannot propagate through the photonic crystal waveguide. The optical switching is in the "OFF" state. The schematic structure of the waveguide–microcavity coupling mechanism with a Lorentzian line shape microcavity mode is shown in Fig. 4.4. The electric field distribution implies the "OFF" state of the optical switching.

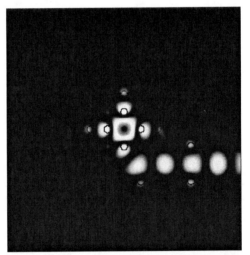

Figure 4.4   Schematic structure of the waveguide-microcavity coupling method with a Lorentzian line-shape microcavity mode. The electric field distribution implies the "OFF" state of the optical switching.

Subsequently, Fan *et al.* proposed an improved photonic crystal configuration to realize all-optical switching effect based on the coupling of waveguide and microcavity [13]. Two identical point defects are introduced in the photonic crystal optical waveguide to provide a partial reflection for the waveguided modes. The schematic structure of the photonic crystal system is shown in Fig. 4.5a. The resonant frequency of the photonic bandgap micro-cavity is ω_0. Without the two point defects in the photonic crystal

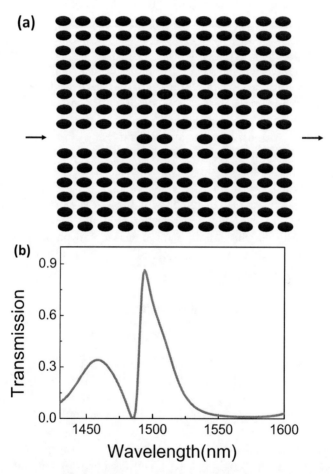

Figure 4.5 Schematic structure of the waveguide-microcavity coupling method. (a) Schematic phonic crystal structure. (b) Transmittance spectrum of the photonic crystal structure.

waveguide, the transmission spectrum of the total photonic crystal system takes on a Lorentzian line shape with the resonant frequency of ω_0. The adding of two point defects in the optical waveguide generates a sharp and asymmetric line shape in the transmittance spectrum of the system, which is also called the Fano line shape, as shown in Fig. 4.5b. The transmittance of the photonic crystal waveguide remains zero at the frequency of ω_0, but increases to 100% at the frequency of ω_t. The value of $\omega_t - \omega_0$ is much less than the line width of the photonic crystal microcavity mode ω_0. Therefore, a small frequency shift can trigger the transition between the "ON" and "OFF" states. At the beginning, the frequency of the probe light can be set equal to the resonant frequency ω_0 of the photonic bandgap microcavity. The transmittance of the probe light is zero and the optical switching is in the state of "OFF". While under the excitation of a pump light, the effective refractive index of the photonic crystal changes. Accordingly, the position of the Fano line shape varies. The frequency of the probe light is in resonance with ω_t, and a 100% transmittance of the probe light can be achieved. Then the optical switching is in the "ON" state.

Just like the optical bistable switching, the all-optical switching based on the waveguide–microcavity coupling depends on the dynamic shift of the resonant frequency of the photonic crystal microcavity. Therefore, large optical nonlinearity and high-Q microcavity are needed.

4.1.5 Waveguide Coupling Method

4.1.5.1 Cross waveguide coupling mechanism

The cross waveguide coupling method is also proposed by Yanik *et al.* in 2003 [14]. Two perpendicular cross waveguides is introduced in a two-dimensional photonic crystal. The intersection region of the cross waveguide consists of a nonlinear photonic crystal microcavity, which supports two resonant dipole modes, the "x" mode and the "y" mode. The schematic structure of the cross waveguide coupling mechanism is shown in Fig. 4.6. Each microcavity mode is even with respect to one waveguide axis and odd with respect to the other waveguide axis. As a result, each waveguide mode only couples to the microcavity mode with the same symmetry, which eliminates the cross-talk between two waveguides [15]. The signal light only couples with the "x" mode and can only propagate in the "x" waveguide, while the control light only couples with the "y" mode and can only propagate in the "y" waveguide. There is a frequency detuning between the probe light and the "x" mode. At the beginning, the probe light cannot propagate through the "x" waveguide due to frequency detuning. The optical switching is in the "OFF" state. When the pump light enters the "y" waveguide, the energy of the pump light can couple into the "y" mode. Owing to the nonlinear optical Kerr effect, the effective refractive index of the photonic crystal microcavity changes. This makes the resonant frequency of "x" mode shift to the frequency of the probe light. Then the probe light can propagate through the "x" waveguide and the optical switching is in the "ON" state.

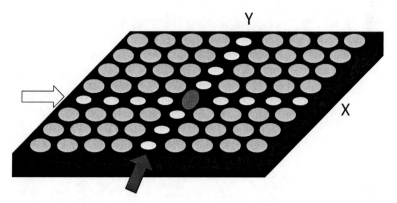

Figure 4.6 Schematic structure of the cross–waveguide coupling method.

4.1.5.1.1 *Parallel waveguide coupling mechanism*

Locatelli and Sharkawy pointed out that all-optical switching effect could be realized on the basis of the coupling of two parallel waveguides 1 and 2 in a two-dimensional nonlinear photonic crystal [16,17]. Each photonic crystal waveguide can provide only one single guided mode. The signal light enters the waveguide "1". When the incident power of the signal light is low, the energy of the signal light in the waveguide "1" can be coupled into the waveguide "2" through the waveguide-coupling region. No signal light output from the waveguide "1". So, the optical switching is in the "OFF" state. With the increase of the incident power of the signal light, the third-order nonlinear optical effect becomes stronger and stronger, which leads to a variation of the effective refractive index of the photonic crystal. The coupling condition cannot be met in the waveguide coupling region because of the effect of nonlinear phase shift. When the intensity of the signal light exceeds a threshold value, all the probe light propagates in the waveguide "1". No signal light outputs from the waveguide "2". Then the optical switching is in the state of "ON".

Soto *et al.* presented an approach to achieving all-optical switching based on a photonic crystal directional coupler [18]. The two-dimensional photonic crystal consists of hexagonal lattice of high dielectric rods embedded in a low dielectric matrix. The photonic crystal directional coupler is composed of two waveguides separated by one row of defect rods. The frequency of the pump light is in resonance with that of the defect modes. At the beginning, the probe light propagating in the waveguide 1 can be coupled into waveguide 2 in the waveguide coupling region, and then output from waveguide 2. No signal is output from waveguide 1. The optical switching is in the state of "OFF". When the pump light is coupled into the defect waveguide between the waveguides 1 and 2, the electric field distribution of the pump light will be strongly confined around the defect structure. A low pump intensity could induce remarkable changes of the refractive index of the defect rods, which varies the coupling coefficients of waveguide 1 and 2. Then the probe light cannot be coupled into the waveguide 2 again, and has to propagate through the waveguide 1 and output from it. Then the optical switching is in the "ON" state. Owing to the strong photon localization effect, low operating threshold pump power can be expected.

The waveguide coupling method requires that the photonic crystal possesses a very large third-order nonlinear optical coefficient. The realization of photonic crystal all-optical switching based on waveguide coupling is not easy in practice.

4.1.6   Photonic Density of State Variation Method

Johnson *et al.* proposed that ultrafast all-optical switching could also be realized based on the changes of photonic density of states in a three-dimensional semiconductor photonic crystal [19]. Under the excitation of a pump laser, a large number of free carriers are generated in semiconductor materials. This results in the variation of the effective refractive index of the semiconductor photonic crystal. Accordingly, the position and the width of the photonic bandgap are changed. At certain frequencies, the photonic density of states in the photonic bandgap can be switched from zero to a very high value with an ultrafast response time. Therefore, the ultrafast optical switching can be realized.

4.1.7 Photonic Flux Switching

Wang *et al.* proposed the concept of photon flux switching in a two-dimensional photonic crystal, which was constructed by periodically placing dielectric nanorods embedded in the matrix of air [20]. When a lattice defect is introduced in the photonic crystal, photon states with certain resonant frequency will be confined at the defect site. This localization effect can result in a sharp transition between the state of the out-of-plane light propagation through the entire photonic crystal and the state in which light propagates only along the defect channels. The wavelength of the probe light is set around the central wavelength of the defect mode. When the refractive index of the lattice defect is higher than the threshold, the probe light can propagate through the photonic crystal along the defect channel. Then the optical switching is in the state of "ON". If the refractive index of the lattice defect is lower than the threshold, the probe light will propagate out of plane and transmit into the free space. Then the optical switching is in the state of "OFF". The changes of the refractive index of the lattice defect can be controlled by different methods, such as chemical, electrical, and optical methods.

4.1.8 Nonlinear Chiral Photonic Crystal Optical Switching

For a photonic crystal all-optical switching, the frequency of the probe light is often different from that of the pump light, so as to eliminate the discrimination problem. Tran proposed an all-optical switching mechanism based on the photonic bandgap shift of nonlinear chiral photonic crystals [21]. The photonic crystal has a kind of material possessing both chirality and optical Kerr nonlinearity. The photonic band structure of the right-circularly polarized light is different from that of the left-circularly polarized light in a chiral photonic crystal. Therefore, the probe and the pump light can have the same frequency. The frequency can be chosen so that the pump light will never transmit even though the probe light at the same frequency can transmit through the photonic crystal. At the beginning, the probe light can propagate through the photonic crystal, and the optical switching is in the state of "ON". Under the excitation of the strong pump light, the effective refractive index of the photonic crystal changes, which makes the photonic bandgap shift. The frequency of the probe light drops in the photonic bandgap and the probe light cannot propagate through the photonic crystal. Then the optical switching is in the state of "OFF". Owing to the chirality, the pump light cannot propagate in the photonic crystal. Therefore, there is no discrimination problem.

4.1.9   Frequency Mixing Method

The frequency mixing method was proposed by Tran in 1996 [22]. Both the frequencies of the pump light ω_1 and probe light ω_2 are set at the photonic bandgap. At the beginning, the probe light cannot propagate through the photonic crystal. The optical switching is in the "OFF" state. When the pump light is switched on, the frequency mixing of the pump and probe lights generates the frequencies of $2\omega_1 - \omega_2$ and $2\omega_1 - \omega_2$, which are outside the photonic bandgap and can propagate through the photonic crystal. The structural parameters of photonic crystal can be carefully designed so that only the pulse with the frequency of $2\omega_1 - \omega_2$ can propagate through the photonic crystal. Then the optical switching is in the "ON" state.

4.1.10   Equifrequency Curve Variation Method

Xiong *et al.* proposed the equifrequency curve variation method for the realization of photonic crystal all-optical switching [23]. The equifrequency surface curves can be adopted to study the propagation properties of light in photonic crystals. They found that the shape of the equifrequency surface curves varied drastically with the refractive index when the frequency and the incident angle of a probe light were selected properly because of the strong dispersion properties of the photonic crystal, and accordingly the strong anisotropy of the photonic band structure. At the beginning, the probe light can propagate through the photonic crystal, and the optical switching is in the "ON" state. According to the third-order nonlinear optical Kerr effect, the effective refractive index of the photonic crystal changes under the excitation of the pump light, which leads to a variation of the photonic band structure. Accordingly, the configuration of the equifrequence curve corresponding to the probe light is changed greatly. The equifrequence curve will be discontinued and the wave vector of the incident light in the photonic crystal drops in the discontinued region of the equifrequency surface curve, also called the partial photonic bandgap. The incident light will be reflected completely by the photonic crystal and cannot propagate through it. Then the optical switching is in the "OFF" state.

For the practical applications of the equifrequency curve variation method, the photonic band structure should be carefully designed, and the structural parameter of the photonic crystal must be carefully selected. One limitation of the equifrequency curve variation method may lie in that this method can operate only for a probe light with a specified frequency and a specified range of incident angle.

Moreover, people have also proposed other methods for the realization of photonic crystal optical switching, such as the one based on the spontaneous emission cancellation in a photonic crystal doped with an ensemble of four-level nanoparticles, the interactions of three-level atoms with a photonic crystal heterostructure, and so on [24,25].

4.2 Photonic Crystal Optical Switching

In the experiment, the mechanisms of photonic bandgap shift, defect mode shift, and the optical bistable switching are widely used to demonstrate photonic crystal optical switching. Various phenomena, including the nonlinear Kerr effect, the electro-optic effect, the magneto-optic effect, and the thermo-optic effect, can be used to modulate the effective refractive index of photonic crystals. Accordingly, there are photonic crystal all-optical switching, electro-optic switching, magneto-optic switching, and thermo-optic switching, respectively.

4.2.1 Photonic Crystal Thermal-Optic Switching

Camargo *et al.* realized a thermo-optic switching in a two-dimensional semiconductor photonic crystal Mach–Zehnder interferometer operating at 1550 nm [26]. The two-dimensional photonic crystal Mach–Zehnder interferometer was fabricated in an $Al_{0.6}Ga_{0.4}As/GaAs/Al_{0.6}Ga_{0.4}As$ heterostructure on a GaAs substrate. The thermo-optic coefficient of the AlGaAs/GaAs photonic crystal heterostructure was $2.5 \times 10^{-4} K^{-1}$ [26]. A heater was used to change the temperature of one of the semiconductor photonic crystal waveguides. Accordingly, the refractive index of the photonic crystal waveguides was changed. The thermally induced variation of the refractive index of AlGaAs/GaAs epitaxial structure changed the phase contrast of the lights propagating in two arms of the photonic crystal Mach–Zehnder interferometer. This resulted in the changes of the transmittance of the probe light. When a π-phase shift was introduced, the maximum transmittance contrast could be obtained. The extinction ratio, extracted from the ratio of the maximum to minimum transmittance value, was –14 dB [26]. The response time of the photonic crystal thermo-optic switching was in the time scales ranging from millisecond to second order [27].

Reese *et al.* reported a nanosecond three-dimensional photonic crystal optical switching by using a colloidal photonic crystal fabricated by self-assembling of poly(*N*-isopropylacrylamide) (PNIPAM) nanogel colloidal nanoparticles [28]. PNIPAM colloidal nanoparticles will undergo a volume phase transition when heated above a low critical solution temperature of about 32°C. Below this critical temperature, the PNIPAM polymer hydrogel is highly swollen in water. Increasing the temperature to above the critical temperature would cause the polymer hydrogel to collapse and squeeze out water, which made the nanogel nanoparticles shrink in volume. Accordingly, the lattice constant was changed. This led to the variation of the position of the Bragg diffraction peak.

The wavelength of probe light was set at the central wavelength of the Bragg diffraction peak. A 1.9 μm, 3 ns pulsed laser was used as the pump light. The pump light was absorbed by the water contained in the sample to heat the photonic crystal. A pump light-induced temperature change from 30 to 35°C made the nanogel nanoparticles shrink. The transmittance contrast of 61% for the probe light was obtained and the response time of the switching was about 100 ns [28].

4.2.2   Photonic Crystal Electro-Optic Switching

Ozaki *et al.* reported a high-speed electro-optic switching based on the tunable defect mode in a one-dimensional photonic crystal containing a nematic liquid crystal defect layer [29]. The one-dimensional photonic crystal consisted of an alternating stack of SiO_2 and TiO_2 layers. A defect layer made of nematic liquid crystal E47 was inserted into the center of the one-dimensional photonic crystal. With the absence of an external electric field, the long axis of the liquid crystal molecular was aligned parallel to the plane of the dielectric layer. The molecule alignment of the nematic liquid crystal in the defect layer varied with the applied voltage, which changed the refractive index of the liquid crystal material. Accordingly, the effective refractive index of the photonic crystal was changed, which made the photonic bandgap and the defect mode shift. Ozaki *et al.* found that the resonant wave-length of the defect mode shifted from 690 nm to 633 nm when the applied voltage changed from zero to 12 V [29]. Then a photonic crystal optical switching could be realized. The response time of the optical switching was in the order of several tens of microseconds.

Recently, Samson *et al.* realized a metallic photonic crystal electro-optic switching by use of the frequency shift of a narrow-band Fano resonance mode induced by changes in the refractive index of an adjacent chalcogenide glass layer [30]. The metallic crystal was composed of square lattice of "$\bar{u}$" shape asymmetric split-ring slits embedded in a 70 nm-thick gold film, covered with a 200 nm-thick amorphous film of gallium lanthanum sulphide (GLS) glass. In such a metallic photonic crystal, the coupling of the trapped plasmon modes with the free-space radiation modes created a narrow, asymmetric, Fano-type reflection and transmission resonance line shape. At the beginning, the reflectivity of a 1450 nm incident light was 60%. Under the excitation of an applied voltage of 45 V, the GLS glass suffered a phase transition from the amorphous state to the crystalline state, which led to a large refractive index change of GLS glass. Accordingly, the reflection peak shifted. Then the reflectivity of the 1450 nm incident light decreased to 20%. A switching efficiency of 40% was achieved. The switching time was about 50 ns, which was determined by the characteristic transition time from amorphous state to crystalline state of the GLS glass.

4.2.3   Photonic Crystal Magneto-Optic Switching

Photonic crystal optical switching can be realized by use of the magneto-optic media based on the magneto-optic effect [31,32]. Wu *et al.* fabricated a magneto-optic $Bi_{0.8}Gd_{0.2}Lu_2Fe_5O_{12}$ film, by use of the liquid-phase-epitaxy method on a gadolinium gallium garnet substrate, which had a Faraday rotation coefficient of 83°/mm [33]. A one-dimensional photonic crystal was fabricated in a 600 nm-thick ridge waveguide of $Bi_{0.8}Gd_{0.2}Lu_2Fe_5O_{12}$ by use of focused ion beam etching method. The one-dimensional photonic crystal was connected with a 900 µm-long $Bi_{0.8}Gd_{0.2}Lu_2Fe_5O_{12}$ waveguide, which was used as a magnetically controlled polarization rotator. When a 1522.8 nm incident light propagated through the $Bi_{0.8}Gd_{0.2}Lu_2Fe_5O_{12}$ waveguide, a 90° polarization rotation could be achieved under the excitation of the longitudinal magnetization. For the magneto-optic photonic crystal, the TE and TM photonic bandgaps were spectrally separated. So, certain frequency of TE light could transmit through the TM bandgap. The 1522.8 nm TE incident light would be changed to be a TM polarization when it propagated through the $Bi_{0.8}Gd_{0.2}Lu_2Fe_5O_{12}$ waveguide under the excitation of a longitudinal magnetization, so that the incident light could propagate through the one-dimensional photonic crystal. Then the optical switching was in the "ON" state. When the external magnetization was switched off, the 1522.8 nm TE incident light would be reflected completely by the one-dimensional photonic crystal. Then the optical switching was in the "OFF" state.

4.2.4   Photonic Crystal All-Optical Switching

For the realization of a photonic crystal all-optical switching, three key characteristics must be considered: the ultrafast switching time, the ultralow operating pump-power, and the high switching efficiency. The switching time can be defined as the time needed for the transition between the "ON" and "OFF" state. The switching efficiency can be defined as the transmittance contrast between the "ON" and "OFF" state. The switching time is mainly determined by the nonlinear time response of the third-order nonlinear optical materials in the photonic crystals. The switching efficiency and the pump power are mainly determined by the magnitude of the third-order nonlinear optical coefficients of the nonlinear optical materials in the photonic crystals. For the conventional nonlinear optical materials, the third-order nonlinear susceptibility is relatively small. This results in a very high operating pump power. The operating pump intensity is in the order of several GW/cm^2. Such a high power consumption has greatly restricted the practical applications of photonic crystal all-optical switching. What is more, the response time slows down greatly when the optical nonlinearity of the materials is resonantly enhanced. This is a tremendous obstacle for the realization of ultrafast and low-power photonic crystal all-optical switching. Tremendous efforts have been made to realize ultrafast, low-power photonic crystal all-optical switching with high switching efficiency.

4.2.4.1   Study of high switching efficiency

High switching efficiency is one of the essential characteristics of photonic crystal all-optical switching, which determines whether or not the optical switching function meets the requirement of practical applications. Semiconductor materials, such as silicon and GaAs, are the very important basis for the development of modern optoelectric industry. Therefore, it is possible to realize integrated photonic circuits and chips based on two-dimensional semiconductor photonic crystals. The organic conjugated polymer materials possess relatively large third-order nonlinear optical susceptibility. Therefore, the conjugated polymer materials are also promising candidates for the realization of photonic crystal all-optical switching with high switching efficiency.

4.2.4.1.1 *Using silicon-based photonic crystals*

Scherbakov *et al.* reported a photonic crystal all-optical switching based on pump light-induced photonic bandgap shift in a high-quality SiO_2 opal infiltrated with vanadium dioxide (VO_2) [34]. The unique property of VO_2 lies in that it can take on a structural phase transition from the semiconductor state at room temperature to the metal state at a temperature higher than 70°C. Accordingly, the dielectric constant of VO_2 changes from 9 to 5.3 [34]. A 532 nm beam (with a pulse width of 10 ns and a repetition rate of 1 kHz) from a Nd:YAG laser system was used to excite the opal. The energy of the laser photons exceeded the energy bandgap of VO_2, which led to a strong one-photon absorption. This triggered the structural phase transition of VO_2 from the semiconductor state to the metal state, and an all-optical switching could be realized. A reflectivity contrast of 40% for the probe laser was obtained under the excitation of 1 GW/cm^2 pump light. The switching time was 100 µs, which originated from the slow relaxation dynamics of photo-excited carriers generated in VO_2.

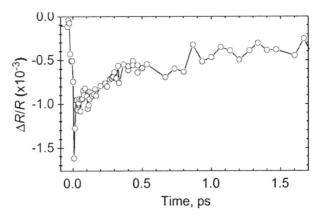

Figure 4.7 Three-dimensional photonic crystal all-optical switching. Reprinted from *J. Luminescience* **108**, D. A. Mazurenko, *et al.*, Ultrafast all-optical switching in a threedimensional photonic crystal, 163–166, Copyright (2004), with permission from Elsevier.

Mazurenko *et al.* reported a three-dimensional silicon photonic crystal all-optical switching in a SiO_2 opal infiltrated with a mixture of nanocrystalline and amorphous silicon [35,36]. The filling factor of silicon was close to 100%. The beam (with a pulse width of 30 fs and a repetition rate of 75 MHz) with a wavelength of 800 nm output from a Ti-sapphire laser was used as the pump laser. Under the excitation of the pump laser, free carriers were generated in silicon, which led to a variation of the refractive index of silicon. This changed the effective refractive index of the photonic crystal and made the photonic bandgap shift. Accordingly, the reflectivity of the probe laser was changed. Then, an all-optical switching could be achieved. The measured reflectivity contrast was 46% for an 800 nm probe laser, as shown in Fig. 4.7 [35]. The pump intensity was 2.1 GW/cm^2. An ultrafast switching time of 30 fs was achieved, which was determined by the pulse duration of the pump laser.

Leonard *et al.* fabricated a two-dimensional triangular lattice silicon photonic crystal [37]. The lattice constant and the diameter of air holes were 500 nm and 412 nm, respectively. A 800 nm beam (with a pulse width of 300 fs and a pulse repetition rate of 250 kHz) from a Ti:sapphire regenerative amplifier was used as the pump laser. The 1.9 μm beam from an optical parameter amplifier system pumped by the Ti:sapphire regenerative amplifier was used as the probe laser. The wavelength of the probe laser dropped in the high-frequency edge of the photonic bandgap. Under the excitation of the pump laser, free carriers were injected into the silicon photonic crystal, which decreased the refractive index of silicon, and accordingly the effective refractive index of the photonic crystal. As a result, the photonic bandgap shifted in the direction of short wavelength and the reflectivity of the probe laser changed. Then an all-optical switching could be realized. The energy density of the pump laser was 1.3 mJ/cm^2. When the probe laser pulse was far away from the pump laser pulse in the time sequence, the reflectivity of the probe laser was only 1%. Then the optical switching was in the "OFF" state. When the probe laser pulse was overlapped with the pump laser pulse in the time sequence, the reflectivity of the probe laser increased, which originated from the blue-shift of the photonic bandgap caused by the pump laser induced free-carrier injection. The reflectivity of the probe laser reached the maximum value of 60% when the probe pulse passed over the pump pulse in the time sequence. Therefore, a high reflectivity of close to 60% was obtained. The switching time was about 400 fs. The maximum reflectivity maintained in the 70 ps time delay between the pump and probe pulses, which originated from the slow relaxation dynamics of the carriers in the silicon governed by the Auger recombination and surface recombination [37].

4.2.4.1.2 *Using polymer photonic crystals*

Zhang's group obtained a 63% transmittance contrast by use of nonlinear three-dimensional polystyrene photonic crystals [38]. They fabricated a three-dimensional organic photonic crystal made of monodispersed polystyrene nanospheres with a diameter of 240 nm by use of the vertical deposition method. The 800 nm beam from a Ti:sapphire laser system (with a pulse duration of 120 fs and a pulse repetition rate of 10 Hz) was used as the pump light. The beam from an optical parameter amplifier pumped by the second harmonic of the same Ti:sapphire laser was used as the probe light. The pump and probe light were collinearly incident upon the polystyrene opal along the Γ–L direction. The optical nonlinearity of polystyrene originated from the delocalization of conjugated π-electrons along the polymer chains, which led to a large third-order nonlinear optical susceptibility and a subpicosecond response time. Under the excitation of a pump laser, the refractive index of polystyrene increased, which led to the shift of the photonic bandgap. If the frequency of the probe light was set at the edge of photonic bandgap, the transmittance of the probe light would vary with the pump intensity. Then a photonic crystal all-optical switching could be realized. The polystyrene photonic crystal all-optical switching effect is shown in Fig. 4.8. The wavelength of the probe light and the pump intensity were 561 nm and 27.5 GW/cm^2, respectively. They also fabricated a two-dimensional polystyrene photonic crystal by use of the focused ion beam etching method [39]. The photonic crystal was composed of regular square arrays of cylindrical air holes embedded in the 300 nm-thick polystyrene slab. The lattice constant and the diameter of air holes were 220 and 180 nm, respectively. The 1.064 μm beam (with a pulse duration of 25 ps and a pulse repetition rate of 10 Hz) from a YAG laser was used as the pump light, while the beam (with a pulse duration of 10 ps and a repetition rate of 10 Hz) from an optical parameter amplifier pumped by the same YAG laser was used as the probe light. The probe light was coupled into the photonic crystal by use of the prism-film coupling method. The probe light propagated through the photonic crystal in the Γ–X direction. The pump light was incident normally to the upper surface plane of the photonic crystal. A delay line was used to adjust the temporal relations between the pump and probe light. The photonic crystal all-optical switching was realized based on the shift of the photonic bandgap induced by the pump light. A high switching efficiency of more than 60% was obtained under the excitation of a 16.7 GW/cm^2 pump light. The switching time was 10 ps, which was determined by the duration of the pump pulse.

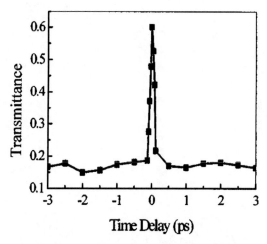

Figure 4.8   Polystyrene opal all-optical switching effect. Reprinted with permission from *Appl. Phys. Lett.*, **86**(15), Y. Liu, *et al.*, Sub-picosecond optical switching in polystyrene opal, Copyright 2005, AIP Publishing LLC.

In 2005, Gong's group achieved a very high transmittance contrast of over 70% by use of the pump light-induced defect mode shift in a two-dimensional polystyrene photonic crystal [40]. The focused ion beam etching method was used to fabricate the periodic patterns of the two-dimensional polystyrene photonic crystal. A line defect with a width of 310 nm was introduced in the center of the two-dimensional photonic crystal. The lattice constant and the radius of air holes were 220 nm and 90 nm, respectively. The patterned area was about 2.5 μm × 100 μm. They adopted a picosecond pump and the probe method to study the all-optical switching effect. The experimental setup is shown in Fig. 4.9a. The 1.064 μm beam (with a pulse duration of 25 ps and a pulse repetition rate of 10 Hz) from a YAG laser was used as the pump light, while the beam (with a pulse duration of 10 ps and a repetition rate of 10 Hz) from an optical parameter amplifier pumped by the same YAG laser was used as the probe light. Both the pump and probe light were TE polarized light with their electric field vector parallel to the polystyrene slab. The probe light, coupled into the photonic crystal by use of the prism-film coupling method, propagated through the photonic crystal in the direction perpendicular to the line defect. The pump light was incident normally to the upper surface plane of the polystyrene photonic crystal. A delay line was used to adjust the temporal relations between the pump and probe light. The probe light transmitted through the photonic crystal was detected by a monochromator, whose output signals were magnified by a photomultiplier before they were input into an oscilloscope. Then, a computer was used to collect and handle the output data from the oscilloscope. The measured transmittance was normalized to that of a reference

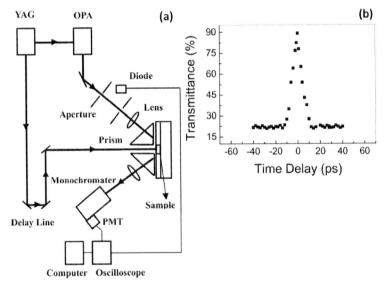

Figure 4.9 Two-dimensional polystyrene photonic crystal all-optical switching. (a) Experimental setup. The thick lines represent optical connections, while thin lines are electronic connections. (b) All-optical switching effect. Reprinted with permission from *Appl. Phys. Lett.*, **87**(23), X. Hu, *et al.*, All-optical switching of defect mode in two-dimensional nonlinear organic photonic crystals, Copyright 2005, AIP Publishing LLC.

polystyrene waveguide without the patterned areas, which is one of the normal methods to study the transmittance properties of two-dimensional photonic crystals. The measured transmittance and central wavelength of the defect mode were 89% and 545 nm, respectively. The quality factor and the line width of the defect mode were about 110 and 5 nm, respectively. According to the nonlinear optical Kerr effect, the refractive index of polystyrene increases under the excitation of a strong pump light with a wavelength of 1.064 μm, which makes the effective refractive index of the two-dimensional photonic crystal increase. As a result, the photonic bandgap and the defect mode shift in the direction of long wavelength. The defect mode shifted about 5 nm when the peak intensity of the pump light was 18.7 GW/cm^2. Then a photonic crystal all-optical switching could be realized. High switching efficiency of 70% and an ultrafast switching time of less than 10 ps were achieved, as shown in Fig. 4.9b.

4.2.4.2   Study of ultrafast switching time

The ultrafast switching time is another basic characteristic of photonic crystal all-optical switching. The ultrafast photonic crystal all-optical switching plays a very important role in the field of ultrahigh speed information processing. The switching time is mainly determined by the nonlinear time response properties of nonlinear optical materials constructing the photonic crystal.

4.2.4.2.1   *Using semiconductor photonic crystals*

Bristow *et al.* fabricated a two-dimensional AlGaAs photonic crystal by use of electron beam lithography and chemically assisted ion beam etching method [41]. The two-dimensional photonic crystal was composed of triangular lattice of air holes embedded in a 400 nm-thick $Al_{0.13}Ga_{0.87}As$ slab. The lattice constant and the filling factor of air were 700 nm and 31%, respectively [41]. To avoid the nonlinear effects of unpatterned waveguides on the nonlinear responses of photonic crystal waveguides, they measured the surface reflectivity of the AlGaAs photonic crystal based on the femtosecond pump and probe method. The surface reflectivity spectrum could provide useful information about the photonic bandgap structure through resonantly coupling the incident probe laser into the guided-wave modes of the AlGaAs photonic crystal waveguide when the phase matching condition was satisfied: $k_g = k_i \sin\theta$, where k_g and k_i are the wave vector of the guided-wave modes of the photonic crystal waveguide, and the incident probe light, respectively. θ is the incident angle [41]. An 840 nm beam from a Ti:sapphire laser system (with a pulse duration of 130 fs and a repetition rate of 1 kHz) was used as the pump light. A white light generated in a thin sapphire crystal excited by a focused 840 nm laser beam output from the femtosecond laser system was used as the probe light. The incident angle was set at 45°. Under the excitation of a 2.3 mJ/cm^2 pump light, the free carriers were generated in AlGaAs, which decreased the refractive index of AlGaAs. This made the photonic bandgap structure of the AlGaAs photonic crystal change and the peak wavelength of the reflection shift from 882 nm to 877 nm. Then a photonic crystal all-optical switching could be realized [41]. A high reflectivity contrast of 30% was achieved. The switching time was about 7.7 ps, which originated from the fast carrier relaxation dynamics of the AlGaAs material.

Hache and Bourgeois realized an all-optical switching in a one-dimensional photonic crystal, constructed by alternatively stacking SiO$_2$ and amorphous Si layers, based on pump light-induced photonic bandgap shift [42]. The thickness was 410 nm for SiO$_2$ layers and 235 nm for Si layers. The center of the photonic bandgap was located at 1400 nm. The femtosecond pump and probe method was used to study the all-optical switching effect. The 1.71 μm idler beam from a femtosecond optical parameter amplifier system (with a pulse duration of 240 fs and a repetition rate of 100 kHz) was used as the pump light. While the 1.51 μm signal beam was used as the probe light, which dropped in the long-wavelength edge of the photonic bandgap of the one-dimensional photonic crystal. A non-collinear geometry was used for the pump and probe laser in their experiment. The probe light transmitted through the one-dimensional photonic crystal was detected by a photodiode detector, whose output was detected by a lock-in amplifier. A computer was used to handle the signals output from the lock-in amplifier. A delay line was used to adjust the temporal relations of the pump and probe lights. Under the excitation of the pump light, free carriers generated in amorphous silicon changed the refractive index of amorphous silicon, and accordingly the effective refractive index of the one-dimensional photonic crystal. Then the photonic bandgap shifted in the short-wavelength and the transmittance of the probe light changed. Under the excitation of an 18 GW/cm^2 pump light, a switching efficiency of 17% was obtained. The response time of the optical switching was about 400 fs, which originated from the ultrafast relaxation dynamics of free carriers in amorphous silicon.

Shimizu and Ishihara fabricated a one-dimensional photonic crystal made of the inorganic-organic layered perovskite semi-conductor material bis-(phenethyl ammonium) tetraiodoplumbate (PEPI) [(C$_6$H$_5$C$_2$H$_4$NH$_3$)$_2$PbI$_4$] [43]. (C$_6$H$_5$C$_2$H$_4$NH$_3$)$_2$PbI$_4$ has a self-organized multiple quantum-well structure, which leads to a large exciton oscillator strength and large optical nonlinearity due to strong dielectric confinement effect [44]. The large optical Stark effect of (C$_6$H$_5$C$_2$H$_4$NH$_3$)$_2$PbI$_4$ results in a remarkable blue shift of the transmittance dips of the photonic crystal under the excitation of a strong pump light. Then a photonic crystal all-optical switching can be realized. The wavelength, the pulse duration, and the intensity of the pump light were 538 nm, 200 fs, and 0.11 mJ/cm^2, respectively. The pulse duration of the probe light was 100 fs. The response time of the optical switching was about 200 fs.

Mazurenko *et al.* achieved a 30 fs ultrafast response time for a three-dimensional Si-based photonic crystal all-optical switching [45,46]. They fabricated a three-dimensional photonic crystal with a face-centered-cubic lattice through self-assembling of SiO$_2$ nanospheres with a diameter of 230 nm. The SiO$_2$ opal was infiltrated with amorphous-nanocrystalline silicon with a filling factor close to 100%. The photonic bandgap was around 800 nm. The opal sample was cut so that the surface was almost parallel to the (111) plane. A 30 fs, 35.5 MHz Ti:sapphire femtosecond laser system was used as the light source. The difference in the propagation of the Bragg diffracted light and the surface reflected light allowed people to distinguish the effects from the photonic crystal structure and from the ordinary surface reflection of the sample [45]. Under the excitation of a pump light, free carriers generated in silicon changed the refractive index of silicon, which made a variation of the effective refractive index of the photonic crystal. As a result, the position and the width of the photonic bandgap were changed. Under the excitation of a 2 GW/cm^2 pump light, a reflectivity contrast up to 1% was achieved for the optical switching. An ultrafast switching time of 30 fs was realized, which originated from the ultrafast relaxation of photogenerated carriers in amorphous silicon.

4.2.4.2.2 *Using polymer photonic crystals*

Recently, Liu *et al.* achieved a 10 fs ultrafast all-optical switching in a polystyrene opal [47]. The polystyrene opal was fabricated by use of the pressure-controlled isothermal heating vertical deposition method made of polystyrene nanospheres with a diameter of 350 nm. The photonic bandgap was located at 800 nm. The femtosecond pump and probe method was used to measure the all-optical switching effect [47]. A 8 fs beam from a Ti: sapphire laser system with a repetition rate of 80 MHz was used as the light source. The wavelength was 785 nm for the pump and probe lights, which was located at the short-wavelength edge of the photonic bandgap. The probe light was incident in the sample in the Γ–L direction. There was a 10° difference between the incident direction of the pump and probe light. At the beginning, the transmittance of the probe light was in the minimum value of 25%. The optical switching was in the "OFF" state. Under the excitation of the pump light, the refractive index of polystyrene increases, which makes the photonic bandgap shift and the transmittance of the probe light increase. Under the excitation of a 20.6 GW/cm^2 pump light, the transmittance of the probe light reached the maximum value, 45%. Then the optical switching was in the "ON" state. An ultrafast switching time of 10 fs was achieved [47].

4.2.4.3 Study of low-power and ultrafast all-optical switching

The operating threshold power and the switching efficiency for a photonic crystal all-optical switching are primarily determined by the magnitude of the third-order nonlinear optical susceptibility of the medium constructing the photonic crystal. Unfortunately, the third-order nonlinear susceptibility of conventional optical materials is relatively small. Moreover, resonant nonlinearity enhancement is accompanied by the sacrifice of the response time. Great efforts have been made to solve this problem. Various approaches have been proposed to realize ultralow power and ultrafast photonic crystal all-optical switching. In general, three approaches are widely adopted: constructing new photonic materials possessing large nonlinear optical coefficients [48], constructing novel photonic crystal structure to enhance the interactions of light and matter [49], and exploring new mechanisms of optical switching [50].

4.2.4.3.1 *Constructing new photonic materials*

Photonic materials having ultrafast response and large nonlinear optical coefficients could find great potential applications not only in the fields of integrated photonic devices but also in the basic research fields of nonlinear optics and nano-photonics. Constructing novel materials is a direct approach to obtaining large nonlinear susceptibility. At the same time, this approach is difficult and arduous due to its complicated synthesis process. Up to now, constructing materials possessing large nonlinear susceptibility and ultrafast response is still a great challenge.

4.2.4.3.1.1 *Chemically synthesizing new nonlinear materials*

The optical nonlinearity of organic conjugated polymer materials originates from the delocalization of conjugated π-electrons along the polymer chains. Therefore, organic conjugated polymer materials have relatively large third-order nonlinear susceptibility and subpicosecond response time [51]. Adopting donor–acceptor substitution is an effective approach to improving the third-order optical nonlinearity of organic conjugated polymer materials [52]. In 2005, May *et al.* synthesized an organic material 1,1,2-tricyano-2-[(4-dimethyl-aminophenyl)ethynyl]ethane (TDMEE), which is a kind of donor-substituted cyanoethynylethene material [53]. The *N,N*-dimethylanilino groups were used as donors, and the cyano groups as acceptors. They found that the third-order molecular polarizability was $\gamma = 53 \times 10^{-48}$ m^5/V^2, which was one order of magnitude larger than that of poly(triacetylene) oligomers. Subsequently, Michinobu *et al.* synthesized a kind of donor–acceptor organic molecules with 4-(dimethylamino)phenyl as donor and 1,1,4,4-tetracyanobuta-1,3-diene acting as acceptor [54]. The third-order molecular polarizability was $\gamma = 12 \times 10^{-48}$ m^5/V^2. Esembeson *et al.* fabricated a nonlinear organic polymer 2-[4-(dimethylamino)phenyl]-3-([4(dimethylamino)phenyl]-ethynyl) buta-1,3-diene-1,1,4,4-tetracarbonitrile (DDMEBT) by use of chemical synthesis method [55]. The third-order nonlinear susceptibility of the DDMEBT film was $\chi^{(3)} = 2 \times 10^{-19}$ m^2/V^2, corresponding to a nonlinear refractive index of $n_2 = 1.7 \times 10^{-13}$ cm^2/W at the off-resonant wavelength. The remarkable advantage of DDMEBT molecules is that DDMEBT molecules have a nonplanar structure, which can reduce the intermolecular interactions and is of great benefit for the fabrication of high-quality film for the practical applications. Recently, Chi *et al.* also reported an ultrafast response time of less than 100 fs and large third-order nonlinear susceptibility of 2.1×10^{-10} esu for a kind of substituted polyacetylene materials [56].

It has been pointed out that large third-order nonlinear susceptibility of the order of 10^{-12} to 10^{-7} esu could be obtained in two kinds of quantum wires: the band insulators of silicon polymers and Peierls insulators of π-conjugated polymers and platinum halides [57–59]. Kishida *et al.* fabricated a kind of one-dimensional Mott-Hubbard insulators, which was a type of quantum wires made of halogen-bridged Ni compounds, including [Ni(cyclohexanediamine)$_2$Br]Br$_2$ and [Ni(cyclohexanediamine)$_2$ Cl]Cl$_2$ [48]. The third-order nonlinear susceptibility was measured to be of the order of 10^{-8} to 10^{-5} esu, which originated from the large dipole moment between the lowest two excited states of the one-dimensional Mott–Hubbard insulators [48]. These materials possess the largest third-order susceptibility compared with the conventional organic and inorganic optical materials.

Optical materials possessing large magneto-optical coefficients may also have great potential applications in the fields of magneto-optic switching. Jin *et al.* fabricated paramagnetic crystals of NaTb(WO$_4$)$_2$ by use of the solid-state reaction method [60]. They found that this material has a large magnetic susceptibility of 8.0×10^{-5} emu/g and ultrafast response time of subpicosecond order. NaTb(WO$_4$)$_2$ is a very promising candidate for the realization of ultrafast and low-power magneto-optic switching.

4.2.4.3.1.2 *Forming composite material structures*

It is very difficult to design and synthesize a new kind of optical material. Forming composite material structures by utilizing the advantages of different kinds of materials is another effective approach to constructing photonic materials with large nonlinear optical susceptibility and ultrafast response. Two kinds of structures are widely adopted: the layered composite structure and the nanocomposite structure [61,62].

Layered composite structures

A layered composite structure is constructed by alternatively stacking subwavelength-thick layers of two kinds of optical materials. According to the effective medium theory, the effective third-order nonlinear susceptibility $\chi_{\text{eff}}^{(3)}$ of the layered composite structure for light polarized perpendicular to the plane of layers can be calculated by the following relation [63]:

$$\chi_{\text{eff}}^{(3)} = f_a \left| \frac{\varepsilon_{\text{eff}}}{\varepsilon_a} \right|^2 \left(\frac{\varepsilon_{\text{eff}}}{\varepsilon_a} \right)^2 \chi_a^{(3)} + f_b \left| \frac{\varepsilon_{\text{eff}}}{\varepsilon_b} \right|^2 \left(\frac{\varepsilon_{\text{eff}}}{\varepsilon_b} \right)^2 \chi_b^{(3)} \qquad (4.1)$$

where f_a and f_b are the filling factors of materials a and b, respectively, ε_a and ε_b are the dielectric constants of materials a and b, respectively, $\chi_a^{(3)}$ and $\chi_b^{(3)}$ are the third-order nonlinear susceptibilities of materials a and b, respectively, and ε_{eff} is the effective dielectric constant of the composite material. The terms $\left| \frac{\varepsilon_{\text{eff}}}{\varepsilon_a} \right|^2 \left(\frac{\varepsilon_{\text{eff}}}{\varepsilon_a} \right)^2$ and $\left| \frac{\varepsilon_{\text{eff}}}{\varepsilon_b} \right|^2 \left(\frac{\varepsilon_{\text{eff}}}{\varepsilon_b} \right)^2$ are the local-field enhancement factors for materials a and b, respectively, which originate from the strong dielectric confinement effect caused by the striking anisotropy of the layered composite. It is very clear that the effective third-order nonlinear susceptibility of the layered composite can be much larger than the nonlinear susceptibility of either the component materials. The physical mechanism of the nonlinearity enhancement can be understood as follows: the electric field distribution of an incident light is not uniform in the component material. Under suitable conditions, the electric field strength within the larger nonlinear component will exceed the spatially averaged electric field strength, which results in the fact that the effective nonlinear susceptibility can exceed that of either the component materials [64]. Moreover, the nonlinear time response properties of layered composite are mainly determined by the time response of the component materials. Ultrafast response time can be expected for the layered composite when the component materials have ultrafast response time.

Noble metal materials, such as gold, silver, and copper, have a large third-order nonlinear susceptibility, which is $10^5 \sim 10^6$ times larger than that of the conventional dielectric materials [65].

However, in the visible and near-infrared range, noble metal materials are almost opaque. This makes that the excellent optical nonlinearity of noble metal materials is not available and feasible for the practical applications when bulk metal materials are used. Scalore *et al.* and Bloemer *et al.* pointed out that a high transmittance can be obtained in a metal-dielectric multilayer structure, also called metallodielectric photonic crystal structure, even if the total thickness of metal layers is much larger than that of the skin depth of the metal material [66,67]. The reason lies in the strong Bragg scattering and interference in the interfaces of metal and dielectric layers. The pass band can be tuned from the visible to the infrared range through adjusting structural parameters, such as the thickness of metal and dielectric layers, and the repetition period. What is more, according to the electromagnetic variation theory, in the low-frequency pass band, also called the dielectric band, the electric field distribution of incident electromagnetic waves will be mainly concentrated in the dielectric layers. High transmittance and low losses can be expected in this photonic crystal structure in the visible and near-infrared range. Therefore, the metal-dielectric multilayer structures may find great potential applications in the field of photonic crystal all-optical switching.

Ma *et al.* reported an Ag/TiO_2 multilayer structure, which was composed of alternating Ag layers and TiO_2 layers. [68]. The number of the repeating periods was 5. The thickness was 310 nm for TiO_2 layers and 25 nm for Ag layers. The nonlinear optical susceptibility of the composite material was measured by use of the femtosecond optical Kerr effect (OKE) technique. A 800 nm laser from a Ti:sapphire laser system (with a pulse duration of 120 fs and a repetition rate of 82 MHz) was used as the light source. The third-order nonlinear susceptibility was determined to be 3.2×10^{-9} esu, and the response time was 2 ps. This indicates that the Ag/TiO_2 multilayer structure possesses excellent optical nonlinearity properties. Lepeshkin *et al.* also reported a Cu/SiO_2 multilayer structure with five periods [69]. The thickness was 16 nm for Cu and 98 nm for SiO_2. The third-order nonlinear optical susceptibility of the Cu/SiO_2 multilayer was one order of magnitude larger than that of the bulk copper.

In the metal-dielectric multilayer structure, also called the metallodielectric photonic crystal, the nodes of the electric field distribution is situated in the metal layers in order to improve the transmittance. In reality, this design also restricts the enhancement of the optical nonlinearity of the total system due to the small electric field distribution in metal layers. In 2009, Du *et al.* presented a novel metal-dielectric heterostructure to obtain very large enhancement of optical nonlinearity [70]. The metal-dielectric heterostructure was composed of truncated all-dielectric one-dimensional photonic crystal connected with a thick metal film. It has been pointed out that an electromagnetic wave can tunnel through a heterostructure composed of a permittivity negative material and a permeability negative material when the impedance matching and the phase mashing conditions are satisfied [71]. The photonic crystal can be considered to be an effective permeability negative material in a frequency region of the photonic bandgap due to the strong dispersion properties [72]. Therefore, the tunneling electromagnetic modes can exist in the interface of a photonic crystal and a thick metal film, which can provide a negative permittivity [73]. The high transmittance and the strong electric field confinement in the metal surface can provide a very large nonlinearity enhancement for the metal-dielectric heterostructure. Du *et al.* found that the third-order nonlinear susceptibility of the metal-dielectric heterostructures could be two orders of magnitude larger than that of the conventional metal-dielectric multilayer structures.

Recently, Scalora *et al.* presented another approach to improving the nonlinearity enhancement of the metal-dielectric multilayer structures [74]. It has been pointed out that the thickness of the dielectric layers of $\lambda/2$ is the better structure parameter for the resonant tunneling of an electromagnetic wave through a metal-dielectric multilayer structure than the $\lambda/4$ dielectric layer [75]. While in the "$\lambda/2$" metal-dielectric multilayer structure, the metal layers act as mirrors, which is not helpful for the confinement of light in the metal layers and the nonlinearity enhancement. The resonant properties of the metal-dielectric multilayer structure can be destroyed by introducing a chirp in the thickness of the dielectric layers. A very simple way to introduce the chirp is to make the thickness of only the external first and last dielectric layers different from that of the inner dielectric layers. Scalora *et al.* found that chirping dielectric layer thickness could dramatically improve the transmittance of the metal-dielectric structure and achieve large electric field inside the metal layers to enhance the optical nonlinearity of the total structure [74].

Nanocomposite structures

Forming nanocomposite structures is another very effective and feasible approach to obtaining photonic materials with large nonlinear susceptibility and ultrafast response time. The nanocomposite material is constructed by nanoparticle inclusions dispersed in a host material. According to the Maxwell Garnett theory, the effective third-order nonlinear susceptibility $\chi_{\text{eff}}^{(3)}$ of the nanocomposite material can be estimated by the following relation [76]:

$$\chi_{\text{eff}}^{(3)} \approx f_a \left| \frac{\varepsilon_{\text{eff}} + 2\varepsilon_b}{\varepsilon_a + 2\varepsilon_b} \right|^2 \left(\frac{\varepsilon_{\text{eff}} + 2\varepsilon_b}{\varepsilon_a + 2\varepsilon_b} \right)^2 \chi_a^{(3)} + f_b \left| \frac{\varepsilon_{\text{eff}} + 2\varepsilon_b}{3\varepsilon_b} \right|^2 \left(\frac{\varepsilon_{\text{eff}} + 2\varepsilon_b}{3\varepsilon_b} \right)^2 \chi_b^{(3)}$$

(4.2)

where f_a and f_b are the filling factors of inclusion a and host material b, respectively, ε_a and ε_b are the dielectric constants of inclusion a and host material b, respectively, and $\chi_a^{(3)}$ and $\chi_b^{(3)}$ are the third-order nonlinear susceptibilities of inclusion a and host material b, respectively. The terms $\left| \frac{\varepsilon_{\text{eff}} + 2\varepsilon_b}{\varepsilon_a + 2\varepsilon_b} \right|^2 \left(\frac{\varepsilon_{\text{eff}} + 2\varepsilon_b}{\varepsilon_a + 2\varepsilon_b} \right)^2$ and $\left| \frac{\varepsilon_{\text{eff}} + 2\varepsilon_b}{3\varepsilon} \right|^2 \left(\frac{\varepsilon_{\text{eff}} + 2\varepsilon_b}{3\varepsilon} \right)^2$ are local-field enhancement factors for inclusion a and host material b, respectively, which originates from the strong dielectric confinement effect. It is very clear that the effective third-order nonlinear susceptibility of the nanocomposite material can be much larger than the nonlinear susceptibility of either the component materials. The physical mechanism of the nonlinearity enhancement can be understood as follows: the electric field distribution of an incident light is not uniformly allotted between the inclusions and the host material due to the difference in the dielectric constant of the two kinds of materials. Under suitable conditions, the electric field strength within the more nonlinear component will exceed the spatially averaged electric field strength, which results in the fact that the effective nonlinear susceptibility can exceed that of either the component materials [76]. Moreover, the nonlinear time response properties are mainly determined by the time response of the component materials. Ultrafast response time can also be expected when the component materials have ultrafast response time.

The nanoparticles of noble metal materials have much larger third-order nonlinear susceptibility than that of their bulk materials and films due to strong quantum confinement effect. Moreover, when the frequency of the incident light is in resonance with the surface plasmon resonance (SPR) frequency of the metal nanoparticles, extremely large nonlinear susceptibility can be expected for metal nanoparticles. At the SPR peak, the term $\varepsilon_a + 2\varepsilon_b = 0$, which results in a very large third-order nonlinear susceptibility for the nanoparticles [77]. Moreover, ultrafast response of subpicosecond order can be achieved due to the ultrafast relaxation dynamics of SPR. Song *et al.* fabricated a $SrBi_2Nb_2O_9$ film containing Ag nanocrystal inclusion by use of the pulsed laser deposition technology [77]. The size of the Ag nanocrystal was about 5 nm. The third-order nonlinear susceptibility was measured to be 8.052×10^{-7} esu for the $Ag:SrBi_2Nb_2O_9$ nanocomposite material by use of the Z-scan technique with an excitation wavelength of 532 nm. Liao *et al.* fabricated a nanocomposite film composed of nanocrystalline Au and ZnO particles by use of the magnetron sputtering technique [78]. The diameter was 40 nm for ZnO nanoparticles and 30 nm for Au nanoparticles. Liao *et al.* measured the third-order nonlinear optical susceptibility of the Au:ZnO nanocomposite by use of a degenerate four-wave mixing method, and found that the $\chi^{(3)}$ was 2×10^{-6} esu under the excitation of a 532 nm, 70 ps laser [78]. Zhang *et al.* reported a nonlinear susceptibility of 4.8×10^{-10} esu and an ultrafast response time of 210 fs for Ag:BaO nanocomposite films at 820 nm [79]. Zhou *et al.* also reported an $Ag:Bi_2O_3$ nanocomposite film with a third-order nonlinear optical susceptibility of 4.1×10^{-10} esu and an ultrafast response time of 550 fs [80]. Tanaka and Saitoh reported a very large of nonlinear refractive index of 10^{-3} cm^2/GW at 1064 nm for a Se-loaded zeolite composite material [81]. Rativa *et al.* reported a lead–germanium oxide composite film having a nonlinear refractive index of 2×10^{-4} cm^2/GW at 800 nm [82]. Araujo *et al.* reported a kind of heavy metal-oxide glasses containing lead and bismuth having a nonlinear refractive index of 10^{-5} cm^2/GW order at 1064 nm and a response time of picosecond order [83].

Semiconductor quantum dots also have large third-order optical nonlinear properties due to strong quantum confinement effect caused by that fact that their size is close to the Bohr radius of excitons [84,85]. Ultrafast response time can be expected for semiconductor quantum dots due to ultrafast exciton relaxation time. It has been reported that the third-order nonlinear optical susceptibility of CdSeS quantum dots was in the order of $10^{-9} \sim 10^{-10}$ esu, while that of the CuS quantum dots was in the order of 10^{-7} esu [86]. The response time of CdSeS quantum dots was in the order of several tens of picoseconds. Recently, Moreels *et al.* reported that PbSe quantum dots had a very large nonlinear refractive index of -4×10^{-2} cm^2/GW and an ultrafast response time of the order of several picoseconds in the near-infrared range [87]. Like metal-dielectric nanocomposite materials, nanocomposite materials composed of semiconductor quantum dots dispersed in a dielectric matrix can also provide very large third-order nonlinear suscepti-bility. Schwerzel and Du reported a nonlinear refractive index of 10^{-4} cm^2/GW order at 532 nm for a nanocomposite material composed of CdS quantum dots dispersed in the polymer matrix [88–90]. Hosono *et al.* fabricated a nanocomposite material composed of amorphous red phosphorous quantum dots with an average diameter of 5 nm dispersed in SiO$_2$ glass by use of the ion implantation technique [91]. Hosono *et al.* measured the third-order nonlinear optical susceptibility of the nanocomposite by use of the degenerate four-wave mixing technique with a 390 nm, 20 ns excimer laser used as the light source. High nonlinear susceptibility of 1×10^{-6} esu was obtained for the P:SiO$_2$ nanocomposite material. Ohtsuka *et al.* also fabricated a nanocomposite material composed of CdTe quantum dots embedded in a SiO$_2$ glass by use of the pulsed laser evaporation method [92]. Ohtsuka *et al.* measured the third-order nonlinear optical susceptibility of the nanocomposite by use of the degenerate four-wave mixing technique with a 5 ps, 4 MHz dye laser pumped by a Nd:YAG laser system used as the light source. The third-order nonlinear susceptibility of the CdTe:SiO$_2$ nanocomposite was estimated to be 4×10^{-7} esu at 580 nm.

4.2.4.3.2 *Exploring new mechanisms of optical switching*

People also take great efforts to explore new mechanisms for the realization of ultrafast and low-power photonic crystal all-optical switching based on various optical phenomena and effects, such as using self-collimating effect [93], anomalous refraction effect [94], slow light effect [95–98], optical solitons [99], coupled-microcavity-induced reflection [100], and so on. Moreover, the double resonance configuration, i.e., the pump and the probe light are resonant with two photonic crystal microcavity modes, respectively, has been adopted to demonstrate photonic crystal all-optical switching [101]. Metamaterials are also used to demonstrate photonic crystal all-optical switching [102]. The coherent control technique could also be used to achieve a low-power photonic crystal optical bistable switching [103]. The coherent control technique could be used to tailor the relative phase between the different spectral components of an incident laser pulse, which can help to enhance the peak energy of the incident laser pulse coupled into a photonic crystal microcavity [104]. As a result, the operating threshold power for an optical bistable switching could be reduced.

4.2.4.3.2.1 *The critical dielectric constant method*

Husakou and Herrmann presented a mechanism of the all-optical switching for a weak light controlling the propagation of a signal light in a metal-dielectric multilayer structure [105]. The effective dielectric constant of a metal-dielectric multilayer structure can be calculated by [105]

$$\varepsilon_{\text{eff}} = \frac{d_{\text{m}}}{a}\varepsilon_{\text{m}} + \frac{d_{\text{d}}}{a}\varepsilon_{\text{d}} \tag{4.3}$$

where d_{m} and d_{d} are the thickness of the metal and dielectric layers, respectively; a is the lattice constant, $a = d_{\text{m}} + d_{\text{d}}$; and ε_{m} and ε_{d} are the dielectric constants of the metal and dielectric layers, respectively. A negative effective dielectric constant, but close to zero, can be achieved by carefully designing the structural parameters of the metal-dielectric multilayer structure. The negative dielectric constant makes the metal-dielectric multilayer structure behave like a bulk metal. At the beginning, the probe light cannot propagate through the metal-dielectric multilayer structure due to strong reflection. A remarkable variation in the effective dielectric constant can be achieved under the excitation of a weak pump light due to the strong third-order optical nonlinearity of the metal-dielectric multilayer structure. This makes a transition of the effective dielectric constant from a negative value to a positive value. As a result, the metal-dielectric multilayer structure will show the dielectric behaviors. As a result, the probe light can propagate through the metal-dielectric multilayer structure with a high transmittance. The switching time is determined not only by the intrinsic time response properties of the nonlinear materials, such as the metal and dielectric medium, but also by the feedback of the metal-dielectric multilayer structure [105]. On the one hand, the electric field distribution in the nonlinear materials will change the refractive index of the nonlinear materials. On the other hand, the variation of the refractive index of the nonlinear materials will change the electric field distribution inside them, which will change the refractive index of the nonlinear materials in turn. Theoretically, an ultrafast switching time of 1.8 ps and low pump intensity of 50 MW/cm^2 can be achieved for an all-optical switching in an Ag/SiO$_2$ multilayer structure.

4.2.4.3.2.2 *Liquid crystal photonic crystal optical switching*

Liquid crystal has a very large third-order nonlinear optical susceptibility. The pump light-induced molecule alignment change and molecule configuration change can result in a remarkable variation of the refractive index of the liquid crystal. This makes it possible to realize a photonic crystal all-optical switching based on liquid crystal material.

Sio *et al.* also presented a mechanism to achieve low-power all-optical switching in a liquid crystal grating structure [106]. The grating structure was constructed by a E7 nematic liquid crystal doped with the nonmesogenic azo-dye methyl-red with a doping concentration of 2%, sandwiched between a glass substrate modified by a polymide layer and a polydimethylsiloxane grating with the period of 4 μm and the slit width of 2 μm. The liquid crystal molecules were aligned along the surface of the glass substrate. A 532 nm beam from a CW diode laser was used as the pump light. A 632.8 nm beam from a He–Ne laser was used as the probe light, which was incident in the grating at the Bragg angle. At the beginning, highly diffracted intensity could be achieved for the probe light due to the strong Bragg diffraction effect. Then the optical switching was in the "ON" state. Under the excitation of the pump light, the *trans-cis* photoisomerization process occurred for the azo-dye, which made the liquid crystal molecules disordered and the effective refractive index of the liquid crystal material close to that of the polydimethylsiloxane. This destroyed the Bragg diffraction effect of the grating structure and the diffracted intensity decreased remarkably. Then the optical switching was in the "OFF" state. The switching time was in the order of several seconds [106]. Sio *et al.* also presented another mechanism to realize low-power all-optical switching in a grating, consisting of films of regularly aligned liquid crystal separated by uniform polymer slices containing azo compounds [107]. Nematic liquid crystal E7, azobenzene derived liquid crystal 1005, and monomer NOA61 were used to fabricate the grating. The thickness and the grating period were 11.4 and 1.6 µm, respectively. A 409 nm beam from an UV diode laser was used as the pump light, while a 632.8 nm beam from a He–Ne laser was used as the probe light. The probe light was incident at the Bragg angle of 11.5°. At the beginning, high-intensity diffracted light could be obtained. Then, the optical switching was in the "ON" state. Under the excitation of the pump light, azo-compounds would undergo a conformational change from the rod-like *trans*-state to the *cis*-state, which made the liquid crystal molecules undergo an isothermal nematic to isotropic phase transition. This made a remarkable variation of the effective refractive index of the composite materials constructing the grating. Accordingly, the diffractive wavelength of the grating changed. The diffracted intensity of the probe light dropped to the minimum value and the optical switching was in the "OFF" state. The operating pump intensity was as low as 74 mW/cm^2. A high switching efficiency of 75% and a switching time of several seconds order were achieved [107].

4.2.4.3.2.3 *Micromechanical optical switching*

Kanamori *et al.* reported a nanomechanical channel drop optical switching by controlling the submicron distance between a silicon photonic crystal microcavity and a silicon photonic crystal waveguide using an ultrasmall electrostatic actuator [108]. The resonant wavelength of the photonic crystal microcavity was 1572.4 nm, which was used as the incident light. At the beginning, 1572.4 nm incident light propagated through the photonic crystal waveguide when the waveguide–microcavity distance was large. Then the optical switching was in the "ON" state. When using a 1.1 V applied voltage driving the actuator, the waveguide–microcavity distance decreased so that the energy of the incident light was coupled into the photonic crystal microcavity. Then the optical switching was in the "OFF" state. Umemori *et al.* also reported a silicon photonic crystal waveguide switching driven by an electrostatic microelectromechanical actuator [109]. The photonic crystal waveguide switching structure consisted of an input waveguide slab, a movable slab acting as the transmission bridge, and an output waveguide. The lattice constant, diameter of air holes, and the thickness of photonic crystal slab were 400, 300, and 205 nm, respectively. The movable slab was controlled by an electrostatic microelectromechanical actuator. When the movable slab was placed in the site of the air gap between the input and output waveguides, the transmission pathway was conductive and the signal light could propagate through the photonic crystal waveguide. Then the optical switching was in the "ON" state. When the movable slab was moved away from the air gap between the input and output waveguide, the transmission pathway was cut off and the signal light could not propagate through the photonic crystal waveguide. Then the optical switching was in the "OFF" state. A switching efficiency of 4 dB was achieved. The switching time was in the order of several seconds.

4.2.4.3.2.4 *Mach–Zehnder type optical switching*

Szymanski *et al.* proposed a scheme to realize all-optical switching based on photonic crystal Mach–Zehnder interferometer [110]. The output intensity of the Mach–Zehnder interferometer could be calculated by the following equation [110]:

$$I = (I_1 + I_2 + 2\sqrt{I_1 I_2} \cos\phi)/2 \qquad (4.4)$$

where I_1 and I_2 are the intensities in arm 1 and arm 2, respectively and ϕ is the phase difference of the two arms. The pump light was focused and incident on only one arm. At the beginning, the output of the Mach–Zehnder interferometer was about zero because of π-phase difference in two arms, resulting in a destructive interference effect. Then the optical switching was in the "OFF" state. Under the excitation of the pump light, a large number of carriers were generated in one arm due to strong two-photon absorption effect, which led to a remarkable variation of the refractive index of the arm, and subsequently the phase difference ϕ. With the increase of the pump intensity, a constructive interference effect was established gradually, and the output intensity of the Mach–

Zehnder interferometer increased to the maximum value. Then the optical switching was in the "ON" state. Owing to the ultrafast surface recombination of photogenerated carriers in air holes, the switching time of several picoseconds order could be achieved in semiconductor photonic crystal.

4.2.4.3.2.5 *Using quasiphase matching technique*

Recently, Padowicz *et al.* presented a mechanism to realize all-optical polarization switching in a quasiperiodic quadratic nonlinear photonic crystal based on the quasiphase matching technique [111]. For the polarization switching, the polarization of the output light depended on the incident intensity. The polarization of the output light was in agreement with that of the incident light without the excitation of the pump light. While under an intense excitation, the polarization of the output light was converted into the orthogonal polarization. In a quasiperiodically poled $LiNbO_3$ photonic crystal, two collinearly cascaded second-order nonlinear processes—the frequency upconversion process followed by a frequency downconversion process—were possible, which could generate a fundamental frequency light with the perpendicular polarization. Padowicz *et al.* found that the operating pump power could be reduced by three orders of magnitude compared with that of the polarization switching based on cascaded cubic optical nonlinearities [111].

4.2.4.3.2.6 *Double-inverse-opal optical switching*

Double inverse opal is a photonic crystal structure consisting of an inverse opal with high dielectric matrix infiltrated with movable dielectric nanospheres in the air pores [112,113]. The photonic bandgap can be tuned by collectively moving the dielectric nanospheres in the air pores. A photonic bandgap can be opened for certain positions of nanospheres in the air pores, while in other positions the photonic bandgap can be completely closed. This is because the position variations of the nanospheres in the air pores remarkably change the spatially dielectric distribution of the photonic crystal structure. Accordingly, the photonic bandgap structure is changed. Therefore, simply by collectively moving the position of nanospheres, a photonic bandgap switching could be realized.

4.2.4.3.2.7   *Using capillary condensation effect*

When injecting a flow of organic vapor close to its saturation point into a porous material, vapor molecules would be absorbed on the surfaces of the pore wells. The collection of the absorbed vapor molecules leads to the formation of a liquid, infiltrating the pores with a size less than the critical diameter. This phenomenon is called the capillary condensation effect [114]. Barthelemy *et al.* fabricated an optical superlattice structure made of porous silicon with pore size close to the diameter of the capillary condensation, the refractive index of which was periodically modulated and a linear gradient of refractive index was added into the modulation [115]. This superlattice structure provides two Wannier–Stark states in the photonic bandgap, and the coupling of these states leads to a high transmittance mode originating from the photon resonant tunneling effect. When the organic vapor is injected, the capillary condensation varies the refractive index distribution of the porous silicon superlattice, which leads to the disappearance of the resonant tunneling mode. The frequency of the probe light is set at the center of the resonant tunneling mode. Then the probe light would be reflected completely by the porous silicon superlattice. The optical switching is in the "OFF" state. While under the excitation of a pump light, laser induced evaporation would recover the original refractive index distribution of the porous silicon superlattice. Accordingly, the probe light can propagate through the porous silicon superlattice. Then the optical switching is in the "ON" state. Barthelemy *et al.* found that an ultralow pump intensity of 600 W/cm^2 could perform the optical switching [115]. The switching time was in the order of several microseconds.

4.2.4.3.3   *Constructing microcavities to enhance interactions of light and matter*

The electric field of the microcavity mode will be enhanced by a scale of Q/V in a photonic crystal microcavity, where Q and V are the quality factor and the mode volume of the photonic crystal microcavity, respectively [116]. High quality factor results in a very narrow line width of the microcavity mode, because of which the magnitude of the frequency shift needed for the realization of all-optical switching will be reduced by a factor of Q. Therefore, a reduction by a factor of V/Q^2 in the operating switching power can be achieved by use of a photonic crystal microcavity with a high quality factor and small mode volume [117]. However, according to the photon localization theory, the microcavity photon lifetime τ can be calculated by $\tau = Q/\omega_0$, where Q and ω_0 are the quality factor and the resonant frequency of the photonic crystal microcavity, respectively [118]. Therefore, the higher the quality factor Q gets, the longer the microcavity photon lifetime τ is. Accordingly, the switching time of the optical switching is slowed down. This may be an obstacle for the practical applications of high quality factor photonic crystal microcavity in the field of ultrafast and low-power all-optical switching. Soljacic *et al.* and Beggs *et al.* also pointed out that extremely low operating energy of even single photon level and a switching time of faster than 100 ps could be achieved in a hybrid system composed of photonic crystal microcavities incorporated with a highly nonlinear ultraslow light medium [119,120]. So, it is a very efficient way of realizing low-power all-optical switching by use of high-Q and small-V photonic crystal microcavity.

4.2.4.3.3.1 One-dimensional photonic crystal

Jin *et al.* realized a low-power all-optical switching in a one-dimensional photonic crystal microcavity containing InAs/GaAs quantum dots [121]. The one-dimensional photonic crystal was composed of a defect layer of InAs/GaAs quantum dots sandwiched in the center of GaAs/Al$_{0.8}$Ga$_{0.2}$As photonic crystal. The resonant wavelength of the microcavity was 1238 nm, which was close to the first excited state of InAs/GaAs quantum dots at 1220 nm. The wavelength of the probe light, a CW laser, was set at around the microcavity mode. A 1238 nm beam from an optical parameter oscillator (with a pulse duration of 130 fs and a repetition rate of 80 MHz) was used as the pump light. The wavelength of the pump light was in resonance with that of the microcavity mode, which led to a strong saturated absorption effect of InAs/GaAs quantum dots. This made the refractive index of InAs/GaAs quantum dots change and the resonant wavelength of the microcavity mode shift in the photonic bandgap. Then the transmittance of the probe light changed and an all-optical switching was realized. They found that a 22 times enhancement of the optical intensity was achieved in the photonic crystal microcavity. Under the excitation of a low pump energy of 10 pJ, up to 100% reflectivity contrast was realized. The switching time was 32 ps. When the wavelength of the pump light was 1298 nm, corresponding to the energy level of the ground state of InAs/GaAs quantum dots, the switching time was increased to 80 ps [121]. This implies that the rapid intersubband relaxation of carriers in the first excited state leads to a faster switching time.

4.2.4.3.3.2 *Two-dimensional photonic crystal microcavity*

Tanabe *et al.* fabricated a two-dimensional high-Q silicon photonic crystal microcavity structure, which consisted of a Fabry–Perot type microcavity, composed of two mirrors consisting of four air holes separated by 5-lattice length, placed in the center of a silicon photonic crystal waveguide [117]. The photonic crystal microcavity provided two resonant modes, mode c and mode s. The resonant wavelength, the line width, the quality factor, and the microcavity photon lifetime were 1530.47 nm, 133 pm, 11500, and 9.3 ps for the mode c, respectively. The resonant wavelength, the line width, the quality factor, and the microcavity photon lifetime were 1568.05 nm, 68 pm, 23000, and 19.1 ps for the mode s, respectively. A 6.4 ps pump laser and a continuous wave probe light were coupled into the photonic crystal microcavity by use of the tapered fiber coupling method. The frequency of the pump light was in resonance with that of the mode c, while the frequency of the probe light was set around the center of the mode s. Under the excitation of the pump light, free carriers were generated in silicon due to strong two-photon absorption effect. This resulted in the variation of the refractive index of silicon and the effective refractive index of the photonic crystal. As a result, the position of the defect mode in the photonic bandgap shifted. An ultrafast switching time of 50 ps, with extremely low switching energy of a few 100 fJ, and switching efficiency of 1% were achieved simultaneously for the photonic crystal switching [117].

Kim *et al.* fabricated a five-cell microcavity in a two-dimensional hexagonal lattice InGaAsP photonic crystal slab containing four $In_{0.76}Ga_{0.24}As_{0.75}P_{0.25}$ quantum wells by use of the electron beam lithography followed by Ar/Cl_2 chemically assisted ion beam etching [121]. The lattice constant and the diameter of air holes were 460 nm and 320 nm, respectively. The central wavelength and the quality factor of the microcavity mode were 1610.84 nm and 25000, respectively. A curved microfiber with a diameter of 1.5 μm and a curvature radius of 50 μm was placed on the surface of the photonic crystal microcavity. Then the 980 nm, 13 ns pump light and the continuous wave probe light could be coupled into the photonic crystal microcavity through evanescent wave coupling method with a coupling efficiency of about 20%. The nonlinear refractive index of the $In_{0.76}Ga_{0.24}As_{0.75}P_{0.25}$ quantum wells was estimated to be -6.84×10^{-1} cm^2/GW at 980 nm due to the strong band-edge resonant enhanced nonlinearity effect [122]. Under the excitation of the 980 nm pump light, the refractive index of $In_{0.76}Ga_{0.24}As_{0.75}P_{0.25}$ quantum wells changed greatly, which led to a great variation of the effective refractive index of the photonic crystal. Then the microcavity mode shifted in the photonic bandgap, which resulted in the variation of the transmittance of the probe light. An ultralow switching energy of less than 75.4 fJ and a switching time of several nanoseconds order were achieved for the photonic crystal all-optical switching [121]. At the beginning, the output power of the probe light was in the high level. While under the excitation of the pump light, the output power of the probe light was switched to the low level. Therefore, a photonic crystal all-optical switching was realized.

The Fano interference, also called the Fano resonance, can be defined as the interference of a discrete energy state with a continuum, resulting in a sharp and asymmetric line shapes [123]. When an optical switching is realized by use of the defect mode shift, smaller magnitude of frequency shift will be needed for Fano resonance due to its sharp and asymmetric line shapes compared with that of the Lorentzian resonance with a symmetric line shape. Therefore, it can be expected that the combination of the sharp, asymmetric line shape of the Fano resonance with the strong photon localization effect of photonic crystal microcavity could remarkably reduce the operating threshold pump power of all-optical switching. Yang *et al.* fabricated silicon photonic crystal microcavity structure composed of a five-cell microcavity side coupled a photonic crystal waveguide with a four lattices distance [124]. The large refractive index contrast of silicon and air in the input and output sides of the photonic crystal waveguide acted as two reflectors, which resulted in a Fano-type resonance of the microcavity mode. The lattice constant, the diameter of air holes, and the thickness of the photonic crystal slab were 420 nm, 243 nm, and 252 nm, respectively. A lensed fiber was used to couple the incident light into the photonic crystal waveguide. The central wavelength and the quality factor of the Fano resonance were 1556.8 nm and 31000, respectively. There was a wavelength detuning of 22 pm between the incident light and the Fano resonance. At the beginning, the transmittance of the incident light was 8%. The optical switching was in the "OFF" state. Under the excitation of the incident light, the refractive index of silicon changed due to strong two-photon absorption induced free-carrier generation, which led to a variation of the effective refractive index of the photonic crystal. As a result, the resonant frequency of the microcavity mode shifted, which made an increase of the transmittance of the incident light. When the incident photon energy was 540 fJ, the transmittance of the incident light changed by 38%. Then the optical switching was in the "ON" state. Therefore, a switching efficiency of 30% was achieved.

4.2.4.3.4 *Low-power organic composite photonic crystal all-optical switching*

Low-power photonic crystal all-optical switching can be realized by using the strong photon confinement effect of photonic crystal microcavities to enhance the nonlinear interaction of light and matter [125,126]. Ultralow operating pump energy of several hundred fJ could be achieved for photonic crystal microcavities with a Q of the order of 10^4 [127]. In order to obtain a high quality factor and small mode volume, materials with a large refractive index and microfabrication precision are usually needed. So, it is not easy to obtain a very high quality factor and very small mode volume simultaneously [128]. Synthesizing new materials with large nonlinear optical susceptibility can also be used to realize low-power photonic crystal all-optical switching. But there still exists a limitation that the larger the nonlinear optical susceptibility is, the slower the nonlinear response time is [129].

The origination of the optical nonlinearity of organic conjugated polymer materials is the delocalization of π-electrons along the polymer chains under the excitation of a pump light, which leads to a relatively large third-order nonlinear optical susceptibility and an ultrafast response time of subpicosecond order. Moreover, when materials with a large optical nonlinearity, including laser dye, noble metal nanoparticles, and semiconductor quantum dots, are dispersed in an organic conjugated matrix, the third-order optical nonlinearity of the organic matrix can be improved greatly. Therefore, composite materials based on organic polymers are promising candidates for the realization of low-power and ultrafast integrated photonic devices. Recently, Gong *et al.* reported a new method based on intermolecular excited-state charge/energy transfer to achieve a large nonlinear susceptibility and an ultrafast response simultaneously in organic composite materials [130–132]. A large third-order nonlinear susceptibility can be realized by near-resonant excitation-enhancing nonlinearity effect. An ultrafast response time can be maintained by fast intermolecular excited-state charge/energy transfer process simultaneously [130–132]. The physical mechanism is simple and clear, while very efficient.

4.2.4.3.4.1 *Polymer composite photonic crystal all-optical switching*

Low-power and ultrafast all-optical switching can be realized in a polymer composite photonic crystal based on the intermolecular excited-state charge/energy transfer method. Gong's group fabricated a polymer composite made of polystyrene doped with laser dye Coumarin 153 (C153) with a doping concentration of 15% [130]. A two-dimensional photonic crystal microcavity, composed of a line defect with a width of 440 nm in the center of square arrays of cylindrical air holes embedded in a 300 nm-thick polymer composite slab, was fabricated by using a focused ion-beam (FIB) etching system. The SEM images of the photonic-bandgap microcavity are shown in Fig. 4.10. The patterned area was about 4 μm × 100 μm. The all-optical switching effect can be measured by the femtosecond pump and probe method, as shown in Fig. 4.11. A beam (with a pulse duration and repetition rate of 120 fs and 76 MHz, respectively) from a tunable Ti:sapphire laser system (Model Mira 900F, Coherent Company) was split into two beams with a ratio of 1:1. One beam was used as the pump light, the frequency of which was doubled by a beta-barium-borate (BBO) crystal. The other beam was used as the probe light. Both the probe and the pump light were transverse-electric (TE) polarized waves with the electric field vector parallel to the polystyrene film. The pump light was focused and normally incident on the upper surface plane of the photonic crystal with a spot size of about 100 μm. Two

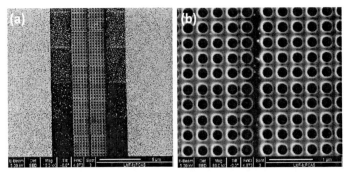

Figure 4.10 SEM images of the organic composite photonic-bandgap microcavity. (a) Large scale image. (b) Small scale image. Reprinted with permission from X. Y. Hu, *et al.*, Picosecond and low-power all-optical switching based on an organic photonic-bandgap microcavity, *Nat. Photon.* **2**, 185–189 (2008). Copyright © 2008, rights managed by Nature Publishing Group.

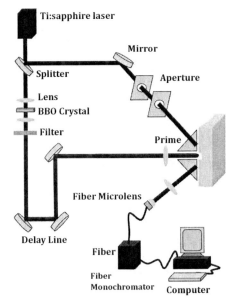

Figure 4.11 Experimental setup for the measurement of all-optical switching effect of organic composite photonic-bandgap microcavity. The thick lines represent optical connections, while thin lines are electronic connections. Reprinted with permission from X. Y. Hu, *et al.*, Picosecond and low-power all-optical switching based on an organic photonic-bandgap microcavity, *Nat. Photon.* **2**, 185–189 (2008). Copyright © 2008, rights managed by Nature Publishing Group.

apertures with a diameter of 1 mm were used to collimate and attenuate the probe light. The evanescent-field coupling method was used to couple the probe light into the polystyrene waveguide. The probe light was incident on the bottom of an input prism placed above the upper surface of the polymer composite waveguide. The angle of incidence of the probe light was adjusted so that the propagation constant of the probe light was equal to that of the guided electromagnetic modes of the polymer composite waveguide. Then the energy of the probe light was coupled into the polystyrene waveguide with the help of the evanescent-field generated in the air gap between the bottom of the prism and the upper surface of the polymer composite waveguide. The probe light propagated through the photonic crystal in the direction perpendicular to the line defect. The probe light coupled out from the output prism was detected by a fiber monochromator (Model HR4000, Ocean Optics Inc.) with a resolution of 0.25 nm, the output signal of which was collected and analyzed by a computer. A delay line was used to adjust the timing of the pump and the probe pulses. The intensity of the probe light propagating through the photonic crystal was weakened greatly because of attenuation and coupling loss. The change of the refractive index of the polymer composite caused by the probe light was in the order of 10^{-5}. So, the influence of the probe light on the nonlinearity of the polymer composite was neglected. The measured transmission spectrum of the photonic crystal microcavity is shown in Fig. 4.12. The resonant wavelength was 788 nm. When a lattice defect is introduced in a perfect photonic crystal, the periodicity of the spatial dielectric distribution is destroyed. According to the photon localization theory, an electromagnetic field with a certain resonant frequency will be strongly confined around the defect site, which leads to the formation of a localized defect mode in the photonic bandgap [133,134]. The electric field distribution decreased exponentially with the distance away from the center of the defect structure. The peak transmittance and the line width of the microcavity resonant mode were 96% and 2.3 nm, respectively. The quality factor of the photonic-bandgap microcavity was 342. The measured all-optical switching effect is shown in Fig. 4.13. The wavelength of the probe and pump light was 788 and 394 nm, respectively. The photon energy of the pump light was 520 fJ. The signal profile shows a fast drop followed by a slow rise. When the pump pulse was far away from the probe pulse in the time sequence, the transmission of the probe light maintained the maximum value of 92%. This indicates that the optical switching is in the "ON" state. The transmission of the probe light began to change when the pump and probe pulse overlap with each other temporally. The reason originates from the pump light-induced microcavity mode shift. The minimum transmission of 10% corresponds to the "OFF" state of the optical switching. A high switching efficiency of more than 80% was reached. The threshold pump intensity was reduced to $110 \, \text{KW/cm}^2$, which is reduced by four orders of magnitude compared with previous reports [38–46]. The large third-order optical nonlinearity attributes to the near-resonant excitation enhancing nonlinearity, and local-field enhancing nonlinearity arising from the non-uniform distribution of the electric field of the pump light in the composite material [135,136]. It is clear that an ultrafast switching time of 1.2 ps was reached.

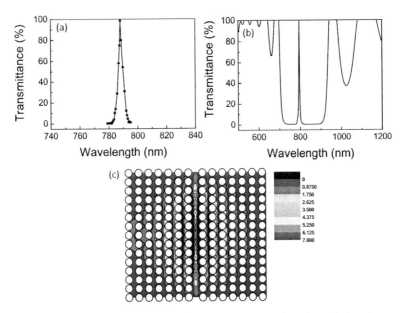

Figure 4.12   Resonant mode of the organic composite photonic-bandgap microcavity. (a) Measured transmittance spectrum. (b) Simulated transmittance spectrum. (c) Electric-field distribution of the microcavity resonant mode. Reprinted with permission from X. Y. Hu, *et al.*, Picosecond and low-power all-optical switching based on an organic photonic-bandgap microcavity, *Nat. Photon.* **2**, 185–189 (2008). Copyright © 2008, rights managed by Nature Publishing Group.

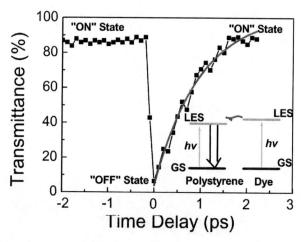

Figure 4.13 All-optical switching effect of the organic composite photonic-bandgap microcavity. The thick line represents the exponentially fitted result. The insert shows the relative energy levels of polystyrene and dye molecule. GS represents ground state. LES is the lowest excited state. Reprinted with permission from X. Y. Hu, *et al.*, Picosecond and low-power all-optical switching based on an organic photonic-bandgap microcavity, *Nat. Photon.* **2**, 185–189 (2008). Copyright © 2008, rights managed by Nature Publishing Group.

Gong *et al.* also fabricated a one-dimensional polymer composite photonic crystal made of poly(3-hexylthiophene) (P3HT) doped with 1-(3-methoxycarbonyl)propyl-1-phenyl-(6,6)C_{61} (PCBM) with a doping concentration of 10%. The PCBM:P3HT photonic crystal consisted of parallel air grooves embedded in a PCBM:P3HT slab. The lattice constant was 760 nm. The width of the air grooves was 290 nm. The length and the width of the patterned area were about 100 and 7 μm, respectively. The wavelength was 793.5 nm for the probe light and 396.75 nm for the pump light. The wavelength of the probe light was in the dielectric bandedge of the PCBM:P3HT photonic crystal, while that of the pump light was in the overlapped region of the linear absorption band of P3HT and PCBM. The threshold pump intensity was as low as 1 MW/cm^2. In the "ON" state, the transmission reached 85%. While in the "OFF" state, the transmission of the probe light reduced to 25%. A high switching efficiency of 60% was obtained. An ultrafast switching time of 58.9 ps was also achieved [137,138].

4.2.4.3.4.2 *Nanocomposite photonic crystal all-optical switching*

Low-power and ultrafast all-optical switching can be realized in a nanocomposite photonic crystal based on the intermolecular excited-state charge/energy transfer method. Poly[2-methoxy-5-(2-ethylhexyloxy)-1,4-phenylenevinylene] (MEH-PPV) is a highly conjugated organic polymer with excellent third-order optical nonlinearity in the near-infrared range [139]. The third-order optical nonlinearity of MEH-PPV can be improved greatly when noble metal nanoparticles are dispersed in a MEH-PPV matrix.

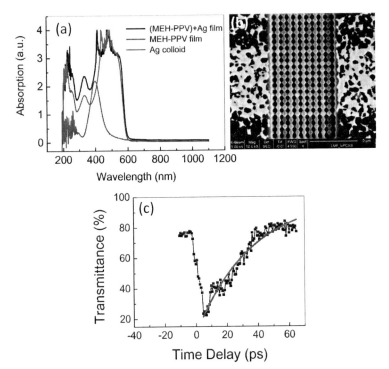

Figure 4.14 All-optical switching effect of two-dimensional nano-Ag:MEH-PPV photonic crystal. (a) Absorption spectra of Ag colloid, pure MEH-PPV film, and nano-Ag:MEH-PPV film. (b) SEM image of nano-Ag:MEH-PPV photonic crystal. (c) Measured all-optical switching effect. The thick red line is the exponentially fitted result. Reprinted with permission from *Appl. Phys. Lett.*, **94**(3), X. Hu, *et al.*, Nano-Ag: polymeric composite material for ultrafast photonic crystal all-optical switching. Copyright 2009, AIP Publishing LLC.

Gong' group fabricated a nanocomposite material consisting of Ag nanoparticles dispersed in an organic polymer matrix made of Poly[2-methoxy-5-(2-ethylhexyloxy)-1,4-phenylenevinylene] (MEH-PPV). The average diameter of Ag nanoparticles, fabricated by a chemical reduction method, was about 15 nm [140]. The absorption spectra of Ag colloid, MEH-PPV film, and nano-Ag:MEH-PPV film are shown in Fig. 4.14a. The SPR frequency of Ag nanoparticles dropped in the absorption bands of MEH-PPV. The SEM image of the two-dimensional nano-Ag:MEH-PPV photonic crystal is depicted in Fig. 4.14b. The photonic crystal was composed of square arrays of cylindrical air holes embedded in a 450 nm-thick nano-Ag:MEH-PPV slab. The lattice constant and the diameter of air holes were 260 nm and 200 nm, respectively [141]. The patterned area was about 3 μm × 100 μm. The all-optical switching effect was measured by use of the femtosecond pump and probe method, as shown in Fig. 4.14c. The wavelength was 799 nm for the probe laser and 399.5 nm for the pump laser. The threshold pump intensity was as low as 0.2 MW/cm^2, which originates from the strong SPR enhancing nonlinearity effect. The switching efficiency reached 65%. An ultrafast switching time of 35 ps was obtained due to fast energy transfer from excited states of MEH-PPV molecules to Ag nanoparticles [142].

References

1. M. Scalora, J. P. Dowling, C. M. Bowden, and M. J. Bloemer, "Optical limiting and switching of ultrashort pulses in nonlinear photonic band gap materials," *Phys. Rev. Lett.* **73**, 1368–1371 (1994).

2. S. Lan, S. Nishikawa, and O. Wada, "Leveraging deep photonic band gaps in photonic crystal impurity bands," *Appl. Phys. Lett.* **78**, 2101–2103.

3. M. Bayindir, B. Temelkuran, and E. Ozbay, "Photonic-crystal-based beam splitters," *Appl. Phys. Lett.* **77**, 3902–3904 (2000).

4. S. Lan and H. Ishikawa, "Coupling of defect pairs and generation of dynamical band gaps in the impurity bands of nonlinear photonic crystals for all-optical switching," *J. Appl. Phys.* **91**, 2573–2577 (2002).

5. P. Tran, "Optical limiting and switching of short pulses by use of a nonlinear photonic bandgap structure with a defect," *J. Opt. Soc. Am. B.* **14**, 2589–2594 (1997).

6. P. R. Villeneuve, D. S. Abrams, S. H. Fan, and J. D. Joannopoulos, "Single-mode waveguide microcavity for fast optical switching," *Opt. Lett.* **21**, 2017–2019 (1996).

7. L. X. Chen, X. Deng, W. Ding, L. Cao, and S. Liu, "Finite-difference time-domain analysis of optical bistability with low threshold in one-dimensional nonlinear photonic crystal with Kerr medium," *Opt. Commun.* **209**, 491–500 (2002).

8. L. X. Chen, X. X. Deng, W. Q. Ding, Y. Zhang, and S. T. Liu, "Low threshold bistable switching by the nonlinear one-dimensional photonic crystal," *Chin. Phys. Lett.* **19**, 798–800.

9. M. Soljacic, M. Ibanescu, S. G. Johnson, Y. Fink, and J. D. Joannopoulos, "Optimal bistable switching in nonlinear photonic crystals," *Phys. Rev. E* **66**, 055601(R) (2002).

10. V. Lousse and J. P. Vigneron, "Bistable behaviour of a photonic crystal nonlinear cavity," *Physica B* **338**, 171–177 (2003).

11. L. X. Chen, D. Kim, Y. L. Song, W. Q. Ding, W. H. Li, and S. T. Liu, "Localization and threshold of bistable switching by gap-edge shifting," *Chin. Phys. Lett.* **20**, 1514–1516.

12. M. F. Yanik, S. H. Fan, and M. Soljacic, "High-contrast all-optical bistable switching in photonic crystal microcavities," *Appl. Phys. Lett.* **83**, 2739–2741 (2003).

13. S. H. Fan, "Sharp asymmetric line shapes in side-coupled waveguide-cavity systems," *Appl. Phys. Lett.* **80**, 908–910 (2002).

14. M. F. Yanik, S. H. Fan, M. Soljacic, and J. D. Joannopoulos, "All-optical transistor action with bistable switching in a photonic crystal cross-waveguide geometry," *Opt. Lett.* **28**, 2506–2508 (2003).

15. S. G. Johnson, C. Manolatou, S. H. Fan, P. R. Villeneuve, J. D. Joannopoulos, and H. A. Haus, "Elimination of cross talk in waveguide intersections," *Opt. Lett.* **23**, 1855–1857 (1998).

16. A. Locatelli, D. Modotto, D. Paloschi, and C. D. Angelis, "All optical switching in ultrashort photonic crystal couplers," *Opt. Commun.* **237**, 97–102 (2004).

17. A. Sharkawy, S. Shi, D. W. Prather, and R. A. Soref, "Electro-optical switching using coupled photonic crystal waveguides," *Opt. Express* **10**, 1048–1059 (2002).

18. F. C. Soto, A. Martinez, J. Garcia, F. Ramos, P. Sanchis, J. Blasco, and J. Marti, "All-optical switching structure based on a photonic crystal directional coupler," *Opt. Express* **12**, 161–167 (2004).

19. P. M. Johnson, A. F. Koenderink, and W. L. Vos, "Ultrafast switching of photonic density of states in photonic crystals," *Phys. Rev. B* **66**, 081102(R).

20. X. Wang, K. Kempa, Z. F. Ren, and B. Kimball, "Rapid photon flux switching in two-dimensional photonic crystals," *Appl. Phys. Lett.* **84**, 1817–1819 (2004).

21. P. Tran, "All-optical switching with a nonlinear chiral photonic crystal structure," *J. Opt. Soc. Am. B* **16**, 70–73 (1999).

22. P. Tran, "Optical switching with a nonlinear photonic crystal: a numerical study," *Opt. Lett.* **21**, 1138–1140 (1996).

23. S. Xiong, and H. Fukshima, "Analysis of light propagation in index-tunable photonic crystals," *J. Appl. Phys.* **94**, 1286–1288 (2003).

24. M. R. Singh, "Switching mechanism due to the spontaneous emission cancellation in photonic band gap materials doped with nano-particles," *Phys. Lett. A* **363**, 177–181 (2007).

25. M. R. Singh, "All-photonic switching in nanophotonic quantum wells," *Phys. Status Solidi A* **206**, 910–914 (2009).

26. E. A. Camargo, H. M. H. Chong, and R. M. D. L. Rue, "2D photonic crystal thermo-optic switch based on AlGaAs/GaAs epitaxial structure," *Opt. Express* **12**, 588–592 (2004).

27. M. Iodice, P. M. Sarro, and M. Bellbcci, "Tansient analysis of a high-speed thermo-optic modulator integrated in all silicon waveguide," *Opt. Eng.* **42**, 169–175 (2003).

28. C. E. Reese, A. V. Mikhonin, M. Kamenjicki, A. Tikhonov, and S. A. Asher, "Nanogel nanosecond photonic crystal optical switching," *J. Am. Chem. Soc.* **126**, 1493–1496 (2004).

29. R. Ozaki, M. Ozaki, and K. Yoshino, "Defect mode switching in one-dimensional photonic crystal with nematic liquid crystal as defect layer," *Jpn. J. Appl. Phys. Part 2*, **42**, L669–L671 (2003).

30. Z. L. Samson, K. F. Macdonald, F. D. Angelis, B. Gholipour, K. Knight, C. C. Huang, E. D. Fabrizio, D. W. Hewak, and N. I. Zheludev, "Metamaterial electro-optic switch of nanoscale thickness," *Appl. Phys. Lett.* **96**, 143105 (2010).

31. C. D. Stanciu, F. Hansteen, A. V. Kimel, A. Kirilyuk, A. Tsukammoto, A. Itoh, and T. Rasing, "All-optical magnetic recording with circularly polarized light," *Phys. Rev. Lett.* **99**, 047601 (2007).

32. A. Shevchenko, M. Korppi, K. Lindfors, M. Heileo, and M. Kaivola, "All-optical reversible switching of local magnetization," *Appl. Phys. Lett.* **91**, 041916 (2007).

33. Z. Wu, M. Levy, V. J. Fratello, and A. M. Merzlikin, "Gyrotropic photonic crystal waveguide switches," *Appl. Phys. Lett.* **96**, 051125 (2010).

34. A. V. Scherbakov, A. V. Akimov, V. G. Golubev, A. S. Kaplyanskii, D. A. Kurdyukov, A. A. Meluchev, and A. B. Pevtsov, "Optical induced Bragg switching in opal-VO$_2$ photonic crystals," *Physica E* **17**, 429–430 (2003).

35. D. A. Mazurenko, A. V. Akimo, A. B. Pevtsov, D. A. Kurdyukov, V. G. Golubev, and J. I. Dijkhuis, "Ultrafast all-optical switching in a three-dimensional photonic crystal," *J. Luminescience* **108**, 163–166 (2004).

36. D. A. Mazurenko, R. Kerst, A. V. Akimo, A. B. Pevtsov, D. A. Kurdyukov, V. G. Golubev, A. V. Selkin, and J. I. Dijkhuis, "Femtosecond Bragg switching in opal-a-nc-Si photonic crystals," *J. Non-Cry. Solids* **338–340**, 215–217 (2004).

37. S. W. Leonard, H. M. Van Driel, J. Schilling, and R. B. Wehrsophn, "Ultrafast band-edge tuning of a two-dimensional silicon photonic crystal via free-carrier injection," *Phys. Rev. B* **66**, 161102(R) (2002).

38. Y. H. Liu, X. Y. Hu, D. X. Zhang, B. Y. Cheng, D. Z. Zhang, and Q. B. Meng, "Ultrafast all-optical switching in polystyrene opal," *Appl. Phys. Lett.* **86**, 151102 (2005).

39. X. Y. Hu, Y. H. Liu, J. Tian, B. Y. Cheng, and D. Z. Zhang, "Ultrafast all-optical switching in two-dimensional organic photonic crystal," *Appl. Phys. Lett.* **86**, 121102 (2005).

40. X. Y. Hu, Q. H. Gong, Y. H. Liu, B. Y. Cheng, and D. Z. Zhang, "All-optical switching of defect mode in two-dimensional nonlinear organic photonic crystals," *Appl. Phys. Lett.* **86**, 231111 (2005).

41. A. D. Bristow, J. P. R. Wells, W. H. Fan, A. M. Fox, M. S. Skolnick, D. M. Whittaker, A. Tahraoui, T. F. Krauss, and J. S. Roberts, "Ultrafast nonlinear response of AlGaAs two-dimensional photonic crystal waveguides," *Appl. Phys. Lett.* **83**, 851–853 (2003).

42. A. Hache and M. Bourgeois, "Ultrafast all-optical switching in a silicon-based photonic crystal," *Appl. Phys. Lett.* **77**, 4089–4091 (2000).

43. M. Shimizu and T. Ishihara, "Subpicosecond transmission change in semiconductor-embedded photonic crystal slab: toward ultrafast optical switching," *Appl. Phys. Lett.* **80**, 2836–2838 (2002).

44. T. Fujita, Y. Sato, T. Kuitani, and T. Ishihara, "Tunable polariton absorption of distributed feedback microcavities at room temperature," *Phys. Rev. B* **57**, 12428–12434 (1998).

45. D. A. Mazurenko, R. Kerst, J. I. Dijkhuis, A. V. Akimov, V. G. Golubev, D. A. Kurdyukov, A. B. Pevtsov, and A. V. Selkin, "Ultrafast optical switching in three-dimensional photonic crystals," *Phys. Rev. Lett.* **91**, 213903 (2003).

46. D. A. Mazurenko, A. V. Akimov, A. B. Pevtsov, D. A. Kurdyukov, V. G. Golubev, and J. I. Dijkhuis, "Ultrafast switching in Si-opals," *Physica E* **17**, 410–413 (2003).

47. Y. Liu, F. Qin, Z. Y. Wei, Q. B. Meng, D. Z. Zhang, and Z. Y. Li, "10 fs ultrafast all-optical switching in polystyrene nonlinear photonic crystals," *Appl. Phys. Lett.* **95**, 131116 (2009).

48. H. Kishida, H. Matsuzaki, H. Okamoto, T. Manabe, M. Yamashita, Y. Taguchi, and Y. Tokura, "Gigantic optical nonlinearity in one-dimensional Mott–Hubbard insulators," *Nature* **405**, 929–932 (2000).

49. O. Wada, "Femtosecond all-optical devices for ultrafast communication and signal processing," *New J. Phys.* **6**, 183 (2004).

50. M. R. Singh and R. H. Lipson, "Optical switching in nonlinear photonic crystals lightly doped with nanostructures," *J. Phys. B: At. Mol. Opt. Phys.* **41**, 015401 (2008).

51. A. Samoc, M. Samoc, M. Woodruff, and B. Luther-Davies, "Tuning the properties of poly(p-phenylenevinylene) for use in all-optical switching," *Opt. Lett.* **20**, 1241–1243 (1995).

52. J. C. May, I. Biaggio, F. Bures, and F. Diederich, "Extended conjugation and donor–acceptor substitution to improve the third-order optical nonlinearity of small molecules," *Appl. Phys. Lett.* **90**, 251106 (2007).

53. J. C. May, J. H. Lim, I. Biaggio, N. N. P. Moonen, T. Michinobu, and F. Diederich, "Highly efficient third-order optical nonlinearities in donor-substituted cyanoethynylethene molecules," *Opt. Lett.* **30**, 3057–3059 (2005).

54. T. Michinobu, J. C. May, J. H. Lim, C. Boudon, J. P. Gisselbrecht, P. Seiler, M. Gross, I. Biaggio, and F. Diederich, "A new class of organic donor–acceptor molecules with large third-order optical nonlinearities," *Chem. Commun.* **6**, 737–739 (2005).

55. B. Esembeson, M. L. Scimeca, T. Michinobu, F. Diederich, and I. Biaggio, "A high-optical quality supramolecular assembly for third-order integrated nonlinear optics," *Adv. Mater.* **20**, 4584–4587 (2008).

56. S. H. Chi, J. M. Hales, C. F. Hernandez, S. Y. Tseng, J. Y. Cho, S. A. Odom, Q. Zhang, S. Barlow, R. R. Schrock, S. R. Marder, B. Kippelen, and J. W.

Perry, "Thick optical-quality films of substituted polyacetylenes with large, ultrafast third-order nonlinearities and application to image correlation," *Adv. Mater.* **20**, 3199–3203 (2008).

57. Y. Iwasa, E. Funatsu, T. Hasegawa, T. Koda, and M. Yamashita, "Nonlinear optical study of quasi-one-dimensional platinum complexes: Two-photon excitonic resonance effect," *Appl. Phys. Lett.* **59**, 2219–2221 (1991).

58. W. S. Fann, S. Benson, J. M. J. Madey, S. Etemad, G. L. Baker, and F. Kajzar, "Spectrum of $\chi^{(3)}$ (-3 $\chi\omega;\omega,\omega,\omega$) in polyacetylene: an application of the free-electron laser in nonlinear optical spectroscopy," *Phys. Rev. Lett.* **62**, 1492–1495 (1989).

59. T. Hasegawa, Y. Iwasa, H. Sunamura, T. Koda, Y. Tokura, H. Tachibana, M. Matsumoto, and S. Abe, "Nonlinear optical spectroscopy on one-dimensional excitons in silicon polymer, polysilane," *Phys. Rev. Lett.* **69**, 668–671 (1992).

60. Z. M. Jin, H. Ma, L. H. Wang, G. H. Ma, F. Y. Guo, and J. Z. Chen, "Ultrafast all-optical magnetic switching in $NaTb(WO_4)_2$," *Appl. Phys. Lett.* **96**, 201108 (2010).

61. S. Q. Chen, W. P. Zang, A. Schulzgen, X. Liu, J. G. Tian, J. V. Moloney, and N. Peyghambarian, "Modeling of Z-scan characteristics for one-dimensional nonlinear photonic bandgap materials," *Opt. Lett.* **34**, 3665–3667 (2009).

62. D. Y. Guan, Z. H. Chen, Y. L. Zhou, K. J. Jin, and G. Z. Yang, "Peaks separation of the nonlinear refraction and nonlinear absorption induced by external electric field," *Appl. Phys. Lett.* **88**, 111911 (2006).

63. R. J. Gehr, G. L. Fescher, R. W. Boyd, and J. E. Sipe, "Nonlinear optical response of layered composite materials," *Phys. Rev. A* **53**, 2792–2798 (1996).

64. G. L. Fischer, R. W. Boyd, R. J. Gehr, S. A. Jenekhe, J. A. Osaheni, J. E. Sipe, and L. A. W. Brophy, "Enhanced nonlinear optical response of composite materials," *Phys. Rev. Lett.* **74**, 1871–1874 (1995).

65. R. S. Bennink, Y. K. Yoon, R. W. Boyd, and J. E. Sipe, "Accessing the optical nonlinearity of metals with metal-dielectric photonic bandgap structures," *Opt. Lett.* **24**, 1416–1418 (1999).

66. M. Scalora, M. J. Bloemer, A. S. Pethel, J. P. Dowling, C. M. Bowden, and A. S. Manka, "Transparent, metallo-dielectric, one-dimensional, photonic band-gap structures," *J. Appl. Phys,* **83**, 2377–2383 (1998).

67. M. J. Bloemer and M. Scalora, "Transmissive properties of Ag/MgF_2 photonic band gaps," *Appl. Phys. Lett.* **72**, 1676–1678 (1998).

68. G. H. Ma and S. H. Tang, "Ultrafast optical nonlinearity enhancement in metallodielectric multiplayer stacks," *Opt. Lett.* **32**, 3435–3437 (2007).

69. N. N. Lepeshkin, A. Schweinsberg, G. Piredda, R. S. Bennink, and R. W. Boyd, "Enhanced nonlinear optical response of one-dimensional metal-dielectric crystals," *Phys. Rev. Lett.* **93**, 123902 (2004).

70. G. Q. Du, H. T. Jiang, Z. S. Wang, and H. Chen, "Optical nonlinearity enhancement in heterostructures with thick metallic film and truncated photonic crystals," *Opt. Lett.* **34**, 578–580 (2009).

71. A. Alu and N. Engheta, "Pairing an epsilon-negative slab with a mu-negative slab: resonance, tunneling and transparency," *IEEE Trans. Antennas Propag.* **51**, 2558–2571 (2003).

72. T. Goto, A. V. Dorofeenko, A. M. Merzlikin, A. V. Baryshev, A. P. Vinogradov, M. Inoue, A. A. Lisyansky, and A. B. Granovsky, "Optical Tamm states in one-dimensional megnetophotonic structures," *Phys. Rev. Lett.* **101**, 113902 (2008).

73. J. Y. Guo, Y. Sun, Y. W. Zhang, H. Q. Li, H. T. Jiang, and H. Chen, "Experimental investigation of interface states in photonic crystal heterostructures," *Phys. Rev. E* **78**, 026607 (2008).

74. M. Scalora, N. Mattiucci, G. Daguanno, M. Larciprete, and M. J. Bloemer, "Nonlinear pulse propagation in one-dimensional metal-dielectric multiplayer stacks: Ultrawide bandwidth optical limiting," *Phys. Rev. E* **73**, 016603 (2006).

75. M. C. Larciprete, C. Sibilia, S. Paoloni, M. Bertolotti, F. Sarto, and M. Scalora, "Accessing the optical limiting properties of metallo-dielectric photonic bandgap structures," *J. Appl. Phys.* **93**, 5013–5017 (2003).

76. J. E. Sipe and R. W. Boyd, "Nonlinear susceptibility of composite optical materials in the Maxwell Garnett model," *Phys. Rev. A* **46**, 1614–1629 (1992).

77. R. Song, D. Y. Guan, L. Ma, and Z. X. Cao, "Exceptionally large third-order optical susceptibility in $Ag:SrBi_2Nb_2O_9$ composite films," *Mater. Lett.* **61**, 1537–1540 (2007).

78. H. B. Liao, W. J. Wen, G. K. L. Wong, and G. Z. Yang, "Optical nonlinearity of nanocrystalline Au/ZnO composite films," *Opt. Lett.* **28**, 1790–1792 (2003).

79. Q. F. Zhang, W. M. Liu, Z. Q. Xue, J. L. Wu, S. F. Wang, D. L. Wang, and Q. H. Gong, "Ultrafast optical Kerr effect of Ag-BaO composite thin films," *Appl. Phys. Lett.* **82**, 958–960 (2003).

80. P. Zhou, G. J. You, Y. G. Li, T. Han, J. Li, S. Y. Wang, L. Y. Chen, Y. Liu, and S. X. Qian, "Linear and ultrafast nonlinear optical response of Ag:Bi_2O_3 composite films," *Appl. Phys. Lett.* **94**, 3876–3878 (2003).

81. K. Tanaka and A. Saitoh, "Optical nonlinearities of Se-loaded zeolite (ZSM-5): a molded nanowire system," *Appl. Phys. Lett.* **94**, 241905 (2009).

82. D. Rativa, R. E. D. Arauji, C. B. D. Arauji, A. S. L. Gomes, and L. R. P. Kassab, "Femtosecond nonlinear optical properties of lead–germanium oxide amorphous films," *Appl. Phys. Lett.* **90**, 231906 (2007).

83. C. B. D. Araujo, E. L. F. Filho, A. Humeau, D. Guichaoua, G. Boudebs, and L. R. P. Kassab, "Picosecond third-order nonlinearity of lead-oxide glasses in the infrared," *Appl. Phys. Lett.* **87**, 221904 (2005).

84. Y. Nosaka, K. Tanaka, and N. Fujii, "Nonlinear optical susceptibility of ultrasmall CdS particles by means of the polarization-discriminated forward degenerate four-wave mixing in a resonant region," *Appl. Phys. Lett.* **62**, 1863–1865 (1993).

85. H. Nakamura, Y. Sugimoto, K. Kanamoto, N. Ikeda, Y. Tanaka, Y. Nakamura, S. Ohkouchi, Y. Watanbe, K. Inoue, H. Ishikawa, and K. Asakawa, "Ultra-fast photonic crystal/quantum dot all-optical switch for future photonic networks," *Opt. Express* **12**, 6606–6614 (2004).

86. V. Klimov, P. H. Bolivar, H. Kurz, V. Karavanskii, and Y. Korkishko, "Linear and nonlinear transmission of Cu_xS quantum dots," *Appl. Phys. Lett.* **67**, 653–655 (1995).

87. I. Moreels, Z. Hens, P. Kockaert, J. Loicq, and D. V. Thourhout, "Spectroscopy of the nonlinear refractive index of colloidal PbSe nanocrystals," *Appl. Phys. Lett.* **89**, 193106 (2006).

88. R. E. Schwerzel, K. B. Spahr, J. P. Kurmer, V. E. Wood, and J. A. Jenkins, "Nanocomposite photonic polymers. 1. third-order nonlinear optical properties of capped cadmium sulfide nanocrystals in an ordered polydiacetylene host," *J. Phys. Chem. A* **102**, 5622–5626 (1998).

89. H. Du, G. Q. Xu, W. S. Chin, L. Huang, and W. Ji, "Synthesis, characterization, and nonlinear optical properties of hybridized CdS-polystyrene nanocomposites," *Chem. Mater.* **14**, 4473–4479 (2002).

90. M. Etienne, A. Biney, A. D. Walser, R. Dorsinville, D. L. V. Bauer, and V. B. Nair, "Third-order nonlinear optical properties of a cadmium sulfide-dendrimer nanocomposite," *Appl. Phys. Lett.* **87**, 181913 (2005).

91. H. Hosono, Y. Abe, Y. L. Lee, T. Tokizaki, and A. Nakamura, "Large third-order optical nonlinearity of nanometer-sized amorphous

semiconductor: phosphorous colloidals formed in SiO_2 glass by ion implantation," *Appl. Phys. Lett.* **61**, 2747–2749 (1992).

92. S. Ohtsuka, T. Koyama, K. Tsunetomo, H. Nagata, and S. Tanaka, "Nonlinear optical properties of CdTe microcrystallites doped glasses fabricated by laser evaporation method," *Appl. Phys. Lett.* **61**, 2953–2954 (1992).

93. Y. L. Zhang, Y. Zhang, and B. J. Li, "Optical switches and logic gates based on self-collimated beams in two-dimensional photonic crystals," *Opt. Express* **15**, 9287–9292 (2007).

94. Y. Y. Wang, J. Y. Chen, and L. W. Chen, "Optical switches based on partial band gap and anomalous refraction in photonic crystals modulated by liquid crystals," *Opt. Express* **15**, 10033–10040 (2007).

95. P. Ma, F. Robin, and H. Jackel, "Realistic photonic bandgap structures for TM-polarized light for all-optical switching," *Opt. Express* **14**, 12794–12802 (2006).

96. S. F. Mingaleev, A. E. Miroshnichenko, and Y. S. Kivshar, "Low-threshold bistability of slow light in photonic crystal waveguides," *Opt. Express* **15**, 12380–12385 (2007).

97. K. Inoue, H. Oda, N. Ikeda, and K. Asakawa, "Enhanced third-order nonlinear effects in slow-light photonic-crystal slab waveguides of line-defect," *Opt. Express* **17**, 7206–7216 (2009).

98. J. B. Khurgin, "Performance of nonlinear photonic crystal devices at high bit rates," *Opt. Lett.* **30**, 643–645 (2005).

99. K. R. Khan, T. X. Wu, D. N. Christodoulides, and G. I. Stegeman, "Soliton switching and multi-frequency generation in a nonlinear photonic crystal fiber coupler," *Opt. Express* **16**, 9417–9428 (2008).

100. S. F. Mingaleev, A. E. Miroshnichenko, and Y. S. Kivshar, "Coupled-resonator-induced reflection in photonic-crystal waveguide structures," *Opt. Express* **16**, 11647–11659 (2008).

101. M. Notomi, T. Tanabe, A. Shinya, E. Kuramochi, H. Taniyama, S. Mitsugi, and M. Morita, "Nonlinear and adiabatic control of high-Q photonic crystal nanocavities," *Opt. Express* **15**, 17458–17481 (2007).

102. B. Kang, J. H. Woo, E. Choi, H. H. Lee, E. S. Kim, J. Kim, T. J. Hwang, Y. S. Park, D. H. Kim, and J. W. Wu, "Optical switching of near infrared light transmission in metamaterial-liquid crystal cell structure," *Opt. Express* **18**, 16492–16498 (2010).

103. S. Sandhu, M. L. Povinelli, and S. H. Fan, "Enhancing optical switching with coherent control," *Appl. Phys. Lett.* **96**, 231108 (2010).

104. H. Rabitz, "Focus on quantum control," *New J. Phys.* **11**, 105030 (2009).

105. A. Husakou and J. Herrmann, "Steplike transmission of light through a metal-dielectric multilayer structure due to an intensity-dependent sign of the effective dielectric constant," *Phys. Rev. Lett.* **99**, 127402 (2007).

106. L. D. Sio, J. G. Cuennet, A. E. Vasdekis, and D. Psaltis, "All-optical switching in an optofluidic polydimethylsiloxane: liquid crystal grating defined by cast-molding," *Appl. Phys. Lett.* **96**, 131112 (2010).

107. L. D. Sio, A. Veltri, C. Umeton, S. Serak, and N. Tabiryan, "All-optical switching of holographic gratings made of polymer-liquid crystal-polymer slices containing azo-compounds," *Appl. Phys. Lett.* **93**, 181115 (2009).

108. Y. Kanamori, K. Takahashi, and K. Hane, "An ultrasmall wavelength-selective channel drop switch using a nanomechanical photonic crystal nanocavity," *Appl. Phys. Lett.* **95**, 171911 (2009).

109. K. I. Umemori, Y. Kanamori, and K. Hane, "Photonic crystal waveguide switch with a microelectromechanical actuator," *Appl. Phys. Lett.* **89**, 021102 (2009).

110. D. M. Szymanski, B. D. Jones, M. S. Skolnick, A. M. Fox, D. Obrien, T. F. Krauss, and J. S. Roberts, "Ultrafast all-optical switching in AlGaAs photonic crystal waveguide interferometers," *Appl. Phys. Lett.* **95**, 141108 (2009).

111. A. G. Padowicz, I. Juwiler, O. Gayer, A. Bahabad, and A. Arie, "All-optical polarization switch in a quadratic nonlinear photonic quasicrystal," *Appl. Phys. Lett.* **94**, 091108 (2009).

112. D. P. Aryal, K. L. Tsakmakidis, C. Jamois, and O. Hess, "Complete and robust bandgap switching in double-inverse-opal photonic crystals," *Appl. Phys. Lett.* **92**, 011109 (2008).

113. T. Ruhl, P. Spahn, C. Hermann, C. Jamois, and O. Hess, "Double-inverse-opal photonic crystals: the route to photonic bandgap switching," *Adv. Funct. Mater.* **16**, 885–890 (2006).

114. P. Barthelemy, M. Ghulinyan, Z. Gaburro, C. Toninelli, L. Pavesi, and D. S. Wiersma, "Optical switching by capillary condensation," *Nat. Photon.* **1**, 172–175 (2007).

115. R. Evans, U. M. B. Marconi, and P. Tarazona, "Fluids in narrow pores: adsorption, capillary condensation, and critical points," *J. Chem. Phys.* **84**, 2376–2399 (1986).

116. M. Belotti, J. F. G. Lopez, S. D. Angelis, M. Galli, I. Maksymov, L. C. Andreani, D. Peyrade, and Y. Chen, "All-optical switching in 2D silicon photonic crystals with low loss waveguides and optical cavities," *Opt. Express* **16**, 11624–11636 (2008).

117. T. Tanabe, M. Notomi, S. Mitsugi, A. Shinya, and E. Kuramochi, "All-optical switches on a silicon chip realized using photonic crystal nanocavities," *Appl. Phys. Lett.* **87**, 151112 (2005).

118. Y. Liu, F. Qin, F. Zhou, and Z. Y. Li, "Ultrafast and low-power photonic crystal all-optical switching with resonant cavities," *J. Appl. Phys.* **106**, 083102 (2009).

119. M. Soljacic, E. Lidorikis, J. D. Joannopoulos, and L. V. Hau, "Ultralow-power all-optical switching," *Appl. Phys. Lett.* **86**, 171101 (2005).

120. D. M. Beggs, T. P. White, L. Ofaolain, and T. F. Krauss, "Ultracompact and low-power optical switch based on silicon photonic crystals," *Opt. Lett.* **33**, 147–149 (2008).

121. C. Y. Jin, O. Kojima, T. Kita, O. Wada, M. Hopkinson, and K. Akahane, "Vertical-geometry all-optical switches based on InAs/GaAs quantum dots in a cavity," *Appl. Phys. Lett.* **95**, 021109 (2009).

122. M. K. Kim, I. K. Hwang, S. H. Kim, H. J. Chang, and Y. H. Lee, "All-optical bistable switching in curved microfiber-coupled photonic crystal resonators," *Appl. Phys. Lett.* **90**, 161118 (2007).

123. U. Fano, "Effects of configuration interaction on intensities and phase shifts," *Phys. Rev.* **124**, 1866–1878 (1961).

124. X. D. Yang, C. Husko, C. W. Wong, M. Yu, and D. L. Kwong, "Observation of femtojoule optical bistability involving Fano resonances in high-Q/V_m silicon photonic crystal nanocavities," *Appl. Phys. Lett.* **91**, 051113 (2007).

125. A. E. Miroshnichenko, E. Brasselet, and Y. S. Kivshar, "All-optical switching and multistability in photonic structures with liquid crystal defects," *Appl. Phys. Lett.* **92**, 253306 (2008).

126. S. Lan and H. Ishikawa, "High-efficiency reflection-type all-optical switch for ultrashort pulses based on a single asymmetrically confined photonic crystal defect," *Opt. Lett.* **27**, 1259–1261 (2002).

127. M. Notomi, A. Shinya, S. Mitsugi, G. Kira, E. Kuramochi, and T. Tanabe, "Optical bistable switching action of Si high-Q photonic-crystal nanocavities," *Opt. Express* **13**, 2678–2687 (2005).

128. S. H. Li and X. H. Cai, "High-contrast all optical bistable switching in coupled nonlinear photonic crystal microcavities," *Appl. Phys. Lett.* **96**, 131114 (2010).

129. P. Bermel, A. Rodriguez, J. D. Joannopoulos, and M. Soljacic, "Tailoring optical nonlinearities via the Purcell effect," *Phys. Rev. Lett.* **99**, 053601 (2007).

130. X. Y. Hu, P. Jiang, C. Y. Ding, H. Yang, and Q. H. Gong, "Picosecond and low-power all-optical switching based on an organic photonic-bandgap microcavity," *Nat. Photon.* **2**, 185–189 (2008).

131. Z. Li, X. Hu, J. Zhang, H. Yang, and Q. Gong, "PCBM:P3HT polymer composites for photonic crystal all-optical switching applications," *J. Phys. D: Appl. Phys.* **43**, 385104 (2010).

132. Y. Zhang, X. Hu, H. Yang, and Q. Gong, "Multi-component nanocomposite for all-optical switching applications," *Appl. Phys. Lett.* **99**, 141113 (2011).

133. L. M. Li and Z. Q. Zhang, "Multiple-scattering approach to finite-sized photonic band-gap materials," *Phys. Rev. B* **58**, 9587–9590 (1998).

134. K. B. Chung and S. H. Kim, "Defect modes in a two-dimensional square-lattice photonic crystal," *Opt. Commun.* **209**, 229–235 (2002).

135. S. A. Kovalenko, J. Ruthmann, and N. P. Ernsting, "Ultrafast strokes shift and excited-state transient absorption of coumarin 153 in solution," *Chem. Phys. Lett.* **271**, 40–50 (1997).

136. J. P. Prineas, J. Y. Zhou, J. Kuhl, H. M. Gibbs, G. Khitrova, S. W. Koch, and A. Knorr, "Ultrafast ac Stark effect switching of the active photonic band gap from Bragg-periodic semiconductor quantum wells," *Appl. Phys. Lett.* **81**, 4332–4334 (2002).

137. E. Lioudakis, I. Alexandron, and A. Othonos, "Ultrafast dynamics of localized and delocalized polaron transitions in P3HT/PCBM blend materials: the effects of PCBM concentration," *Nanoscale Res. Lett.* **4**, 1475–1480 (2009).

138. S. Trotzky, T. Hoyer, W. Tuszynski, C. Lienau, and J. Parisi, "Femtosecond up-conversion technique for probing the charge transfer in a P3HT: PCBM blend via photoluminescence quenching," *J. Phys. D: Appl. Phys.* **42**, 055105 (2009).

139. G. H. Ma, L. J. Guo, J. Mi, Y. Liu, S. X. Qian, J. H. Liu, G. F. He, Y. F. Li, and R. Q. Wang, "Investigations of third-order nonlinear optical response of poly(p-phenylenevinylene) derivatives by femtosecond optical Kerr effect," *Physica B* **305**, 147–154 (2001).

140. K. C. Grabar, R. G. Freeman, M. B. Hommer, and M. J. Natan, "Preparation and characterization of Au colloid monolayers," *Anal. Chem.* **67**, 735–748 (1995).

141. X. Y. Hu, P. Jiang, C. Xin, H. Yang, and Q. H. Gong, "Nano-Ag:polymeric composite material for ultrafast photonic crystal all-optical switching," *Appl. Phys. Lett.* **94**, 031103 (2009).

142. K. M. Gaab and C. J. Bardeen, "Wavelength and temperature dependence of the femtosecond pump-probe anisotropies in the conjugated polymer MEH-PPV: implications for energy-transfer dynamics," *J. Phys. Chem. B* **108**, 4619–4626 (2004).

Questions

4.1 How can an all-optical switching be realized in a two-dimensional photonic crystal based on the photonic bandgap shift or defect mode shift?

4.2 Please discuss how the nonlinear coefficient of the component material and the quality factor of the photonic crystal microcavity influence the realization of the low-power photonic crystal all-optical switching.

Chapter 5

Tunable Photonic Crystal Filter

Optical filters can selectively permit the propagation of the signal light with a certain frequency, and forbid the propagation of other signal lights with different frequencies. Moreover, optical filters with a size of nano/micrometer scale play a very important role in the integrated photonic circuits, optical interconnection networks, and wavelength division multiplexing systems. Photonic crystals are perfect basis for the realization of future integrated photonic devices due to their unique properties of controlling the propagation states of photons. Therefore, much attention has been paid to the realization of optical filter based on photonic crystals. The optical channel or pass band of a tunable photonic crystal can be tuned by adjusting the external parameters. Therefore, in practice, tunable photonic crystal filter can find wider and greater applications.

5.1 Configuration of Photonic Crystal Filter

The photonic bandgap effect and the photon localization effect are the essential basis for the realization of photonic crystal filter. Owing to the strong Bragg scattering and strong interference effect, photonic crystal can prohibit the transmission of electro-magnetic waves whose frequencies drop in the photonic bandgap.

Photonic Crystals: Principles and Applications
Qihuang Gong and Xiaoyong Hu
Copyright © 2014 Pan Stanford Publishing Pte. Ltd.
ISBN 978-981-4267-30-4 (Hardcover), 978-981-4364-83-6 (eBook)
www.panstanford.com

When a structural defect is introduced in a perfect photonic crystal, electromagnetic waves with a certain resonant frequencies will be confined around the defect site. Accordingly, defect states with very high transmittance will appear in the photonic bandgap. The defect modes with a high transmittance and a high quality factor can form perfect filtering channels. Moreover, a narrow transmission window can be formed in the photonic bandgap when the defect structure is adequately designed. The narrow pass band window can also form a narrow-band photonic crystal filter.

5.1.1 Single-Channel Photonic Crystal Filter

The single-channel photonic crystal filter has one optical channel, only permitting the propagation of a signal light with a certain resonant frequency. Therefore, single-channel photonic crystal filter can be constructed by use of a point defect, a microcavity, or a line defect in a photonic crystal [1,2]. By carefully designing the defect structures, a single-channel photonic crystal filter with a high frequency resolution and a high transmittance can be realized. The basic characteristics of single-channel photonic crystal filters are high wavelength resolution (corresponding to a high quality factor of the optical channel) and high transmittance contrast between the optical channel and the photonic bandgap.

5.1.1.1 The point defect type

Introducing a simple point defect in a photonic crystal structure can be used to construct a single-channel photonic crystal filter [3]. Li *et al.* proposed a very simple approach to achieve a single-channel filter by use of an optical waveguide etched with a row of air holes in it, which is placed periodically along the optical waveguide [4]. This one row of air holes could also be considered a one-dimensional photonic crystal, because the refractive index was periodically modulated along the optical waveguide. Perfect photonic bandgap can be obtained. A point defect is introduced in the center of the one-dimensional photonic crystal by enlarging (or reducing) the distance between two air holes. Then a defect mode will appear in the photonic bandgap, which forms the optical channel for the photonic crystal filter. This is the very simple and feasible approach to realizing photonic crystal filter. However, there is only one row of air holes in the photonic crystal. Accordingly, the structural period of the one-dimensional photonic crystal is small. This leads to a relatively weak photon confinement effect. So, the optical channel has a low quality factor, and accordingly, a bad wavelength resolution. The schematic structure of a silicon photonic crystal filter formed by etching eight air holes in a silicon waveguide is shown in Fig. 5.1 [4]. The lattice constant and the diameter of air holes were 360 and 200 nm, respectively.

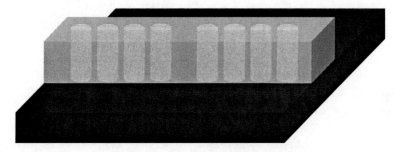

Figure 5.1   Schematic structure of single-channel photonic crystal filter with eight air holes etched in a silicon waveguide.

5.1.1.2 The surface-emitting type

Based on the coupling of a point defect and a waveguide in a photonic crystal, the signal light with a specific frequency in the optical waveguide can be trapped into the point defect site, which can perform the functions of single-channel photonic crystal filter [5,6]. When the photonic crystal structure is carefully designed so that the signal light is emitted into the free space from the point defect, a surface-emitting-type photonic crystal filter can be realized. For the surface-emitting photonic crystal filter with a single point defect, the maximum output efficiency is only 50% when the condition of $Q_{in} = Q_v$ is satisfied, where Q_{in} is the in-plane quality factor, which is determined by the coupling between the defect mode and the waveguide modes. Q_v is the vertical directional quality factor, which is determined by the coupling between the defect mode and the free space [7]. The reason lies in that there exist a great deal of forward and backward propagation signal light in the waveguide. In order to improve the output efficiency, two microcavities with a degenerate resonant frequency and different symmetry with respect to the plane perpendicular to the waveguide should be adopted. Then, both the forward and backward propagation signal light in the waveguide can be destructively interfered. When the condition of $Q_{in} = Q_v$ is satisfied, a very high output efficiency of up to 100% could be reached theoretically [7]. Moreover, if a beam of wide-band signal light propagates in the free space, the special signal light with a required frequency can be selected out by the point defect through resonantly trapping effect, and subsequently coupled into the optical waveguide [8]. Therefore, the surface-emitting photonic crystal filter can find versatile applications in practice.

The typical structure of the surface-emitting photonic crystal filter consists of a single point defect made of an enlarged air hole, which also called the acceptor-type defect, side-coupled with a photonic crystal waveguide, as shown in Fig. 5.2 [9]. When a wide-band signal light propagates in the photonic crystal waveguide, the required signal light component can be coupled into the defect mode when its frequency is in resonance with that of the defect mode, and then emitted to the free space. The resonant frequency of the defect mode determines which signal light can be filtered out from the bus waveguide and emitted into the free space. The emission frequency can be tuned by modulating the size of the defect structure and the distance between the point defect and waveguide. On one hand, a red shift of the filtered frequency can be achieved by enlarging the diameter of the air hole constructing the point defect. The reason lies in that the effective refractive index of the defect structure decreases with the increase of the air hole diameter. According to the electromagnetic variational theory, high-frequency electromagnetic modes will concentrate their energy in the low dielectric region. Therefore, the defect structure can only confine electromagnetic modes with the higher frequency. On the other hand, a red shift of the filtered frequency can also be achieved by enlarging the distance between the point defect and the waveguide. The electric field distribution of the defect mode exponentially decays with the distance away from the center of the defect structure. Therefore, with the increase of the distance between the single defect and the waveguide, very weak electric field distribution of the defect mode can reach that of the waveguide modes. This not only leads to a very weak coupling of the defect mode and the waveguide mode, but also decreases the effective refractive index of the defect structure [9]. As a result, the resonant frequency of the defect mode shifts in the high-frequency direction. However, the low effective refractive index of the acceptor-type point defect results in a weak photon confinement effect. Accordingly, the quality factor of the optical channel of the photonic crystal filter is relatively small. This leads to a low wavelength resolution.

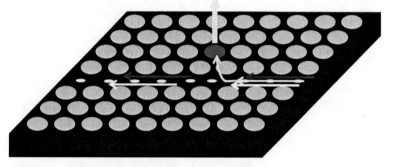

Figure 5.2   Schematic structure of surface-emitting single-channel photonic crystal filter.

To improve the wavelength resolution of the photonic crystal filter, the defect structure with a high quality factor has to be adopted. The donor-type defect is composed of several missing air holes, which leads to a relatively large effective refractive index of the defect structure. As a result, the photonic confinement effect of the donor-type defect is much stronger than that of the acceptor-type defect. When a donor-type defect composed of three missing air holes in line was used, the quality factor and the line width of the optical channel were 2600 and 0.6 nm for the silicon photonic crystal filter, respectively. While for an L4 donor-type defect composed of four missing air holes in line was used, a high quality factor of 3200 and a narrower line width of 0.2 nm could be obtained [10,11]. Even higher quality factor of 7300 can be reached by adopting the defect structure composed of two donor-type defects providing symmetric and antisymmetric modes with a degenerate resonant frequency [12].

5.1.1.3 The multilayer type

When a defect layer is inserted into a one-dimensional photonic crystal, a defect mode will appear in the photonic bandgap, which forms the perfect filtering channel. Then a multiplayer-type photonic crystal filter can be realized. Not only dielectric materials but also metamaterials can be used to construct the photonic crystal filter.

Lee *et al.* reported a single-channel one-dimensional Si/SiO_2 photonic crystal filter at the wavelength of 1.55 μm [13]. The thickness was 95.3 nm for silicon layers and 234.9 nm for SiO_2 layers, respectively. A defect layer made of 222.6 nm-thick silicon layer was inserted in the center of the multilayer structure. A defect mode with the resonant wavelength of 1.55 μm appeared in the photonic bandgap, which forms the optical channel of the photonic crystal filter. The line width of the defect mode, also called the full width at half maximum (FWHM), was 2.1 nm, which indicates that the photonic crystal filter possesses good wavelength resolution. The quality factor of the optical channel was 775.

The multilayer structure formed by alternatively placing layers of negative-permittivity material and negative-permeability material also have perfect photonic bandgap, also called zero effective phase gap, which originates from the interactions of evanescent waves in the single-negative materials, not from the Bragg scattering effect [14]. When a defect layer made of positive refractive index material is inserted in the center of the multilayer structure, a defect mode will appear in the photonic bandgap, which can form the optical channel of photonic crystal filter.

5.1.1.4 The channel-drop type

The channel-drop photonic crystal filter has two basic configurations. The first basic configuration is formed by one (or several) photonic crystal microcavity sandwiched between two parallel photonic crystal optical waveguides [15]. The second basic configuration is formed by two closely and parallel placed photonic crystal waveguides. If a beam of wide-band signal light enters the bus waveguide, the special signal light with a certain resonant frequency can be transferred into the drop waveguide via the photonic crystal microcavity between them based on the photon tunneling effect (for the first configuration), or via the coupling of these two waveguides directly (for the second configuration). Therefore, the channel-drop photonic crystal filter may find great potential applications in the wavelength division multiplexing systems.

5.1.1.4.1 Basic configurations

5.1.1.4.1.1 *Using microcavity-waveguide coupling*

This kind of channel-drop photonic crystal filter consists of one (or several) microcavity side-coupled with two optical waveguides, as shown in Fig. 5.3 [16]. The structure parameters of two microcavities can be carefully designed so that the resonant frequencies of two microcavity modes are equal to each other. The maximum efficiency of transferring a signal light from the bus waveguide to the drop waveguide via the microcavities can be obtained when the following condition is met [17]

$$kd = \left(n + \frac{1}{2}\right)\pi \qquad (5.1)$$

where k is the wave vector of the operating frequency, d is the distance between two photonic crystal microcavities, n is an integer. If the resonant condition of equation 5.1 is satisfied, almost all the signal light whose frequency is in resonance with that of the microcavity modes can be transferred from the bus waveguide to the drop waveguide and subsequently output from the drop waveguide. Fan *et al.* pointed out that by carefully designing the structural parameters of one single microcavity that supports two degenerate resonant modes with opposite symmetries, high transfer efficiency could also be achieved [18].

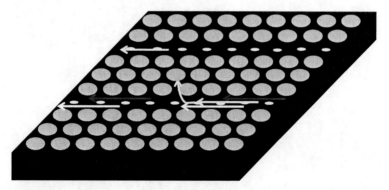

Figure 5.3 Schematic structure of channel-drop type photonic crystal filter.

5.1.1.4.1.2 *Using parallel waveguide coupling*

This kind of channel-drop photonic crystal filter consists of two closely and parallel placed photonic crystal waveguides. If a periodic corrugation is introduced in the bus waveguide, the waveguiding and the diffraction phenomenon coexist and complete with each other [20]. When the phase difference Δk between the fundamental mode and a high-order mode satisfies the condition $\Delta k = m(2\pi/a)$, where m and a are the integer and the period of the corrugation, respectively, the fundamental mode will convert to the high-order mode [21]. Therefore, when a beam of wide-band signal light enters the bus waveguide, the fundamental mode of a special signal light will convert to a high-order mode. Then this high-order mode will be transferred to the drop waveguide through the resonant tunneling effect. In the drop waveguide, the high-order mode was converted into the fundamental mode again and the original certain signal light is filter out from the bus waveguide [22].

The contra-directional dropping can also be realized in the channel drop photonic crystal filter. Figure 5.4 shows the schematic structure of the channel-drop photonic crystal filter based on the contra-directional coupling of two photonic crystal waveguides [19]. The bus and drop waveguides are separated by one row of air holes. The length of the coupling region is carefully designed so that only the required signal light in the bus waveguide can be coupled into the drop waveguide. One output port of the drop waveguide is closed by the photonic crystal region. Owing to the

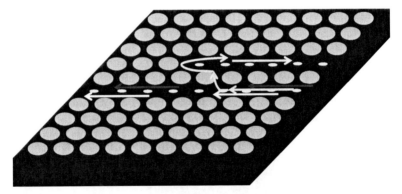

Figure 5.4   Schematic structure of a contra-directional coupling, channel-drop type photonic crystal filter.

strong Bragg scattering effect of photonic crystal in the closed part of the drop waveguide, the signal light transferred into the drop waveguide can only propagate in the contra-direction compared with the electromagnetic wave in the bus waveguide.

5.1.1.4.2 *Enhancing the dropping efficiency*

It has been pointed out that the maximum efficiency of extracting the light trapped in a point defect into a nearby optical waveguide is 25%, which greatly restricts the practical applications of the channel-drop type photonic crystal filter [23]. Takano *et al.* proposed an approach to achieving a very high dropping efficiency by use of a photonic crystal heterostructure [24]. The photonic crystal heterostructure consists of two photonic crystal slabs with different lattice constant, connected with each other directly. The interface of the heterostructure can act as a mirror. When the frequency of a special signal light propagating in the bus waveguide was in resonance with that of the photonic crystal microcavity, the energy of the signal light will be trapped by the microcavity, and subsequently coupled into the drop waveguide through the photon resonant tunneling effect. By properly designing the structural parameters, the reflected light at the heterostructure interface can make a destructive interference in the bus waveguide and a constructive interference in the drop waveguide. This can greatly enlarge the transfer efficiency from the bus waveguide to the drop waveguide via the photonic crystal microcavity. Ren *et al.* founded that adopting an additional wavelength-selective reflection microcavity side-coupled with the bus waveguide could also be used to improve the dropping efficiency of the channel-drop photonic crystal filter [25]. The channel-drop photonic crystal filter contains two microcavities, one for resonant tunneling-based signal light trapping and transferring, the other for wavelength-selective reflection feedback in the bus waveguide. Owing to the wavelength-selective reflection microcavity, the reflected light at the heterostructure interface can make a destructive interference in the bus waveguide and a constructive interference in the output waveguide. This also can greatly enlarge the transfer efficiency from the bus waveguide to the drop waveguide. Theoretically, the dropping efficiency close to 100% can be achieved by use of these two approaches.

5.1.1.5   The surface-wave type

The surface wave can exist in the surface of a terminated photonic crystal, the resonant frequency of which being in the photonic bandgap [26]. When a reflection object is placed in other side, a surface-wave microcavity can be formed. The surface-wave microcavity can provide resonant electromagnetic modes confined in the surface of the photonic crystal [27]. Adopting surface wave microcavity can also realize the function of an optical filter. Zhang *et al.* reported a single-channel filter based on two-dimensional photonic crystal surface-mode cavity in amorphous silicon-on-silica structure [28]. The photonic crystal was composed of a triangular lattice of air holes in 280 nm-thick silicon film, with a lattice constant of 380 nm and a diameter of air holes of 228 nm, respectively. The microcavity was constructed by enlarging the diameter of six air holes in the center of edge of the photonic crystal so that the microcavity could provide the surface-wave mode with a resonant wavelength of 1580 nm. The air gap between the photonic crystal edge and the 425 nm wide optical waveguide was 200 nm. The 1580 nm signal light propagating in the waveguide can be trapped and confined in the surface-wave microcavity. Therefore, a single-channel photonic crystal filter based on surface-wave microcavity can be realized.

5.1.2   Multichannel Photonic Crystal Filter

The multichannel photonic crystal filter has two or more optical channels, permitting the propagation of several signal lights with different resonant frequencies at one time. The central frequency of one optical channel is different from others. Each optical channel only selects a certain frequency signal light. Therefore, several defect structures have to be introduced in a perfect photonic crystal to form the optical channels of the photonic crystal filter.

5.1.2.1   Microcavity-waveguide coupling type

A single photonic crystal microcavity side-coupled with a bus waveguide can be used to extract a specific signal light from the bus waveguide, which can perform the function of a single-channel photonic crystal filter. Accordingly, several photonic crystal microcavities with different resonant frequencies side-coupled with a bus waveguide can extract different signal light from the bus waveguide. Therefore, a multichannel photonic crystal filter can be realized. There are several configurations, such as the surface emitting type, the splitting-beam type, and the in-plane channel-drop type.

5.1.2.1.1 *The surface emitting type*

Noda *et al.* reported a surface-emitting multichannel photonic crystal filter based on the coupling of several single defects side-coupled with an optical waveguide in a two-dimensional photonic crystal slab [29,30]. The schematic structure of the multichannel photonic crystal filter is shown in Fig. 5.5. They fabricated a two-dimensional triangular-lattice InGaAsP photonic crystal slab with a thickness of 250 nm by use of the electron-beam lithography and reactive ion etching technique. Two single defects side-coupled with a linear waveguide were introduced in the photonic crystal slab. The resonant wavelength was 1545 nm for defect 1 and 1566 nm for defect 2. The quality factor of the defect mode was about 400. When a beam of wide-band signal light propagated in the optical waveguide, the 1545 nm signal light and the 1566 nm signal light would be trapped by the defects 1 and 2, respectively, and then emitted to the free space.

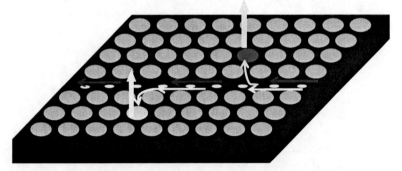

Figure 5.5   Schematic structure of surface-emitting multichannel photonic crystal filter.

5.1.2.1.2 *The splitting-beam type*

When a point defect is introduced in a photonic crystal waveguide, a defect mode with a certain resonant frequency can be formed. Only the signal light whose frequency is in resonance with that of the defect mode can be confined in the point defect site and subsequently transmitted through the optical waveguide based on the resonant tunneling effect [31]. Therefore, the photonic crystal waveguide containing a point defect can be used as an element to construct the optical channel for multichannel filters. Jin *et al.* fabricated a Y-type photonic crystal waveguide system, each output waveguide having a point defect in it, and adopted it to realize multichannel filter function [32]. When a beam of wide-band signal light enters the bus waveguide, each drop waveguide only selects the signal light whose frequency is resonant with that of the point mode in it.

Kim *et al.* proposed a five-channel photonic crystal filter having five photonic crystal waveguides, each waveguide having a point defect in it, as shown in Fig. 5.6 [33]. The lattice constant and the radius of the two-dimensional square lattice photonic crystal are a and r, respectively. The point defect consisted of five air holes with the center one having a changed radius, $r = 0.1a$, $0.08a$, $0.065a$, $0.05a$, and 0 for defects A, B, C, D, and E, respectively. The dielectric constant was 11.65. The waveguide A was used as the bus waveguide. Five defect modes with high transmittance close to 100% appeared in the photonic crystal bandgap, which forms the optical channels for the photonic crystal filter. Moreover, defect A in the bus waveguide provided a reflection feedback, which helped to improve the transfer efficiency from the bus waveguide to the drop waveguide up to 96% for the five channels [33].

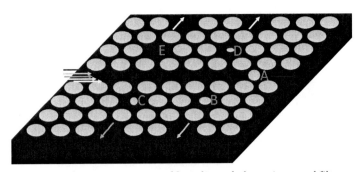

Figure 5.6 Schematic structure of five-channel photonic crystal filter.

5.1.2.1.3 *The in-plane channel-drop type*

Based on the microcavity and waveguide coupling, the in-plane multichannel optical filters consisting of several microcavities and waveguides in two-dimensional photonic crystal slabs can be realized [34]. To improve the dropping efficiency of the in-plane channel-drop filter, closing the bus and drop waveguides in the microcavity-waveguide coupling region is a very effective approach, which can act as perfect reflection mirrors, resulting in a constructive interference in the drop waveguides and a destructive interference in the bus waveguide [35]. Ren *et al.* reported a two-channel filter based on closed-waveguides and microcavity coupling in a two-dimensional triangular lattice silicon photonic crystal, as shown in Fig. 5.7 [36]. The lattice constant and the diameter of air holes were 430 nm and 290 nm, respectively. Waveguides 1 and 2 acted as drop waveguides, while waveguide 3 was the bus waveguide. Two microcavities, C1 and C2, were formed by three missing air holes. The resonant wavelength was 1529.5 nm for C1 and 1531 nm for C2. The distance between the ends of microcavity and the waveguide was 1290 nm. When a beam of wide-band signal light enters the bus waveguide, only the signal light whose frequency is resonant with the microcavity mode is trapped by the microcavity and transferred to the drop waveguide due to resonant tunneling effect. Accordingly, the 1529.5 nm signal light transmits from the drop waveguide 1, while

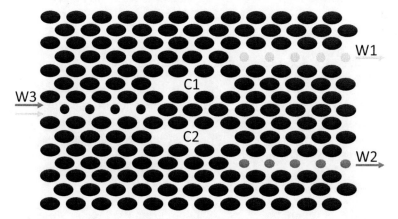

Figure 5.7 Schematic structure of two-channel silicon photonic crystal filter.

the 1531 nm signal light transmits from the drop waveguide 2. Therefore, the wavelength difference and the quality factor of two optical channels are only 1.5 nm and over 1000, which ensures high wavelength resolution. Adopting photonic crystal ring resonators instead of point-defect microcavities is another effective approach to reaching high dropping efficiency [37], because ring resonators have much larger size than that of the point-defect microcavity, which can not only sustain multiple-microcavity modes but also benefit the optimal coupling of microcavity and waveguide.

5.1.2.2 Perturbed photonic crystal type

Lei *et al.* proposed a mechanism to realize high-quality multichannel filter in a perturbed one-dimensional photonic crystals [38]. The perfect one-dimensional photonic crystal consists of a periodic placement of the unit element made of high- and low-dielectric layers. The perturbed one-dimensional photonic crystal consists of a lot of blocks, composed of randomly repeated stacks of a number of unit elements, separated by spacers. The spacers are made of the same low-dielectric material with a random thickness. According to the photon localization theory, most of the electromagnetic waves will be localized in the photonic crystal due to the strong disorders. There still exist some extended-state electromagnetic modes, which form high transmittance modes with high quality factor through the resonant tunneling effect. These resonant-tunneling modes can form perfect optical channels with high wavelength resolution for the photonic crystal filter.

5.1.2.3 Photonic crystal heterostructure type

The photonic crystal heterostructure, which is composed of two or more different photonic crystal components, can also be used to realize multichannel filter functions. There are one- and two-dimensional photonic crystals heterostructures that can be used to construct multichannel filter.

5.1.2.3.1 *For one-dimensional case*

The simplest photonic crystal heterostructure has a symmetric configuration of $(AB)^n B^m (BA)^n$, where A and B are two kinds of different dielectric materials and m and n are the numbers of repeating elements [39]. The symmetric photonic crystal hetero-structure also possesses perfect photonic bandgap properties due to spatially quasiperiodic dielectric distribution of the hetero-structure [40]. The B^m layer in the center of the heterostructure can be regarded as a defect layer. Multiple perfect transmittance peaks will appear in the photonic bandgap when the value of m is larger than 2, which originates from the internal symmetry of the heterostructure and the modulation of optical thickness of the heterostructure [39]. These transmittance peaks form the optical channels for the photonic crystal filter. Moreover, the number and the center wavelength of the optical channels can be tuned by adjusting the m value of the defect layer. With the increment of m value, the number of the optical channel increases, with results from the larger defect size that can sustain more defect modes.

Qiao *et al.* presented a mechanism to realize multichannel optical filter based on a one-dimensional photonic crystal quantum well [41]. The photonic crystal quantum well has a configuration of $(AB)^5(CD)^n(AB)^5$, where AB and CD represent two different one-dimensional photonic crystals. The CD photonic crystal can be regarded as a defect layer of the photonic crystal quantum well structure. If both AB and CD photonic crystals can sustain the propagation of an electromagnetic wave, the electromagnetic wave can be an extended-state mode. If only one photonic crystal can sustain the propagation of an electromagnetic wave, the electro-magnetic wave will be a localized-state mode [41]. The structural parameters of the photonic crystal quantum well are properly designed so that the second pass band of the CD photonic crystal drops in the second photonic bandgap of the AB photonic crystal. So, localized-state modes will appear in the photonic crystal quantum well. When the value of $n = 1$, only one high-Q transmission mode appears in the photonic bandgap of the photonic crystal quantum well, which forms the optical channel for the photonic crystal filter. With the increase of the n value, multiple optical channels can be formed. The increase of the n value corresponds to the increment of the thickness of the defect layer. A larger size of the defect layer can sustain multiple defect modes. Therefore, a multichannel photonic crystal filter can be realized.

Lee *et al.* reported a multichannel transmission filter based on a one-dimensional Si/SiO$_2$ photonic crystal heterostructure with a configuration of $(AB)^2(CD)^m(AB)^2$, where A, B, C, and D represent 93.4 nm-thick Si layer, 230.6 nm-thick SiO$_2$ layer, 186.8 nm-thick Si layer, and 461.2 nm-thick SiO$_2$ layer, respectively, and m is the repeating number of photonic crystal CD [42]. The photonic crystal CD can be considered a defect layer inserted in the center of a one-dimensional photonic crystal. Several defect modes with a high transmittance and a high quality factor appeared in the photonic bandgap, which forms the optical channels of the photonic crystal filter. The number of the defect modes was $2m$. This is because with the increase of the m value, the size of the defect layer increases, owing to which the defect layer sustains more resonant electromagnetic modes. Therefore, by simply adjusting the repeating number m of photonic crystal CD, a multichannel photonic crystal filter with tunable optical channel can be achieved.

5.1.2.3.2 *For two-dimensional case*

Two-dimensional photonic crystal heterostructure multichannel filters can be formed by simply connecting several filter units with the structure proportional to an optimized basic unit and operating at a different wavelength [43,44]. The remarkable advantages of a two-dimensional photonic crystal heterostructure are the excellent suitability for the integration applications, the willful increase in optical channels, and the greatly improved dropping efficiency by use of the destructive interference to eliminate undesirable outputs due to the reflection at the heterostructure interface [45].

The typical case of a multichannel filter based on two-dimensional photonic crystal heterostructure is shown in Fig. 5.8a [46]. The filter device is composed of several photonic crystal slabs, PC1, PC2, PC3, ... Each photonic crystal unit has a proportional lattice constant and a different operating wavelength. When a beam of wide-band signal light propagates in the bus waveguide, the signal light whose frequency is resonant with that of the microcavity mode will be trapped by the corresponding point-defect microcavity, and then transferred to the drop waveguide due to the resonant tunneling effect. The heterostructure interfaces between neighbor photonic crystal slabs can act as reflection mirrors [47]. A part of the transferred signal light through the microcavity propagates in the direction of hetero-structure interfaces and is reflected, resulting in a destructive interference. This enhances the energy propagation in the output waveguide, and a high dropping efficiency up to 100% can be achieved theoretically.

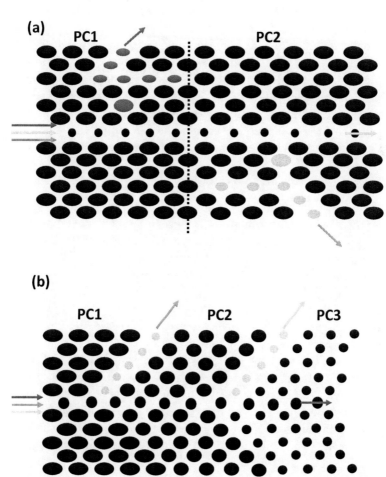

Figure 5.8   Schematic structure of a two-dimensional photonic crystal heterostructure filter. (a) Using dielectric material. (b) Using metal material.

Florous *et al.* proposed an approach to achieving a multi-channel filter in a photonic crystal heterostructure, which consists of three two-dimensional metal photonic crystals connected in line, as shown in Fig. 5.8b [48]. The metal photonic crystal is composed of a triangular lattice of cylindrical metallic rods embedded in air. These metallic photonic crystals have different rod diameters, but the same lattice constant. Three line defects with different width are introduced in the photonic crystal hetero-structure. The mechanism of confining electromagnetic waves in the line defect originates from confining the photons in the air core of line defects due to the total external reflection occurring in the interface between the metallic photonic crystal and air [49]. The line defects with different widths can provide electromagnetic waves with different resonant frequencies. Therefore, different signal light in the bus waveguide can be trapped by different line defects, and a multichannel filter can be realized.

5.1.2.4 Branchy defect type

Increasing the number of defect structures in a photonic crystal is the traditional method to achieve a multichannel photonic crystal filter [50]. Li *et al.* reported a multichannel filter based on a branchy defect structure in a microstrip photonic crystal in the microwave range [51]. The microstrip photonic crystal consists of a row of circles periodically etched on a 50 Ω microstrip line. The lattice constant and the diameter of circles were 20 mm and 4.5 mm, respectively. A defect was introduced in the center of periodic structure by enlarging the distance of two neighboring air holes. The length of the microstrip line in the defect region determined the resonant frequency of the defect mode. Microstrip lines with different lengths provide different defect modes. Therefore, by adding several microstrip lines with different lengths around the defect region, multi-defect modes can be achieved. This makes it possible to realize the multichannel photonic crystal filter.

5.1.2.5 Quasiperiodic photonic crystal type

Quasiperiodic photonic crystals still possess perfect photonic bandgap properties due to their spatially quasiperiodic distribution of dielectric function. On the other hand, they possess strong intrinsic disorder degree, which forms defect modes in the photonic bandgap. Therefore, multichannel filter can also be realized by use of quasiperiodic photonic crystals.

Kee *et al.* proposed that a quasiperiodically poled lithium niobate with a Fibonacci sequence can be used to realize multi-channel photonic crystal filter [52]. A Fibonacci sequence can be constructed based on the rule $S_{i+2} = S_i + S_{i+1}$, with S_1 = A and S_2 = B, respectively. A and B are two building elements. For a quasi-periodically poled lithium niobate with a Fibonacci sequence, A and B represent the positive ferroelectric domain and the negative ferroelectric domain, respectively. The central wavelengths of the optical channels are given by [52]

$$\lambda_m = \frac{(n_e - n_o)a}{2m - 1} \tag{5.2}$$

where m is an integer, a is the period of the photonic crystal, n_e and n_o are the extraordinary and ordinary refractive index of lithium niobate, respectively.

5.1.3 Narrow-Band Photonic Crystal Filter

The narrow-band photonic crystal filter can permit the output of a certain continuous frequency range of signal light at one time. Therefore, a narrow pass band window has to be formed in the photonic bandgap. Several mechanisms can be adopted to obtain a narrow pass band in the photonic bandgap structure, such as using the coupling of several identical defect structures, cascaded identical resonant grating strictures, coupling of surface-wave modes in a multilayer structure, and so on [53–55].

5.1.3.1 The resonant grating type

Jacob *et al.* proposed a mechanism to realize a flat-topped narrow-band photonic crystal filter by use of cascaded identical resonant grating structure with a cascaded π out of phase in each filtering elements [56]. Each filtering element is composed of the antireflective layer, the waveguide layer, the grating layer, and the separating layer in sequence. They found that the π out of phase arrangement could result in a line shape with a broader peaks and narrower bases compared with that of a single filtering element. The reason lies in that the π phase shift can nearly cancel the off resonance of two filtering element, which makes a flat-topped and narrow-band reflective window.

5.1.3.2 The multiple defects type

The traditional method to form a narrow-band filter is to introduce many identical defects in a perfect photonic crystal to construct a multiple-microcavity structure. The coupling of the neighboring microcavities makes the eigenfrequency split and form a continuous transmittance band in the photonic bandgap [57]. In a two-dimensional photonic crystal, introducing an array of identical defects could yield a narrow transmission band in the photonic bandgap, which can form a narrow-band filter [58]. Similarly, successively inserting defects in a photonic crystal waveguide along the direction parallel to the waveguide can also realize a narrow-band filter [59].

The multilayer structure having multiple defect layers can also be used to construct narrow-band photonic crystal filter. Troitski pointed out that when the following conditions are satisfied, a narrow transmission band can be achieved [60]:

$$\Phi = (2m + 1)\pi \quad \text{and} \quad \frac{d\Phi}{d\omega} = 0 \qquad (5.3)$$

where Φ is the total phase shift per round trip in a photonic crystal microcavity, ω is the frequency of the incident light. Based on Troitski's concept, Chen *et al.* realized a flat-topped and narrow-band one-dimensional GaAs/AlO photonic crystal filter having multiple defect structures, fabricated by use of the molecular beam epitaxy and oxidation technique [61]. The thickness of GaAs and AlO was 160.5 and 131.3 nm, respectively. The thickness of the GaAs layer in the center of the photonic crystal was enlarged to 515 nm, which resulted in a one-dimensional microcavity. They also introduced two additional defect layers in the lateral photonic crystal mirrors to induce an anomalous dispersion to meet the conditions of equation 5.3. A flat-topped transmittance band centered at 1600 nm was achieved. The width of the pass band was 65 nm.

5.1.3.3 The metallodielectric type

Metal materials possess a relatively small real part of the dielectric constant, which leads to a high dielectric contrast between metal and dielectric materials. Moreover, owing to the perfect reflectivity of metal, metallodielectric photonic crystals are perfect basis for the reflectance-type optical filters [62,63]. Jaksic *et al.* [64] fabricated a one-dimensional metallodielectric photonic crystal optical bandpass filter for the ultraviolet range. The metallo-dielectric photonic crystal was composed of Ag/SiO_2 multilayer fabricated by the radiofrequent sputtering method. Satisfactory suppression of undesirable visible and infrared parts of the spectrum could be achieved even for a small number of periodic units. Owing to the resonant tunneling through the metal layers, the one-dimensional metallodielectric photonic crystal combines certain properties of bulk metal with a high transparency in the desired wavelength range [64]. The important characteristic, compared with the Fabry–Perot-based structures, is that no higher-order transmittance maximum exists and the only one peak in spectrum from the ultraviolet to microwave range is that at the desired frequency [62]. The metallodielectric photonic crystal bandpass filter is composed of two Ag/SiO_2 pairs and an additional SiO_2 layer with a half-thickness compared to that in the multilayer structure, which serves as an antireflection coating. The thickness is 26 nm for Ag and 40 nm for SiO_2. The central wavelength and the width of the pass band are about 340 nm and 100 nm, respectively. The average transmittance of the pass band is 47%. Jaksic *et al.* also found that an increase in the dielectric layer thickness causes a red shift of the central wavelength of the pass band, and the decrease causes a blue one. An increase in the silver layer thickness will result in a decrease of the transmittance of the pass band. With an increase of the number of the periodic units, the width of the pass band will be narrowed and the long-wavelength transmittance of the pass band will be decreased, which results from the increase of the reflection of metal layer [64].

5.1.3.4   Photonic crystal superprism type

For a photonic crystal, the effective refractive index near the Brillouin zone edge is strong frequency dependent, i.e., possessing a violent dispersion effect [65]. For example, the superprism effect can exist in a photonic crystal, which allows a wide-angle deflection of a laser beam by a slight change of frequency and incident angle [66]. Baba and Matsumoto proposed that narrow-band optical filters could also be achieved by use of the superprism effect in a photonic crystal [66]. By carefully designing the structural parameters of a photonic crystal, the signal light, the frequencies of which drop in a certain frequency range, will be strongly deflected by the photonic crystal, which can act as a narrow-band photonic crystal filter.

5.1.3.5   The channel-drop type

The channel-drop type photonic crystal filter consists of two parallel photonic crystal optical waveguides separated by a defect structure. When the defect structure is properly designed so that the defect can provide narrow-band resonant modes, the signal light whose frequency dropping in the resonant frequency range of the defect modes can be transferred from one optical waveguide to the other optical waveguide via the defect structure based on the photon resonant tunneling effect [67]. Then a narrow-band photonic crystal filter can be achieved. Several identical micro-cavities in line can form a coupled-microcavity waveguide, which provides narrow-band resonant modes. So, a coupled-microcavity waveguide can trap a narrow-band of electromagnetic modes from a photonic crystal optical waveguide. Therefore, the coupling of a bus waveguide and a coupled-microcavity waveguide could also be used to construct narrow-band photonic crystal filter, as shown in Fig. 5.9 [68]. Bayindir and Ozbay proposed a two-dimensional square-lattice photonic crystal having a plane waveguide and two coupled-microcavity waveguides with different radius of defect air holes, $0.124a$ for r_c and $0.062a$ for r_D, respectively [68], where a is the lattice constant. The refractive index of the dielectric matrix is 3.1. It is very clear that channel C could trap the signal light from the bus waveguide in the frequency range from $0.311\ c/a$ to

0.339 c/a, while channel D could trap the signal light from the bus waveguide in the frequency range from 0.353 c/a to 0.383 c/a.

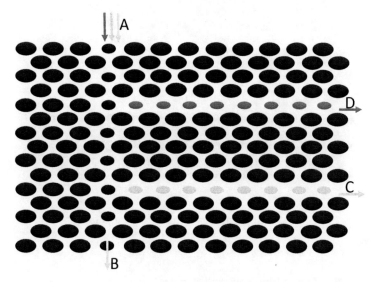

Figure 5.9   Schematic structure of narrow-band photonic crystal filter constructed by coupled-microcavity waveguides and a bus waveguide.

The cutoff frequency of the fundamental guiding mode of a photonic crystal waveguide varies with the structure parameters of the waveguide. When the diameter of air holes in two inner sides of the photonic crystal waveguide is enlarged, cutoff frequency of the fundamental guiding mode will take on a blue shift [69]. Niemi *et al.* proposed a unique narrow-band filter using a planar photonic crystal waveguide with consecutive sections of large border holes [70]. Enlarging the diameter of the air holes at the inner sides in consecutive sections along the photonic crystal waveguide will successively prevent a narrow band of guided modes near the low-frequency transmission bandedge from propagating [70]. The blocked part of the signal light can be transferred into a drop waveguide through resonant tunneling effect. Therefore, a multiple narrow-band photonic crystal filter can be realized. The schematic structure of the narrow-band photonic crystal filter is shown in Fig. 5.10.

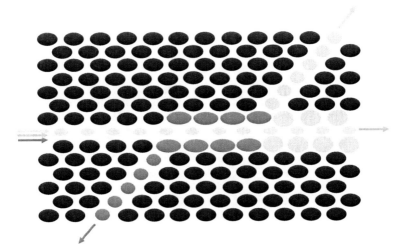

Figure 5.10   Schematic structure of a narrow-band photonic crystal filter realized by a waveguide with enlarged air holes at inner-sides.

5.1.3.6   The guided resonance type

Two-dimensional photonic crystal slabs possess the unique properties of guided resonances. The guided resonances, originating from the guided modes of a uniform dielectric slab, are a kind of electromagnetic modes strongly confined within the dielectric slab. When the phase matching condition is met, the guided resonances can couple into the radiation modes [71,72]. A photonic crystal slab can provide even and odd guided resonances with respect to the plane parallel to the surface of the slab, and set at the center of the slab. Through properly designing the structural parameters of the photonic crystal slab, the frequencies of the even and odd resonances can be equal. Suh and Fan found that by setting even and odd resonances with the equal frequency, and in the vicinity of the frequency of the signal light, a narrow-band photonic crystal filter can be realized when the signal light was incident in the direction perpendicular to the slab surface [73]. Owing to the finite thickness of the photonic crystal slab, the signal light could directly transmit through the photonic crystal slab due to the Fabry–Perot resonant effect. On the other hand, the photonic crystal structure can be considered period arrays of microcavities, which couple with each other through evanescent field. The coupling of normally incident signal light with these microcavity arrays results in a radiation into the free space. As a result, the Fano interference of the directly and indirectly transmitted light through the photonic crystal slab results in a narrow-band filter [74,75]. The guided resonance phenomenon exists not only in a two-dimensional photonic crystal slab but also in a slab grating structure, which can be considered a one-dimensional photonic crystal [76,77].

When the frequency ω and the decay rate γ of the even and odd resonant modes are equal to each other, i.e., $\omega_{even} = \omega_{odd}$, and $\gamma_{even} = \gamma_{odd}$, the transmittance of the normally incident signal light through the photonic crystal slab can be calculated by [78]

$$T = \frac{(\omega - \omega_0)^4}{(\omega - \omega_0)^4 + 4\gamma^4} \tag{5.4}$$

where $\omega_0 = (\omega_{even} + \omega_{odd})/2$. It is very clear that there exists a narrow frequency range in the vicinity of ω_0 whose reflectivity is near 100%. Therefore, a reflection type narrow-band photonic crystal filter can be achieved simply by adopting a single photonic crystal slab.

5.2 Tunable Photonic Crystal Filter

For traditional photonic crystal filter, the filtering properties, including the resonant frequency of the optical channel, the wavelength resolution, and the width of the transmission band, cannot be tuned. Tuning the frequencies of optical channels on a large scale can be achieved by modifying the structural geometry of photonic crystal microcavities [79–81], engineering the waveguide–cavity coupling via changing the configuration of the photonic crystal filter [82], or directly adjusting the permittivity of defect structure [83], which remarkably changes the photonic bandgap and the defect state structure. The first approach to finely tune the resonant frequencies of optical channels can be achieved by removing the ultrathin layer of material of the photonic crystal microcavity via a wet chemical digital etching technique, which will change the thickness of photonic crystal microcavities and accordingly the effective refractive index of photonic crystal microcavities [84]. The second approach is to locally infiltrate the photonic crystal microcavities with liquid, such as water, alcohol, and so on, via the nanofluidic technique [85]. However, in practical applications, it is strongly required that the tuning of the filtering properties be possible according to the changes of external conditions. The refractive index of third-order nonlinear optical materials can be adjusted by use of pump light, electric field, magnetic field, or even thermal radiation [86]. Various approaches have been proposed to demonstrate the tunable photonic crystal filter, such as the use of intense laser, electric field, magnetic field, mechanical force, and so on [87–89].

5.2.1 The Electric Field Tuning Method

Under the excitation of an external electric field, the orientation of the liquid crystal molecules changes when the applied voltage is larger than the critical voltage V_c, which leads to a remarkable change in the refractive index of liquid crystal materials. Accordingly, the photonic bandgap structure of a photonic crystal containing liquid crystal will vary with the applied voltage, which results in a variation of the filtering properties. Therefore, if liquid crystal is one of the components that construct the photonic crystal filter, the filtering properties can be tuned by use of an external electric field [90,91].

5.2.2 The Mechanical Tuning Method

Microelectrical-mechanical systems provide a powerful approach to adjust the properties of optical devices. If a mechanical force (or strain stress) can change the structural parameters of a photonic crystal, the photonic bandgap structure of the photonic crystal will vary with the applied mechanical force (or strain stress). So, the filtering properties can be tuned by use of an external mechanical force (or strain stress) [92].

Drysdale *et al.* reported a mechanically tunable metallic photonic crystal filter for terahertz frequency applications [93]. The filter was constructed by stacking two gold-coated silicon plates comprising two orthogonal linear grids with integral mounting lugs separated by an air gap of 30 μm. The lattice constant, the width, and the thickness of the metallic lugs were 1.8 mm, 450 μm, and 500 μm, respectively. The incident direction of the electromagnetic waves was normal to the surface of the metallic plate. The unique properties of metallic photonic crystals lie in that they possess a plasmon-like photonic bandgap extending from zero to the cutoff frequency [94]. When a structural defect was introduced in the center of the metallic plate, a transmittance peak appeared in the photonic bandgap just above the cutoff frequency, which forms the optical channel of single-channel photonic crystal filter. The resonant frequency of the filtering channel can be calculated by [95]

$$f = \frac{1}{2\pi\sqrt{LC}} \tag{5.5}$$

where L is the inductance and C is the capacitance. A lateral shift of the metallic plate will affect the inductance and the capacitance. Accordingly, the central frequency of the filtering channel will change with the increase of the lateral shift of the metallic plate. When the magnitude of the lateral shift increased from zero to the value equal to the width of the metallic lugs, 450 μm, the effective size of the defect structure increased, which resulted in an increment of the capacitance. As a result, the central frequency of the filtering channel decreased. When the magnitude of the lateral shift was larger than the width of the metallic lugs, the effective size of the defect structure decreased, which resulted in a decrease of the capacitance and an increase in the central frequency of the filtering channel. Drysdale *et al.* obtained a 3.5 GHz tuning range of the frequency of the filtering channel.

Suh *et al.* presented a mechanism to realize displacement-sensitive photonic crystal filter based on the guide resonance effect [96,97]. The photonic crystal filter consisted of two photonic crystal slabs separated by an air gap, each one made of square lattice of air holes embedded in a high-dielectric slab. The dielectric constant was 12. The thickness of each photonic crystal slab was a. The radius of air holes and the thickness of the photonic crystal slab were $0.55a$ and $0.4a$, respectively. The operation mechanism is based on the guided resonance effect in the slab, which forms a strong reflectivity in the vicinity of the resonant frequency for a single photonic crystal slab for the normally incident light. Stacking two photonic crystal slabs makes a wide frequency range of reflection band from 0.515 to 0.555 (c/a) with a high transmittance peak within the band, which forms the optical channel of the photonic crystal filter. The reason originates from the Fabry–Perot microcavity effect. The resonant frequency of the optical channel varies with the spacing between two photonic crystal slabs, as shown in Fig. 5.11.

Figure 5.11 Schematic structure of mechanically tunable filter based on guided resonance of two photonic crystal slabs.

5.2.3 The Thermal Tuning Method

If the refractive index of the material constructing the photonic crystal filter varies with the circumstance temperature, a thermally tunable photonic crystal filter can be achieved. Schuller *et al.* reported a thermally tunable multichannel semiconductor photonic crystal filter infiltrated with liquid crystals [98]. They fabricated a two-dimensional triangular lattice InGaAsP photonic crystal by use of electron-beam lithography and reactive ion etching technique. The lattice constant and the air filling factor were 340 nm and 40%, respectively. The photonic crystal microcavity was constructed by two photonic crystal reflectors, made of five rows of air holes, separated by 25 rows of air holes in a photonic crystal waveguide. The liquid crystal E7 was infiltrated in the air holes. The critical temperature of E7 liquid crystal was 60°C. Above this temperature, the transition from the nematic to the isotropic phase occurs. This makes the resonant frequencies of the microcavity modes shift. The photonic crystal microcavity sustained 16 TE resonant modes and 8 TM resonant modes. When the temperature changed from 15°C to 60°C, the central wavelength of 1516 nm mode shifts to 1517 nm.

Intonti *et al.* reported a thermally tunable single-channel photonic crystal filter by controllable removal of locally infiltrated water via changing the temperature [99]. They fabricated a two-dimensional triangular lattice GaAs photonic crystal containing a point defect. The lattice constant and the diameter of air holes were 311 and 500 nm, respectively. The point defect was formed by an enlarged air hole with a diameter of 600 nm. The air holes were infiltrated with water. The central wavelength and the quality factor of the defect mode were 1252 nm and over 10,000, respectively. This formed the perfect optical channel for the photonic crystal filter. The temperature changes were achieved under excitation of a laser beam. This controlled the evaporation rate of water and led to a variation of the effective refractive index of the point defect. Accordingly, a tunable photonic crystal filter could be achieved. The authors achieved a tunable range of 20 nm for the central wavelength of the optical channel.

For perovskite ferroelectric materials, such as $KTaO_3$ and $CaTiO_3$, their permittivity is temperature dependent under the condition of above the critical temperature [100]. Therefore, a thermal tunable photonic crystal filter can be reached by use of these materials. Nemec *et al.* fabricated a one-dimensional photonic crystal made of SiO_2 and air layers with a $KTaO_3$ defect layer [101]. The resonant frequency of the optical channel changed from 220 GHz at 170 K to 267 GHz at 290 K, reaching a large relative tunability of 20%.

5.2.4 The Acousto-Optic Tuning Method

The acousto-optic effect can result in a variation in the refractive index of the material or the shape change of the material, which leads to the change of the photonic bandgap structure when the acousto-optic materials are used to construct the photonic crystals. Therefore, the acousto-optic effect could also be adopted to realize the tunable photonic crystal filter.

Hong *et al.* reported a tunable photonic crystal fiber filter based on the acousto-optic effect [102]. The photonic crystal fiber filter was composed of a two-mode photonic crystal fiber attached with an acoustic horn, a piezoelectric disk, and an acoustic damper. The excited acoustic wave made the photonic crystal fiber bend. Therefore, when a beam of wide-band signal light is coupled into the photonic crystal fiber, the coupling between the fundamental core mode and the high-order core mode can be removed by bending the photonic crystal fiber based on the acousto-optic effect, and a tunable single-channel optical filter can be realized [102]. A tunable range from 700 nm to 1700 nm with a response time of 100 μs was achieved. The bandwidth of the optical channel changed from 4.5 to 9.5 nm.

5.2.5 The All-Optical Tuning Method

According to the nonlinear optical Kerr effect, the refractive index of the third-order nonlinear materials will change under the excitation of a pump light. Under the excitation of a laser beam, the refractive index of photosensitive materials will change due to the molecular configuration changes or component changes originating from the photochemical interactions. Therefore, all-optical tunable photonic crystal filter can be realized by use of these kinds of materials [103,104].

5.2.5.1 Single-channel organic photonic crystal filter

Because of their outstanding advantages over inorganic materials, such as large third-order nonlinearity, ultrafast time response, easy structure tailor, and superior waveguide properties, organic conjugated polymers are considered to be one of the ideal candidates for future integrated photonic devices [105]. Among these polymers, polystyrene is one of the most promising polymer materials for its high environmental and thermal stability, high optical damage threshold, and excellent film quality. Gong *et al.* adopted the polystyrene powder with a normal molecular weight of 8,000,000 (Fluka Chemie Company, Switzerland) to fabricate two-dimensional organic photonic crystal filter in the visible range [106]. Polystyrene powder was dissolved in toluene with a weight ratio of 1:140. Spin coating method was used to fabricate 300 nm-thick polystyrene film slab on silica substrates, which were pre-cleaned carefully. A focused ion beam etching system (Model DB235, FEI Company, USA) was employed to prepare the periodical patterns of two-dimensional photonic crystal filter. The Ga^+ ion beam generated by a Canion ion gun was connected to the ultrahigh vacuum chamber, where the sample was placed. A spot current of 30 pA was obtained from a weak emission current of 1 µA at 25 keV. A weak current can reduce the sample damage and the Gaussian wings of the ion beam, so that the quality of air holes can be guaranteed [107]. The sample of the two-dimensional photonic crystal filter is composed of regular square arrays of cylindrical air holes embedded in the background medium of polystyrene slab, as shown in Fig. 5.12. The radius of the air hole and the lattice constant are 90 nm and 220 nm, respectively. The width of the line defect is 310 nm. The total etched area is about 2.5 µm × 100 µm. The highly ordered and periodical structures of the patterns of air holes indicate the perfect quality of the two-dimensional photonic crystal filter. A defect mode with the center wavelength of 545 nm appeared in the photonic bandgap, which forms the optical channel for the photonic crystal filter. The quality factor and the line width of the optical channel were 230 and 2.1 nm, respectively, which indicates that the photonic crystal possesses excellent filtering properties.

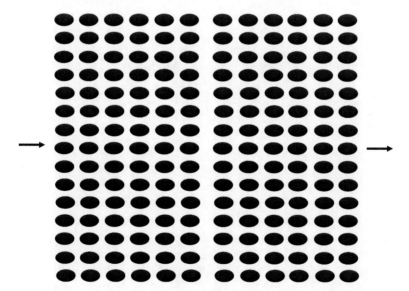

Figure 5.12   Schematic structure of a single-channel organic photonic crystal filter in the visible range.

The tunability of the photonic crystal filter can be achieved by changing the width of the line defect. The central wavelength of optical channel increases with the increment of the width of line defect. The reason lies in the fact that the defect structure determines the properties of two-dimensional photonic crystal filter. The line defect adds extra high dielectric materials in the photonic crystal where they should not be there. This donor-type defect, or called the dielectric defect, can support the defect modes in the photonic bandgap whose frequency is near the bottom of air band [108]. However, according to the electromagnetic variational theorem, the low-frequency modes concentrate their energy in the high-dielectric-constant regions, while the high-frequency modes concentrate their energy in the low dielectric constant regions [108]. Therefore, as the width of the line defect increases, the resonant frequency of the defect modes decreases, because the electromagnetic field are concentrated more and more in a high dielectric region. The closer the optical channel is to the center of photonic bandgap, the narrower the line width of optical channel is. The defect mode with 560 nm central wavelength is near the center of photonic bandgap. Accordingly, the line width of the optical channel with 560 nm central wavelength has the minimal value, 2 nm. The defect mode with 534 nm central wavelength is far away from the center of photonic bandgap, and it has the largest line width, 3.1 nm. When a defect mode is close to the center of photonic bandgap the photon confinement effect becomes more and more strong [109]. While the spectral width of defect mode shrinks with the increase of photon confinement. When the optical channel approaches the center of photonic bandgap, the quality factor increases. The quality factor of the optical channel with 534 nm-central wavelength is 172, while that of the optical channel with 560 nm-central wavelength is 280.

The femtosecond pump and probe method can be used to study the all-optical tunability of the photonic crystal filter. Gong *et al.* also fabricated a high-quality photonic crystal filter in the near infrared range [110]. The central wavelength, the peak transmittance, and the line width of the optical channel were 791 nm, more than 92%, and 1.6 nm, respectively. The quality factor of the microcavity was 500, which indicates that the photonic crystal filter has an excellent wavelength resolution.

The position of the optical channel in the photonic bandgap continuously shifted in the long-wavelength direction with the increase of pump intensity. The reason lies in the intensity-dependent refractive index of polystyrene. Polystyrene has a positive nonlinear refractive index. According to the nonlinear Kerr effect, the refractive index of polystyrene increases with the enhancement of pump intensity, which leads to the increment of the effective refractive index of photonic crystal and the red shift of the central wavelength of the optical channel. The central wavelength of the optical channel shifted 3.1 nm under the excitation of 9.4 GW/cm^2 pump intensity. The response time of the photonic crystal filter can be extracted from the transmittance changes of a 791 nm probe light as functions of the time delay between pump and probe pulses. The position of the optical channel in the photonic bandgap transmittance changes occurred only when the pump light was switched on. The subsequent changes were negligible after the pump light disappeared. The half width of the signal envelope, ~120 fs, was in proximity to the pulse duration of the pump light. Moreover, the signal profile almost showed a symmetrical distribution around the zero time delay. This evidence indicates that the response time of the photonic crystal filter is faster than the experimental time resolution and the pulse-width limited nonlinear response of polystyrene is faster than 120 fs.

5.2.5.2 Multichannel organic photonic crystal filter

When several identical defects are introduced in a two-dimensional photonic crystal, a multichannel optical filter can be reached based on the coupling of the defect modes. Gong *et al.* fabricated a multichannel filter constructed by three identical line defects separated by two rows of air holes in the center of a two-dimensional square lattice polystyrene photonic crystal [111]. The radius of the air holes and the lattice constant were 90 nm and 220 nm, respectively. The width of each line defect was 310 nm, as shown in Fig. 5.13. The two-dimensional photonic crystal filter has three optical channels, whose central wavelengths are 495.5 nm for channel 1, 549 nm for channel 2, and 621.5 nm for channel 3, respectively. The transmittance of the three channels was higher

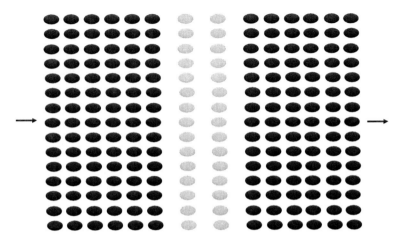

Figure 5.13 Schematic structure of multichannel organic photonic crystal filter.

than 85% and the transmittance contrast between the optical channel and the stop band was over 70%. These are the basics for optical filter. Moreover, the closer the optical channel is to the photonic bandgap center, the higher the quality factor is. The quality factor was estimated to be 190 for channel 2, 83 for channel 1, and 103 for channel 3. The quality factor of a defect mode can be estimated as $Q \propto \exp(L/\zeta)$, where L is the system length and ζ is the localization length [112]. The electric field distributions of the three optical channels are depicted in Fig. 5.14. It is very clear that when the position of the defect mode approaches the center of the photonic bandgap, the photon confinement effect becomes more and more violent. Kee's calculation indicated that the stronger the photon confinement is, the shorter the localization length is [113]. Compared with channel 1 and channel 3, the electromagnetic confinement of channel 2 is the strongest, which leads to the lowest value of localization length. As a result, the strong localization effect results in a high quality factor of channel 2. It was also observed that the stronger the confinement of the defect mode is, the narrower the line width of the optical channel is. Channel 2 had the smallest line width, 2 nm, in the three channels. The reason is that the spectral width of the optical channel shrinks with the increase of photon confinement, which leads to a narrow line width and superior wavelength resolution [114].

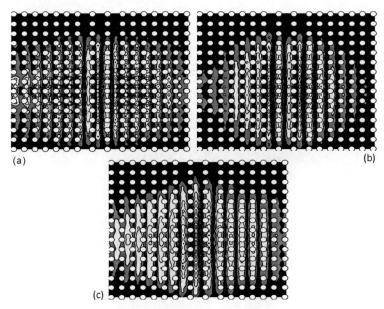

Figure 5.14   Electric field distribution of optical channels of three channel polystyrene photonic crystal filter. (a) For channel 1. (b) For channel 2. (c) For channel 3.

The all-optical tunability of the photonic crystal filter was studied by use of the picosecond pump and probe method. The red shift in the position of optical channels was continuous with the increase of pump intensity, due to the intensity-dependent refractive index of polystyrene. Accordingly, the optical channel in the photonic bandgap shifted in the same direction. Under an excitation of 15.9 GW/cm^2 pump intensity, channels 1 and 3 shifted about 4 nm and 6 nm, respectively, which is related to the different magnitudes of the shift of the long-wavelength edge and the short-wavelength edge of the photonic bandgap. According to the electromagnetic variation theorem, the long-wavelength band is more sensitive to the dielectric constant changes of high dielectric materials compared with the short-wavelength band. As a result, under the same pump intensity, the shift magnitude of the long-wavelength edge of the photonic bandgap would be larger than that of the short-wavelength edge. Accordingly, the shift magnitude of the defect modes near the long-wavelength edge would be larger than those at the short-wavelength edge. The response time of the photonic crystal filter can be extracted from the transmittance changes of a 549 nm probe light as a function of the time delay between pump and probe pulses. The measurements were performed as follows. First, the time delay was varied from –25 ps to 0. Then, a time delay of 0 was maintained. It was very clear that the transmittance changes occurred only when the pump light was switched on. The subsequent changes were negligible after the pump light disappeared. The transient process reflects the time response progress of the ultrafast transmittance changes of the probe light, from which the time response of the optical filter can be extracted. The falling time of the transient process is 10 ps, which approximates the duration of the pump pulse. It is well known that the nonlinear response of polystyrene is in subpico-second order [115]. This shows that the time response of the photonic crystal filter was limited by the experimental time resolution. Therefore, the response time of the two-dimensional photonic crystal filter was estimated to be less than 10 ps.

5.2.5.3 Narrow-band organic photonic crystal filter

Gong *et al.* fabricated a narrow-band optical filter based on the coupling of four identical organic photonic crystal waveguides [116]. The photonic crystal filter was constructed by four waveguides with a width of 310 nm in the center of regular square arrays of cylindrical air holes embedded in the background matrix of a 300 nm-thick polystyrene slab, as shown in Fig. 5.15. The radius of the air holes and the lattice constant were 90 nm and 220 nm, respectively. Owing to the limitations of the operation frequency range of the laser system, the transmittance spectrum of the photonic crystal filter could not be measured when the wavelength was lower than 430 nm. The central wavelength and the estimated bandwidth of the pass band of the photonic crystal filter were 496 nm and 112 nm, respectively. Owing to the coupling of four photonic crystal waveguides, a transmittance band could be formed in the photonic bandgap. This resulted in the formation of the wide pass band of the photonic crystal filter.

The pass band transmittance changed slightly, which indicated that the pass band possessed a flat top. The average transmittance of the pass band was more than 80% and the transmittance contrast between the pass band and the stop band was higher than 60%. The high transmittance and the steep roll-off of the pass band imply that the photonic crystal filter possesses excellent filtering properties.

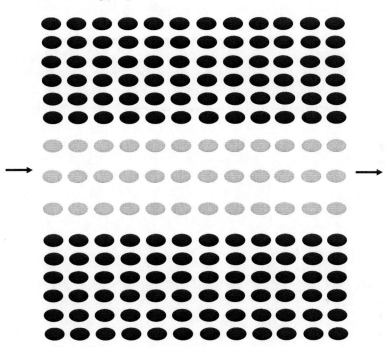

Figure 5.15   Schematic structure of narrow-band organic photonic crystal filter.

The picosecond pump and probe method was used to study the all-optical tunability of the photonic crystal filter. The shift magnitude of the pass band increases with the increment of the pump intensity. According to the nonlinear Kerr effect, the positive value of the third-order nonlinear susceptibility of polystyrene results in the increase of the effective refractive index of the photonic crystal under the excitation of the pump light, which makes the pass band of the photonic crystal filter shift in the long-wavelength direction. The maximal shift was 4.2 nm under

14.7 GW/cm^2 pump intensity, which was in agreement with the calculated results. The average transmittance and the bandwidth of the pass band changed slightly under the excitation of the pump light, which shows that the photonic crystal filter possesses excellent tunability. The time response of the photonic crystal filter can be extracted from the transmittance changes of the probe light as functions of the time delay between the pump and probe pulse. The wavelength of the probe light and the pump intensity were 551 nm and 14.7 GW/cm^2, respectively. It is very clear that the transmittance changed only when the pump and probe pulse overlapped with each other. The maximal transmittance was obtained for zero time delay, with two pulses overlapping completely in the temporal domain. Moreover, the half width of the signal envelope, ~10 ps, was in proximity to the pulse duration of the pump light. The signal profile showed an almost symmetrical distribution around the zero time delay. This evidence showed that the time response of the tunable photonic crystal filter was faster than the experimental time resolution. Therefore, the measured time response of the tunable photonic crystal filter, ~10 ps, is limited by the pulse duration of the pump light.

References

1. J. Zi, J. Wan and C. Zhang, "Large frequency range of negligible transmission in one-dimensional photonic quantum well structures," *Appl. Phys. Lett.* **73**, 2084–2086 (1998).

2. J. N. Munday and W. M. Robertson, "Slow electromagnetic pulse propagation through a narrow transmission band in a coaxial photonic crystal," *Appl. Phys. Lett.* **83**, 1053–1055 (2003).

3. J. C. Chen, H. A. Haus, S. H. Fan, P. R. Villeneuve, and J. D. Joannopoulos, "Optical filters from photonic bandgap air bridges," *J. Lightwave Technol.* **14**, 2575–2580 (1996).

4. M. Li, K. Mori, M. Ishizuka, X. B. Liu, Y. Sugimoto, N. Ikeda, and K. Asakawa, "Photonic bandpass filter for 1550 nm fabricated by femtosecond direct laser ablation," *Appl. Phys. Lett.* **83**, 216–218 (2003).

5. E. Miyai and S. Noda, "Structural dependence of coupling between a two-dimensional photonic crystal and a wire waveguide," *J. Opt. Soc. Am. B* **21**, 67–72 (2004).

6. M. Bayindir and E. Ozbay, "Dropping of electromagnetic waves through localized modes in three-dimensional photonic bandgap structures," *Appl. Phys. Lett.* **81**, 4514–4516 (2002).

7. B. K. Min, J. E. Kim, and H. Y. Park, "High-efficiency surface-emitting channel drop filters in two-dimensional photonic crystal slabs," *Appl. Phys. Lett.* **86**, 011106 (2005).

8. T. Asano, B. S. Song, Y. Tanaka, and S. Noda, "Investigation of a channel-add/drop-filtering device using acceptor-type point defects in a two-dimensional photonic-crystal slab," *Appl. Phys. Lett.* **83**, 407–409 (2003).

9. A. Chutinan, M. Mochizuki, M. Imada, and S. Noda, "Surface-emitting channel drop filters using single defects in two-dimensional photonic crystal slabs," *Appl. Phys. Lett.* **79**, 2690–2692 (1994).

10. Y. Akahane, M. Mochizuki, T. Asano, Y. Tanaka, and S. Noda, "Design of a channel drop filter by using a donor-type cavity with high-quality factor in a two-dimensional photonic crystal slab," *Appl. Phys. Lett.* **82**, 1341–1343 (2003).

11. Y. Akahane, T. Asano, B. S. Song, and S. Noda, "Investigation of high-Q channel drop filters using donor-type defects in two-dimensional photonic crystal slabs," *Appl. Phys. Lett.* **83**, 1512–1514 (2003).

12. K. H. Hwang and G. H. Song, "Design of a high-Q channel add-drop multiplexer based on the two-dimensional photonic-crystal membrane structure," *Opt. Express* **13**, 1948–1957 (2005).

13. H. Y. Lee, H. Makino, and T. Yao, "Si-based omnidirectional reflector and transmission filter optimized at a wavelength of 1.55 μm," *Appl. Phys. Lett.* **81**, 4502–4504 (2002).

14. H. T. Jiang, H. Chen, H. Q. Li, Y. W. Zhang, and S. Y. Zhu, "Compact high-Q filters based on one-dimensional photonic crystals containing single-negative materials," *J. Appl. Phys.* **98**, 013101 (2005).

15. H. Takano, Y. Akahane, T. Asano, and S. Noda, "In-plane-type channel drop filter in a two-dimensional photonic crystal slab," *Appl. Phys. Lett.* **84**, 2226–2228 (2004).

16. M. Qiu, and B. Jaskorzynska, "Design of a channel drop filter in a two-dimensional triangular photonic crystal," *Appl. Phys. Lett.* **83**, 1074–1076 (2003).

17. S. H. Fan, P. R. Villeneuve, J. D. Joannopoulos, M. J. Khan, C. Manolatou, and H. A. Haus, "Theoretical analysis of channel drop tunneling processes," *Phys. Rev. B* **59**, 15882–15892 (1999).

18. S. H. Fan, P. R. Villeneuve, J. D. Joannopoulos, and H. A. Haus, "Channel drop filters in photonic crystals," *Opt. Express* **3**, 4–11 (1998).

19. M. Qiu, M. Mulot, M. Swillo, S. Anand, B. Jaskorzynska, A. Karlsson, M. Kamp, and A. Forchel, "Photonic crystal optical filter based on contra-directional waveguide coupling," *Appl. Phys. Lett.* **83**, 5121–5123 (2003).

20. S. Olivier, H. Benisty, C. Weisbuch, C. Smith, T. Krauss, and R. Houdre, "Coupled-mode theory and propagation losses in photonic crystal waveguide," *Opt. Express* **11**, 1490–1496 (2003).

21. S. Olivier, M. Rattier, H. Benisty, C. Weisbuch, C. J. M. Smith, R. M. D. L. Rue, T. F. Krauss, U. Oesterle, and R. Houdre, "Mini-stopbands of a one-dimensional system: the channel waveguide in a two-dimensional photonic crystal," *Phys. Rev. B* **63**, 113311 (2001).

22. S. Olivier, C. Weisbuch, and H. Benisty, "Compact and fault-tolerant photonic crystal add-drop filter," *Opt. Lett.* **28**, 2246–2248 (2003).

23. S. Fan, P. R. Villeneuve, and J. D. Joannopoulos, "Channel drop tunneling through localized states," *Phys. Rev. Lett.* **80**, 960 (1998).

24. H. Takano, B. S. Song, T. Asano, and S. Noda, "Highly efficient in-plane channel drop filter in a two-dimensional heterophotonic crystal," *Appl. Phys. Lett.* **86**, 241101 (2005).

25. H. L. Ren, C. Jiang, W. S. Hu, M. Y. Gao, and J. Y. Wang, "Photonic crystal channel drop filter with a wavelength-selective reflection microcavity," *Opt. Express* **14**, 2446–2458 (2006).

26. J. K. Yang, S. H. Kim, G. H. Kim, H. G. Park, Y. H. Lee, and S. B. Kim, "Slab-edge modes in two-dimensional photonic crystals," *Appl. Phys. Lett.* **84**, 3016–3018 (2004).

27. S. Xaio and M. Qiu, "Surface-mode microcavity," *Appl. Phys. Lett.* **87**, 111102 (2005).

28. Z. Y. Zhang, M. Dainese, L. Wosinski, S. Xiao, M. Qiu, M. Swillo, and U. Andersson, "Optical filter based on two-dimensional photonic crystal surface-mode cavity in amorphous silicon-on-silica structure," *Appl. Phys. Lett.* **90**, 041108 (2007).

29. S. Noda, A. Chutinan, and M. Imada, "Trapping and emission of photons by a single defect in a photonic bandgap structure," *Nature* **407**, 608–610 (2000).

30. M. Imada, S. Noda, A. Chutinan, M. Mochizuki, and T. Tanaka, "Channel drop filter using a single defect in a 2-D photonic crystal slab waveguide," *J. Lightwave Technol.* **20**, 873–878 (2002).

31. M. Agio, E. Lidorikis, and C. M. Soukoulis, "Impurity modes in a two-dimensional photonic crystal: coupling efficiency and Q factor," *J. Opt. Soc. Am. B* **17**, 2037–2042 (2000).

32. C. J. Jin, S. Z. Han, X. D. Meng, B. Y. Cheng, and D. Z. Zhang, "Demultiplexer using directly resonant tunneling between point defects and waveguides in a photonic crystal," *J. Appl. Phys.* **91**, 4771–4773 (2002).

33. S. Kim, I. Park, H. Lim, and C. S. Kee, "Highly efficient photonic crystal-based multichannel drop filters of three-port system with reflection feedback," *Opt. Express* **12**, 5518–5525 (2004).

34. W. Jiang and R. T. Chen, "Multichannel optical add-drop processes in symmetrical waveguide-resonator system," *Phys. Rev. Lett.* **91**, 213901 (2003).

35. Z. Y. Zhang and M. Qiu, "Coupled-mode analysis of a resonant channel drop filter using waveguides with mirror boundaries," *J. Opt. Soc. Am. B* **23**, 104–113 (2006).

36. C. Ren, J. Tian, S. Feng, H. H. Tao, Y. H. Liu, K. Ren, Z. Y. Li, B. Y. Cheng, and D. Z. Zhang, "High resolution three-port filter in two dimensional photonic crystal slabs," *Opt. Express* **14**, 10014–10020 (2006).

37. Z. Qiang, W. D. Zhang, and R. A. Soref, "Optical add-drop filters based on photonic crystal ring resonators," *Opt. Express* **15**, 1823–1831 (2007).

38. X. Y. Lei, H. Li, F. Ding, W. Y. Zhang, and N. B. Ming, "Novel application of a perturbed photonic crystal: high-quality filter," *Appl. Phys. Lett.* **71**, 2889–2891 (1991).

39. Z. Wang, R. W. Peng, F. Qiu, X. Q. Huang, M. Wang, A. Hu, S. S. Jiang, and D. Feng, "Selectable-frequency and tunable-Q perfect transmissions of electromagnetic waves in dielectric heterostructures," *Appl. Phys. Lett.* **84**, 3969–3971 (2000).

40. Z. S. Wang, L. Wang, Y. G. Wu, L. Y. Chen, X. S. Chen, and W. Lu, "Multiple channeled phenomena in heterostructures with defects mode," *Appl. Phys. Lett.* **84**, 1629–1631 (2004).

41. F. Qiao, C. Zhang, J. Wan, and J. Zi, "Photonic quantum-well structures: multiple channeled filtering phenomena," *Appl. Phys. Lett.* **77**, 3698–3700 (2000).

42. H. Y. Lee, S. J. Cho, G. Y. Nam, W. H. Lee, T. Baba, H. Makino, M. W. Cho, and T. Yao, "Multiple-wavelength-transmission filters based on Si-SiO$_2$ one-dimensional photonic crystals," *J. Appl. Phys.* **97**, 103111 (2005).

43. B. S. Song, S. Noda, and T. Asano, "Photonic devices based on in-plane hetero photonic crystals," *Science* **300**, 1537 (2003).

44. A. Shinya, S. Mitsugi, E. Kuramochi, and M. Notomi, "Ultrasmall multi-port channel drop filter in two-dimensional photonic crystal on silicon-on-insulator substrate," *Opt. Express* **14**, 12394–12400 (2006).

45. B. S. Song, T. Asano, and S. Noda, "Role of interfaces in heterophotonic crystals for manipulation of photons," *Phys. Rev. B* **71**, 195101 (2005).

46. H. Takano, B. S. Song, T. Asano, and S. Noda, "Highly efficient multi-channel drop filter in a two-dimensional hetero photonic crystal," *Opt. Express* **14**, 3491–3496 (2006).

47. B. S. Song, T. Asano, Y. Akahane, Y. Tanaka, and S. Noda, "Transmission and reflection characteristics of in-plane hetero-photonic crystals," *Appl. Phys. Lett.* **85**, 4591–4593 (2004).

48. N. L. Florous, K. Saitoh, M. Koshiba, and M. Skorobogatiy, "Low-temperature-sensitivity heterostructure photonic-crystal wavelength-selective filter based on ultralow-refractive-index metamaterials," *Appl. Phys. Lett.* **88**, 121107 (2006).

49. B. T. Schwartz and R. Piestun, "Total external reflection from metamaterials with ultralow refractive index," *J. Opt. Soc. Am. B* **20**, 2448–2453 (2003).

50. C. S. Kee, I. Park, and H. Lim, "Photonic crystal multi-channel drop filters based on microstrip lines," *J. Phys. D: Appl. Phys.* **39**, 2932–2934 (2006).

51. Y. H. Li, H. T. Jiang, L. He, H. Q. Li, Y. W. Zhang, and H. Chen, "Multichanneled filter based on a branchy defect in microstrip photonic crystal," *Appl. Phys. Lett.* **88**, 081106 (2006).

52. C. S. Kee, J. M. Lee, and Y. L. Lee, "Multiwavelength Sòlc filters based on $\chi^{(2)}$ nonlinear quasiperiodic photonic crystals with Fibonacci sequences," *Appl. Phys. Lett.* **91**, 251110 (2007).

53. Z. M. Jiang, B. Shi, D. T. Zhao, J. Liu, and X. Wang, "Silicon-based photonic crystal heterostructure," *Appl. Phys. Lett.* **79**, 3395–3397 (2001).

54. B. Z. Steinberg, A. Boag, and R. Lisitsin, "Sensitivity analysis of narrowband photonic crystal filters and waveguides to structure variations and inaccuracy," *J. Opt. Soc. Am. A* **20**, 138–146 (2003).

55. F. Villa, and J. A. G. Armenta, "Photonic crystal to photonic crystal surface modes: narrow-bandpass filters," *Opt. Express* **12**, 2338–2355 (2004).

56. D. K. Jacob, S. C. Dunn, and M. G. Moharam, "Flat-top narrow-band spectral response obtained from cascaded resonant grating reflection filters," *Appl. Opt.* **41**, 1241–1245 (2002).

57. D. Park, S. Kim, I. Park, and H. Lim, "Higher order optical resonant filters based on coupled defect resonators in photonic crystals," *J. Lightwave Technol.* **23**, 1923–1928 (2005).

58. W. Nakagawa, P. C. Sun, C. H. Chen, and Y. Fainman, "Wide-field-of-view narrow-band spectral filters based on photonic crystal nanocavities," *Opt. Lett.* **27**, 191–193 (2002).

59. R. Costa, A. Melloni, and M. Martinelli, "Bandpass resonant filters in photonic-crystal waveguides," *IEEE Photon. Technol. Lett.* **15**, 401–403 (2003).

60. Y. V. Troitski, "Dispersion-free, multiple-beam interferometer," *Appl. Opt.* **34**, 4717–4722 (1995).

61. C. H. Chen, K. Tetz, W. Nakagawa, and Y. Fainman, "Wide-field-of-view GaAs/AlO one-dimensional photonic crystal filter," *Appl. Opt.* **44**, 1503–1511 (1995).

62. J. A. Oswald, B. I. Wu, K. A. McIntosh, L. J. Mahoney, and S. Verghese, "Dual-band infrared metallodielectric photonic crystal filters," *Appl. Phys. Lett.* **77**, 2098–2100 (2000).

63. A. Kao, K. A. McIntosh, O. B. Mcmahon, R. Atkins, and S. Verghese, "Calculated and measured transmittance of metallodielectric photonic crystals incorporating flat metal elements," *Appl. Phys. Lett.* **73**, 145–147 (1998).

64. Z. Jaksic, M. Maksimovic, and M. Sarajlic, Silver-silica transparent metal structures as bandpass filters for the ultraviolet range. *J. Opt. A: Pure Appl. Opt.* **7**, 51–55 (2005).

65. Y. H. Ye, D. Y. Jeong, T. S. Mayer, and Q. M. Zhang, "Finite-size effect on highly dispersive photonic crystal optical components," *Appl. Phys. Lett.* **82**, 2380–2382 (2003).

66. T. Baba and T. Matsumoto, "Resolution of photonic crystal superprism," *Appl. Phys. Lett.* **81**, 2325–2327 (2002).

67. Y. Sugimoto, Y. Tanaka, N. Ikeda, T. Yang, H. Nakamura, K. Asakawa, K. Inoue, T. Maruyama, K. Miyashita, K. Ishida, and Y. Watanabe, "Design, fabrication, and characterization of coupling-strength-controlled directional coupler based on two-dimensional photonic crystal slab waveguides," *Appl. Phys. Lett.* **83**, 3236–3238 (2003).

68. M. Bayindir and E. Ozbay, "Band-dropping via coupled photonic crystal waveguides," *Opt. Express* **10**, 1279–1284 (2009).

69. M. Notomi, A. Shinya, S. Mitsugi, E. Kuramochi, and H. Y. Ryu, "Waveguides, resonators and their coupled elements in photonic crystal slabs," *Opt. Express* **12**, 1551–1561 (2004).

70. T. Niemi, L. H. Frandsen, K. K. Hede, A. Harpoth, P. I. Borel, and M. Kristensen, "Wavelength-division demultiplexing using photonic crystal waveguides," *IEEE Photon. Technol. Lett.* **18**, 226–228 (2006).

71. S. H. Fan and J. D. Joannopoulos, "Analysis of guided resonances in photonic crystal slabs," *Phys. Rev. B* **65**, 235112 (2002).

72. F. Yang, G. Yen, and B. T. Cunningham, "Integrated 2D photonic crystal stack filter fabricated using nanoreplica molding," *Opt. Express* **18**, 11846–11858 (2010).

73. W. Suh and S. H. Fan, "All-pass transmission or flattop reflection filters using a single photonic crystal slab," *Appl. Phys. Lett.* **84**, 4905–4907 (2004).

74. C. C. Lin, Z. L. Lu, S. Y. Shi, G. Jiin, and D. W. Prather, "Experimentally demonstrated filters based on guided resonance of photonic-crystal films," *Appl. Phys. Lett.* **87**, 091102 (2005).

75. E. H. Cho, H. S. Kim, B. H. Cheong, P. Oleg, W. Xianyua, J. S. Sohn, D. J. Ma, H. Y. Choi, N. C. Park, and Y. P. Park, "Two-dimensional photonic crystal color filter development," *Opt. Express* **17**, 8621–8629 (2009).

76. Y. Ding and R. Magnusson, "Resonant leaky-mode spectral-band engineering and device applications," *Opt. Express* **12**, 5661–5674 (2004).

77. N. Ganesh and B. T. Cunningham, "Photonic-crystal near-ultraviolet reflectance filters fabricated by nanoreplica molding," *Appl. Phys. Lett.* **88**, 071110 (2006).

78. B. H. Cheong, O. N. Prudnikov, E. Cho, H. S. Kim, J. Yu, Y. S. Cho, H. Y. Choi, and S. T. Shin, "High angular tolerant color filter using subwavelength grating," *Appl. Phys. Lett.* **94**, 213104 (2009).

79. G. Subramania, S. Y. Lin, J. R. Wendt, and J. M. Rivera, "Tuning the microcavity resonant wavelength in a two-dimensional photonic crystal by modifying the cavity geometry," *Appl. Phys. Lett.* **83**, 4491 (2003).

80. B. Kuhlow, G. Przyrembel, S. Schluter, W. Furst, R. Steingruber, and C. Weimann, "Photonic crystal microcavities in SOI photonic wires for WDM filter applications," *J. Lightwave Technol.* **25**, 421–431 (2007).

81. Z. Y. Li, H. Y. Sang, L. L. Lin, and K. M. Ho, "Evanescent-wave-assisted wideband continuous tunability in photonic crystal channel-drop filter," *Phys. Rev. B* **72**, 035103 (2005).

82. L. L. Lin, Z. Y. Li, and B. Lin, "Engineering waveguide-cavity resonant side coupling in a dynamically tunable ultracompact photonic crystal filter," *Phys. Rev. B* **72**, 165330 (2005).

83. D. Stieler, A. Barsic, G. Tuttle, M. Li, and K. M. Ho, "Effects of defect permittivity on resonant frequency and mode shape in the three-dimensional woodpile photonic crystal," *J. Appl. Phys.* **105**, 103109 (2009).

84. K. Hennessy, A. Badolato, A. Tamboli, P. M. Petroff, E. Hu, M. Atature, J. Dreiser, and A. Imamoglu, "Tuning photonic crystal nanocavity modes by wet chemical digital etching," *Appl. Phys. Lett.* **87**, 021108 (2005).

85. S. Vignolini, F. Riboli, D. S. Wiersma, L. Balet, L. H. Li, M. Francardi, A. Gerardino, A. Fiore, M. Gurioli, and F. Intonti, "Nanofluidic control of coupled photonic crystal resonators," *Appl. Phys. Lett.* **96**, 141114 (2010).

86. D. M. Pustal, A. Sharkawy, S. Shi, and D. W. Prather, "Tunable photonic crystal microcavities," *Appl. Opt.* **41**, 5574–5579 (2002).

87. X. F. Chen, J. H. Shi, Y. P. Chen, Y. M. Zhu, Y. X. Xia, and Y. L. Chen, "Electro-optic Sölc-type wavelength filter in periodically poled lithium niobate," *Opt. Lett.* **28**, 2115–2117 (2003).

88. Y. M. Zhu, X. F. Chen, J. H. Shi, Y. P. Chen, Y. X. Xia, and Y. L. Chen, "Wide-range tunable wavelength filter in periodically poled lithium niobate," *Opt. Commun.* **228**, 139–143 (2003).

89. J. H. Shi, J. H. Wang, L. J. Chen, X. F. Chen, and Y. X. Xia, "Tunable Sölc-type filter in periodically poled $LiNbO_3$ by UV-light illumination," *Opt. Express* **14**, 6279–6284 (2006).

90. Y. K. Ha, Y. C. Yang, J. E. Kim, H. Y. Park, C. S. Kee, H. Lim, and J. C. Lee, "Tunable omnidirectional reflection bands and defect modes of a one-dimensional photonic bandgap structure with liquid crystals," *Appl. Phys. Lett.* **79**, 15–17 (2001).

91. I. D. Villar, I. R. Matias, F. J. Arregui, and R. O. Claus, "Analysis of one-dimensional photonic bandgap structures with a liquid crystal defect towards development of fiber-optic tunable wavelength filters," *Opt. Express* **11**, 430–436 (2003).

92. T. D. Drysdale, I. S. Gregory, C. Baker, E. H. Linfield, W. R. Tribe, and D. R. S. Cumming, "Transmittance of a tunable filter at terahertz frequencies," *Appl. Phys. Lett.* **85**, 5173–5175 (2004).

93. T. D. Drysdale, R. J. Blaikie, and D. R. S. Cumming, "Calculated and measured transmittance of a tunable metallic photonic crystal filter for terahertz frequencies," *Appl. Phys. Lett.* **83**, 5362–5364 (2003).

94. F. Gadot, A. D. Lustrac, J. M. Lourtioz, T. Brillat, A. Ammouche, and E. Akmansoy, "High-transmission defect modes in two-dimensional metallic photonic crystals," *J. Appl. Phys.* **85**, 8499–8501 (1999).

95. D. F. Sievenpiper, E. Yablonovitch, J. N. Winn, S. Fan, P. R. Villeneuve, and J. D. Joannopoulos, "3D metallo-dielectric photonic crystal with strong capacitive coupling between metallic islands," *Phys. Rev. Lett.* **80**, 2829–2832 (1998).

96. W. Suh, M. F. Yanik, O. Solgaard, and S. H. Fan, "Displacement-sensitive photonic crystal structures based on guided resonance in photonic crystal slabs," *Appl. Phys. Lett.* **82**, 1999–2001 (2003).

97. W. Suh and S. H. Fan, "Mechanically switchable photonic crystal filter with either all-pass transmission or flat-top reflection characteristics," *Opt. Lett.* **28**, 1763–1765 (2003).

98. C. Schuller, J. P. Reithmaier, J. Zimmermann, M. Kamp, A. Forchel, and S. Anand, "Polarization-dependent optical properties of planar photonic crystals infiltrated with liquid crystals," *Appl. Phys. Lett.* **87**, 121105 (2005).

99. K. S. Hong, H. C. Park, B. Y. Kim, I. K. Hwang, W. Jin, J. Ju, and D. I. Yeom, "100 nm tunable acousto-optic filter based on photonic crystal fiber," *Appl. Phys. Lett.* **92**, 031110 (2008).

100. C. Ang, A. S. Bhalla, and L. E. Cross, "Dielectric behaviour of paraelectric $KTaO_3$, $CaTiO_3$, and $(Ln_{1/2}Na_{1/2})TiO_3$ under a dc electric field," *Phys. Rev. B* **64**, 184104 (2001).

101. H. Nemec, L. Duvillaret, F. Garet, P. Kuzel, P. Xavier, J. Richard, and D. Rauly, "Thermally tunable filter for terahertz range based on a one-dimensional photonic crystal with a defect," *J. Appl. Phys.* **96**, 4072–4075 (2004).

102. F. Intonti, S. Vignolini, F. Riboli, M. Zani, D. S. Wiersma, L. Balet, L. H. Li, M. Francardi, A. Gerardino, A. Fiore, and M. Gurioli, "Tuning of photonic crystal cavities by controlled removal of locally infiltrated water," *Appl. Phys. Lett.* **95**, 173112 (2009).

103. N. C. Panoiu, M. Bahl, and R. M. Osgoodjr, "All-optical tunability of a nonlinear photonic crystal channel drop filter," *Opt. Express* **12**, 1605–1610 (2004).

104. D. Sridharan, E. Waks, G. Solomon, and J. T. Fourkas, "Reversible tuning of photonic crystal cavities using photochromic thin films," *Appl. Phys. Lett.* **96**, 153303 (2010).

105. L. Yang, R. Dorsinville, Q. Z. Wang, P. X. Ye, R. R. Alfano, R. Zamboni, and C. Taliani, "Excited-state nonlinearity in polythiophene thin films investigated by the Z-scan technique," *Opt. Lett.* **17**, 323–325 (1992).

106. X. Y. Hu, Q. H. Gong, Y. H. Liu, B. Y. Cheng, and D. Z. Zhang, "Fabrication of two-dimensional organic photonic crystal filter," *Appl. Phys. B* **81**, 779–781 (2005).

107. Z. L. Wang, Q. Wang, H. J. Li, J. J. Li, P. Xu, Q. Luo, A. Z. Jin, H. F. Yang, and C. Z. Gu, "The field emission properties of high aspect ratio diamond nanocone arrays fabricated by focused ion beam milling," *Sci. Technol. Adv. Mater.* **6**, 799–803 (2005).

108. J. D. Joannopoulous, P. R. Villeneuve, and S. Fan, "Photonic crystals: putting a new twist on light," *Nature* **386**, 143–149 (1997).

109. M. Bayindir, B. Temelkuran, and E. Ozbay, "Tight-binding description of the coupled defect modes in three-dimensional photonic crystals," *Phys. Rev. Lett.* **84**, 2140–2143 (2000).

110. X. Y. Hu, P. Jiang, H. Yang, and Q. H. Gong, "All-optical tunable photonic bandgap microcavities with a femtosecond time response," *Opt. Lett.* **31**, 2777–2779 (2006).

111. X. Y. Hu, Q. H. Gong, Y. H. Liu, B. Y. Cheng, and D. Z. Zhang, "Ultrafast tunable filter in two-dimensional organic photonic crystal," *Opt. Lett.* **31**, 371–373 (2006).

112. B. Liu, A. Yamilov, and H. Cao, "Effect of Kerr nonlinearity on defect lasing modes in weakly disordered photonic crystals," *Appl. Phys. Lett.* **83(6)**, 1092–1094 (2003).

113. C. S. Kee, H. Lim, and J. Lee, "Coupling characteristics of localized photons in two-dimensional photonic crystals," *Phys. Rev. B* **67**, 073103 (2003).

114. K. Guven and E. Ozbay, "Coupling and phase analysis of cavity structures in two-dimensional photonic crystals," *Phys. Rev. B* **71(8)**, 085108 (2005).

115. M. A. Bader, H. M. Keller, and G. Marowsky, "Polymer-based waveguides and optical switching," *Opt. Mater.* **9**, 334–341 (1998).

116. X. Y. Hu, P. Jiang, C. Ding, and Q. H. Gong, "All-optical tunable narrow-band organic photonic crystal filters," *Appl. Phys. B* **87**, 255–258 (2007).

Questions

5.1 Please discuss the difference, including the advantages and the disadvantages, between the different methods to realize the multiple-channel photonic crystal filter.

5.2 Please discuss the physical mechanism of realizing all-optical tunable photonic crystal filter.

Chapter 6

Photonic Crystal Laser

As an essential integrated photonic device, photonic crystal laser plays a very important role in the fields of all-optical computing, optical communication, and ultrahigh speed information processing. Photonic crystal possesses unique photonic bandgap properties, which can effectively prohibit the spontaneous radiation, which makes it possible to achieve a thresholdless laser [1]. The dream of thresholdless laser has inspired people to make great efforts in the fields of low-power photonic crystal laser. The simplest photonic crystal laser uses two one-dimensional photonic crystals as high reflection mirrors to construct a microcavity. When the active material was filled in the microcavity structure, a photonic crystal laser can be reached [2]. This approach often leads to a large size of the photonic crystal laser. The micro/nano laser could be realized in two-dimensional photonic crystals based on the photonic bandedge effect or defect states [3]. In the photonic bandgap, the photon state density is zero. The value of photon state density rapidly increased from zero to a large magnitude in the photonic bandedges and the center of defect modes. Moreover, the group velocity v_g of electromagnetic waves almost approaches zero in the photonic bandedges and the center of the defect modes, which enhances the lifetime of photons and interactions of light and matter greatly. These two factors indicate

Photonic Crystals: Principles and Applications
Qihuang Gong and Xiaoyong Hu
Copyright © 2014 Pan Stanford Publishing Pte. Ltd.
ISBN 978-981-4267-30-4 (Hardcover), 978-981-4364-83-6 (eBook)
www.panstanford.com

that the Bloch waves in the photonic crystal can provide a strong and intrinsic feedback mechanism in the photonic bandedge and defect states for the realization of laser oscillation [4].

When a gain medium is infiltrated in the photonic crystal, the spontaneous radiation rate and the spectral shape of the spontaneous radiation of the gain medium will be modified by a factor proportional to the local photonic density of modes [5]. On one hand, a sharp peak in the photonic density of modes will enhance the spontaneous radiation rate of gain medium. On the other hand, the gain is proportional to $1/v_g$ for the electromagnetic wave modes with a peak photonic density of modes, where v_g is the group velocity [5]. The electromagnetic modes in the photonic bandedge and the defect states possess very small group velocity and very large local photonic density of states. Strong gain can be obtained for these electromagnetic modes in the photonic bandedge and the defect states. When the gain exceeds the total losses, photonic crystal laser can be realized [6]. When a photonic crystal waveguide is placed close to the photonic crystal microcavity laser, the laser emission can be guided to propagate in the photonic crystal waveguide if the optimum coupling condition is satisfied [7]. This makes it possible to integrate the photonic crystal microcavity laser into the integrated photonic circuits to be used as a signal resource.

6.1 Photonic Crystal Laser

Two essential factors to realize laser emission are the active medium to provide strong optical gain, and the resonator to provide strong optical feedback. Organic laser dye, highly π-conjugated organic polymer, semiconductor quantum well material, and semiconductor quantum dot can all be used as the gain medium [8–11]. Accordingly, the wavelength of the laser oscillation ranging from visible to terahertz range can be realized [12]. The violent changes of photon density of states and the zero gap velocity of electromagnetic modes in the photonic bandedges and the defect modes can offer intense optical feedback. This makes it possible to achieve the laser oscillation base on the photonic bandedge or defect modes in the photonic bandgap [13]. Photonic crystal microcavities with a high quality factor can enhance the interactions of light and the gain medium, which can remarkably reduce the threshold pump power for the laser oscillation. In a defect-free photonic crystal with larger refractive index contrast, the photonic bandedge near the Brillouin zone edge becomes flat, which makes the group velocity of the electromagnetic modes very small. This also reduces the threshold pump power [14]. Photonic crystals can be fabricated by using various materials, such as dielectric materials, ferroelectric materials, metallic materials, and liquid crystals. Accordingly, the laser emission can be achieved in one-, two-, or three-dimensional photonic crystals made of these materials [15,16]. Quasiperiodic photonic crystals have long-range rotational symmetry and quasiperiodic dielectric distribution, which leads to the formation of a photonic bandgap. Therefore, laser emission could also be realized not only in periodic structure photonic crystals but also in the quasiperiodic photonic crystals [17].

6.1.1 Dielectric Photonic Crystal Laser

6.1.1.1 Based on photonic bandgap effect

By use of the extremely small group velocity and long photonic lifetime of the electromagnetic modes in the photonic bandedge and the defect states, laser oscillation can be achieved when the gain exceeds the total losses. Various photonic bandedge modes and different types of photonic crystal microcavities could be adopted to provide laser oscillation [18–20].

6.1.1.1.1 *Photonic bandedge laser*

For the photonic bandedge laser in a one-, two-, or three-dimensional photonic crystal, the active material is often infiltrated in the photonic crystals. The extremely small group velocity and the shape variation of the photonic density of state provide a strong feedback for the laser oscillation [21]. A photonic crystal bandedge laser array can be realized by the laser holography method or the traditional microfabrication method, which can output several laser emissions with different wavelengths [22].

Every photonic bandgap has a dielectric bandedge and an air bandedge. Superficially, it is possible to achieve a laser oscillation mode at the dielectric bandedge and another laser oscillation mode with a different wavelength at the air bandedge simultaneously under excitation of a pump light. But in practice, the laser oscillation mode with the wavelength at the dielectric bandedge is easier to be obtained than that at the air bandedge. The reason lies in that the flatness of the dielectric bandedges is different from that of the air bandedge in the photonic band structure of a photonic crystal. The slope of the air bandedge is larger than that of the dielectric bandedge, because of which the dielectric bandedge is flatter than the air bandedge. Large flatness of the dielectric bandedge results in a much lower group velocity of the electromagnetic modes and stronger feedback [23]. So, the laser oscillation is much easier to be established at the dielectric bandedge than at the air bandedge. Not only is it the fundamental photonic bandgap, but also the high-order photonic bandgap could be adopted to realize photonic bandedge laser [24–26].

Using one-dimensional photonic crystal

The traditional configuration of the vertical cavity surface emitting laser (VCSEL) consists of an active material layer sandwiched between two distributed Bragg reflection (DBR) layers. When the top DBR layer was etched with periodic arrays of air holes, a photonic crystal VCSEL could be achieved. Song *et al.* fabricated a photonic crystal VCSEL using semiconductor materials [27]. The DBR layers were made of 24 $Al_{0.15}Ga_{0.85}As/Al_{0.95}Ga_{0.05}As$ pairs. The top 17 pairs were etched with a single-defect two-dimensional photonic crystal with a lattice constant and diameter of air holes of 5 and 3 μm, respectively. When the pump current was larger than the threshold current of 250 mA, laser emission with a wavelength of 848 nm could be achieved. The function of the defect was to guide laser mode.

Using two-dimensional photonic crystal

Anodic porous alumina is composed of an array of uniform air holes embedded in the alumina matrix, which can be fabricated by the anodization of aluminum matrix. The anodic porous alumina constructs a perfect two-dimensional photonic crystal. Masuda *et al.* fabricated a triangular lattice anodic porous alumina with a lattice constant of 200 nm and a filling factor of 0.3 for alumina, respectively [28]. A layer of dendrimer-encapsulated fluorescence dye pyrromethene 597 with a thickness of 1 μm was covered on the surface of the two-dimensional photonic crystal. The photonic bandgap covered the wavelength range from 450 to 560 nm. The fluorescence band of the pyrromethene 597 overlapped with the photonic bandgap, which ensured that effective feedback could be achieved in the photonic bandedge. A 532 nm laser beam from a Nd:YAG laser system (with a pulse duration of 8 ns and a repetition rate of 10 Hz) was used as the pump light. The pump light was incident in the direction parallel to the air holes. When the energy of the pump light was larger than the threshold energy of 0.12 μJ, remarkable laser emission with a wavelength of 547 nm could be obtained in the Γ–X direction of the photonic crystal, as shown in Fig. 6.1. The 547 nm laser mode was situated at the long wavelength edge of the photonic bandgap. The line width of the laser mode was only 0.08 nm.

At low temperature, the surface recombination rate of free carriers in a semiconductor material can be decreased greatly, which will effectively reduce the surface recombination losses for the laser oscillation. Ryu *et al.* reported a low-threshold photonic bandedge laser in free-standing semiconductor photonic crystal slab at a temperature of 80 K [29]. The triangular lattice of air holes was patterned in a free-standing slab made of InGaAsP quantum well material with a thickness of 200 nm. The lattice constant and the diameter of air holes were 1000 and 700 nm, respectively. The InGaAsP quantum well material was used as the gain medium. A 980 nm laser with a pulse duration of 10 ns was used as the pump light. They found that when the lattice constant was less than 600 nm, an in-plane laser emission could be obtained, which originated from the photonic bandedge below the light line, while a vertical emitting laser could be achieved when the lattice constant was larger than 600 nm, which originated from the high symmetric Γ point above the light line. A vertical emitting laser with a wavelength of 1466 nm could be reached when the power of the pump light was larger than the threshold power of 0.035 mW.

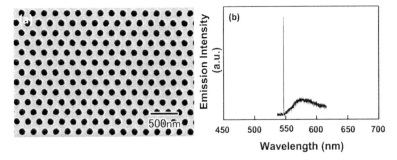

Figure 6.1   Anodic porous alumina photonic crystal laser. (a) SEM image of the photonic crystal. (b) Spectrum of laser emission mode under excitation of different pump light exceeding the threshold energy. Reprinted with permission from *Adv. Mater.*, **18**, H. Masuda, *et al.*, Lasing from two-dimensional photonic crystals using anodic porous alumina, 213–216. Copyright © 2006 WILEY-VCH Verlag GmbH & Co. KGaA, Weinheim.

Notomi *et al.* reported the photonic bandedge laser in a two-dimensional photonic quasicrystal with a Penrose lattice [30]. The photonic quasicrystal consisted of 150 nm deep air holes with a quasiperiodic lattice of 10-fold symmetry in a SiO_2 substrate, fabricated by use of the electro-beam lithography and reaction ion beam etching technology. The photonic quasicrystal was covered by a 300 nm-thick organic layer made of laser dye DCM doped with Alq_3. The lattice constant was 560 nm. A 337 nm, 0.6 ns pulsed laser was used as the pump light. When the power of the pump light was larger than the threshold pump power, remarkable laser emission with a wavelength of 617.73 nm could be obtained. The wavelength of 617.73 nm was located in the photonic bandedge of the photonic quasicrystal.

Using three-dimensional photonic crystal

Monodispersed mesoporous silica spheres are nano-scale silica spheres having radially aligned fine mesopores, which can also be used to fabricate three-dimensional photonic crystal [31]. Yamada *et al.* realized a colloidal photonic crystal laser made of monodispersed mesoporous silica spheres with a diameter of 197 nm [32]. They immersed the colloidal photonic crystal into an aqueous solution of Rhodamine 610, a kind of organic laser dye, and Rhodamine 610 was introduced in the air pores of the mesoporous silica spheres. The central wavelength of the photonic bandgap was about 600 nm. While the fluorescence peak of Rhodamine 610 was around 580 nm, which was dropped near the short-wavelength edge of the photonic bandgap of the colloidal photonic crystal. A 532 nm beam from a YAG laser (with a pulse duration of 8 ns and a pulse repetition rate of 10 Hz) was used as the pump light. The pump light was incident in the direction of 10° away from the normal direction of the surface of the colloidal photonic crystal, as shown in Fig. 6.2a. When the pump

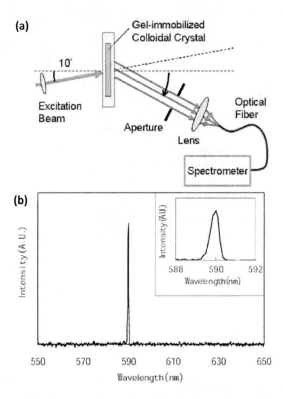

Figure 6.2 Colloidal photonic crystal laser made of monodispersed mesoporous silica spheres doped with Rhodamine 610. (a) Experimental setup. (b) Laser spectrum. Insert is the enlarged spectrum of the peak of laser mode. Reprinted with permission from *Adv. Mater.*, **21**, H. Yamada, *et al.*, Colloidal-crystal laser using monidispersed mesoporous silica spheres, 4134–4138. Copyright © 2009 WILEY-VCH Verlag GmbH & Co. KGaA, Weinheim.

intensity was larger than the threshold intensity of 6 MW/cm^2, remarkable laser emission could be obtained with the wavelength of 590 nm, which originates from the strong optical feedback provided by the photonic bandedge. The laser spectrum was shown in Fig. 6.2b.

6.1.1.1.2 *Photonic crystal microcavity laser*

Microcavities can provide a strong feedback for the realization of laser oscillation. The microcavity resonant mode was confined in plane by the photonic bandgap effect, and in the vertical direction by the total internal reflection effect. Under excitation of over the threshold pump power, a single (or coupled several) photonic bandgap microcavity, or even a weakly disordered photonic crystal can sustain strong laser oscillation when the gain medium is infiltrated in the microcavity regions [33–35]. Moreover, when the microcavity resonant excitation condition is satisfied, i.e., the frequency of the pump light is in resonance with the microcavity mode instead of the laser mode, the threshold pump power could be reduced greatly due to the enhanced absorption originating from the microcavity resonant effect [36,37]. Photonic crystal microcavity can be constructed not only by a point defect structure formed by removing one or several air holes [38–41] but also by a line defect waveguide [42]. The line defect waveguide can provide defect modes, also called waveguiding modes, located in the photonic bandgap. Calculation indicates that in the edge of the zero-order waveguiding modes, the group velocity of the electromagnetic wave is almost zero and a zigzag-shaped standing wave can be formed due to the strong Bragg scattering, which makes it possible to achieve laser oscillation in the line defect waveguide [43].

Point defect microcavity

Park *et al.* reported a two-dimensional semiconductor photonic crystal laser based on monopole-mode photonic bandgap microcavity [44]. A triangular lattice photonic crystal was fabricated in the InGaAsP slab with a thickness of 200 nm. The InGaAsP quantum well material was used as the gain medium. The lattice constant and the diameter of air holes were 570 and 188 nm, respectively. The photonic bandgap microcavity was constructed by removing one air hole in the center of the photonic crystal with six neighboring air holes with a reduced diameter of 125 nm. The schematic structure of the photonic bandgap microcavity is shown in Fig. 6.3. The quality factor of the photonic bandgap microcavity was about 1900. A 980 nm laser with a pulse duration of 10 ns was used as the pump laser. When the pump power was higher than the threshold power of 0.3 mW, a 1546 nm laser emission can be obtained. The line width of the laser mode was 0.8 nm.

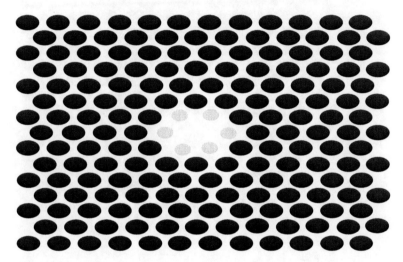

Figure 6.3   Schematic structure of semiconductor photonic crystal microcavity laser.

Painter *et al.* also reported a two-dimensional photonic crystal single point defect microcavity laser containing the InGaAsP quantum well material as the gain medium [45]. The schematic structure of the photonic crystal microcavity laser is shown in Fig. 6.4. The single-defect microcavity was constructed by removing one air hole in the center of a triangular lattice photonic crystal, with the diameter of neighboring two air holes enlarged. The neighboring enlarged two holes were used to destroy the degeneracy of the dipole microcavity mode. The thickness of the photonic crystal, the lattice constant, the diameter of normal and enlarged air holes were 220, 515, 180, and 240 nm, respectively. When the power of the pump light was larger than the threshold

pump power, a laser emission with a wavelength of 1504 nm
could be obtained. Ee *et al.* fabricated a photonic crystal microcavity
by enlarging the distance of two air holes in the center of a
two-dimensional square lattice InGaAsP photonic crystal with a
thickness of 200 nm [46]. To achieve a stronger photon confinement
effect, the surrounding air holes were also enlarged, as shown in
Fig. 6.5. The lattice constant, normal air hole radius, enlarged

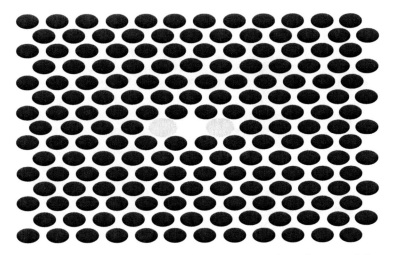

Figure 6.4   Schematic structure of photonic crystal single point defect
microcavity laser.

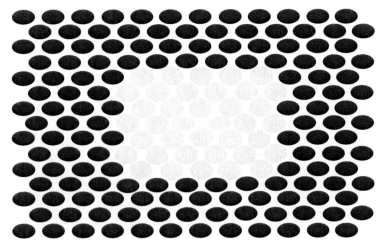

Figure 6.5   Schematic structure of photonic crystal point defect microcavity
laser.

air hole radius were 520, 171.6, 192.4 nm, respectively. The enlarged lattice was 794 nm. The InGaAsP quantum well materials embedded in the photonic crystal slab was used as the gain medium. A 980 nm, 10 ns laser was used as the pump light. When the excitation power of the pump light was larger than the threshold pump light of 130 μW, remarkable emission with a wavelength of 1511 nm could be obtained.

Ring microcavity

Kim *et al.* reported a semiconductor photonic crystal ring laser using the InGaAsP quantum well material as the gain medium [47]. The photonic crystal ring microcavity was composed of six waveguides connected with each other, forming six 120° bends in the center of a triangular lattice InGaAsP photonic crystal slab. The InGaAsP quantum well material was used as the gain medium. The schematic structure of the photonic crystal laser is shown in Fig. 6.6. The lattice constant and the diameter of air holes were 570 and 410 nm, respectively. A 980 nm laser diode with a pulse duration of 10 ns was used as the pump light. When the power

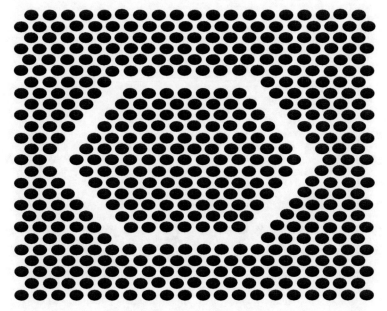

Figure 6.6   Schematic structure of semiconductor photonic crystal ring microcavity laser.

of the pump light was larger than the threshold power of 3 mW, remarkable laser emission with a wavelength of 1625 nm can be obtained. The line width of the laser mode was 0.8 nm. This kind of photonic crystal ring microcavity will sustain a pair of double degenerate modes having twofold symmetry, just like that of the whispering-gallery modes [47]. The laser mode can be switched to a neighbor mode through varying the optical pump condition. While for the practical applications of photonic crystal laser, the stable and single laser mode oscillation is preferred. Point defects could be added into the ring microcavity to achieve a single-mode laser oscillation. Alija *et al.* reported a single-mode two-dimensional InGaAsP photonic crystal coupled-cavity ring laser [48]. Six point defects made of one missing air hole were placed in the six corners of a hexagonal photonic crystal ring microcavity. The six waveguides of the ring microcavity were formed by eight missing air holes in line. The schematic structure of the photonic crystal ring microcavity is shown in Fig. 6.7. The point defect could sustain a single-defect mode.

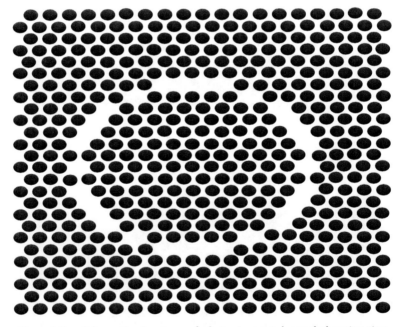

Figure 6.7 Schematic structure of photonic crystal coupled-cavity ring laser.

So, it is feasible to adopt the point defect to select a single mode for the laser oscillation in the multiple modes of the ring microcavity. The thickness of the photonic crystal, lattice constant, and the diameter of air holes were 237, 450, and 297 nm, respectively. A 780 nm pulse was used as the pump light. When the power of the pump plight was larger than the threshold pump power of 0.3 mW, single-mode laser emission with a wavelength of 1500 nm was achieved.

Photonic crystal heterostructure microcavity

Photonic crystal heterostructure could sustain high quality defect modes. The photon confinement originates from strong reflection in the interface between two different photonic crystal structures, which makes the connection region a nanocavity. Shih *et al.* reported a high-quality-factor InGaAsP photonic crystal hetero-structure microcavity laser [49]. The heterostructure microcavity was constructed by increasing the lattice constant in the central region of the photonic crystal waveguide formed by removing three rows of air holes. The schematic structure of the photonic crystal heterostructure microcavity is shown in Fig. 6.8.

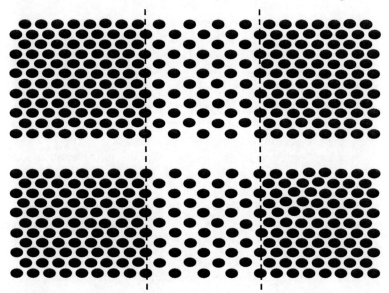

Figure 6.8   Schematic structure of photonic crystal heterostructure microcavity laser.

The thickness of the InGaAsP film, the lattice constant of the normal photonic crystal and the microcavity region were 240, 430, and 441 nm, respectively. The resonant modes sustained by the microcavity region are bound state modes with resonant frequencies near the bottom of the band of the guided modes [49]. A 850 nm, 8 ns pulsed laser was used as the pump light. Two laser modes with a wavelength of 1444 and 1577 nm could be achieved when the excitation power of the pump light was larger than the threshold pump light. The threshold pump power was 3.5 mW for the 1444 nm laser mode and 4.2 mW for the 1577 nm laser mode, respectively. The quality factor was in the order of more than 10^4 for two laser modes.

Whispering-gallery mode type microcavity

Two-dimensional quasiperiodic photonic crystal possesses perfect photonic bandgap effect due to quasiperiodic dielectric distribution. When a structural defect is introduced in the quasi-periodic photonic crystal, defect modes will appear in the photonic bandgap. More often than not, the defect mode of a two-dimensional quasiperiodic photonic crystal was confined not only by the photonic bandgap effect but also by the boundary reflection effect. So, the defect mode is really a kind of the whispering gallery mode [50,51]. If gain medium is infiltrated in the microcavity region, laser emission can be achieved when the excitation power is larger than the threshold pump power due to the strong feedback provided by the microcavity.

Compared with the periodic structure photonic crystal microcavity, the quasiperiodic photonic crystal microcavities possess stronger photon confinement effect due to the fact that the photon confinement originating from the joint action of the Bragg scattering and the interference reflection. This leads to a much lower threshold pump power for the laser oscillation [51]. Nozaki and Baba reported a semiconductor quasiperiodic photonic crystal microcavity laser with low threshold power [52]. The 12-fold quasiperiodic photonic crystal was fabricated in a 240 nm-thick GaInAsP quantum well material, which was also used as the gain medium. The lattice constant and the diameter of air holes were 610 nm and 435 nm, respectively. The microcavity was constructed by removing seven air holes in the center of the photonic crystal, as shown in Fig. 6.9. A 980 nm pulse light was used as the pump light. When the power of the pump light was larger than 0.8 mW, remarkable laser emission with a wavelength of 1600 nm could be obtained at room temperature. The laser wavelength was in agreement with that of the defect mode. On the other hand, when the quasiperiodic photonic crystal was excited outside the microcavity region, a laser emission with a wavelength of 1600 nm can also be obtained. This mode is weakly localized, but extending over the whole photonic crystal due to the self-similarity effect of the quasiperiodic photonic crystal [50]. This kind of electromagnetic mode does not exist in the periodic photonic crystal.

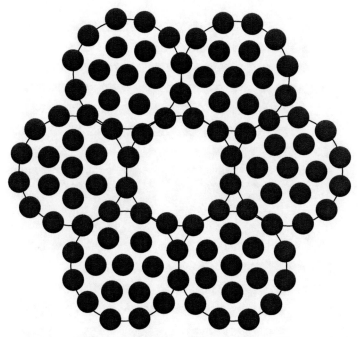

Figure 6.9   Schematic structure of quasiperiodic semiconductor photonic crystal microcavity laser.

Bragg reflection microcavity

An electromagnetic wave will be completely reflected by a photonic crystal when its frequency drops in the photonic bandgap. Therefore, in practice, one-dimensional and three-dimensional photonic crystals can be adopted as perfect reflection mirrors. When a layer of gain medium is placed between two photonic crystal mirrors, a photonic crystal microcavity laser could be constructed.

One-Dimensional Case: Yoon *et al.* reported a one-dimensional organic/inorganic hybrid photonic crystal defect mode laser with a dye-doped defect layer [53]. The one-dimensional photonic crystal was composed of alternating layers of TiO$_2$ nanoparticles and polymethylmethacrylate (PMMA) with a defect layer made of organic dye DCM-doped PMMA. The thickness of defect layer, TiO$_2$ layer, and PMMA layer was 1540 nm, 88 nm, and 99 nm, respectively. The schematic structure of the photonic crystal laser is shown in Fig. 6.10. There exist two defect modes in the photonic

Figure 6.10 Schematic structure of one-dimensional organic/inorganic hybrid photonic crystal laser.

bandgap with the wavelength of 582 and 620 nm, respectively. A 532 nm 8 ns intense laser was used as the pump light. When the intensity of the pump light was larger than the threshold of the pump intensity of the 582 nm mode, but less than that of the 620 nm mode, strong laser emission with a wavelength of 582 nm could be obtained.

The vertical-cavity surface-emitting lasers are inherently one-dimensional photonic crystal microcavity lasers, composed of active material layer sandwiched between two DBR mirrors [54]. One-dimensional and two-dimensional photonic crystals can also be integrated into vertical-cavity surface-emitting lasers to control the laser emission properties, such as the mode shape, polarization, divergence angle, and phase [55–59]. By etching proper photonic crystal microcavity structures into the top DBR layer, the single or multiple laser mode operation can be achieved for the vertical-cavity surface-emitting lasers. Moreover, when two or several point defects are introduced in the photonic crystal structure etched into the top DBR layer, the coherent coupling between the point defect microcavities laser modes can be realized by modifying the refractive index of the coupling region through varying the air hole parameters [60]. Even coupled photonic crystal heterostructures can be adopted to modulate the laser emission properties of vertical-cavity surface-emitting lasers [61]. Levallois *et al.* reported a long wavelength vertical-cavity surface-emitting laser using an electro-optic index modulator [62]. The photonic crystal laser was composed of a gain medium layer made of InGaAs/InGaAsP quantum well material and a nanopolymer dispersed liquid crystal layer sandwiched between two distributed Bragg reflectors. The schematic structure of the photonic crystal laser is shown in Fig. 6.11. A 1.064 μm intense pulse laser was used as the pump light. When the intensity of the pump light was larger than the threshold pump intensity, laser emission with a wavelength of 1561.6 nm could be obtained. While the refractive index of the nanopolymer dispersed liquid crystal decreased under excitation of an applied voltage due to electric field–induced reorientation of liquid crystal molecules. This makes a blue shift of resonant wavelength defect mode. Accordingly, the wavelength of the laser oscillation decreased. When the applied voltage increased from 0 to 170 V, the wavelength of the laser emission changed from 1561.6 nm to 1551.8 nm. A tuning range of 10 nm could be reached.

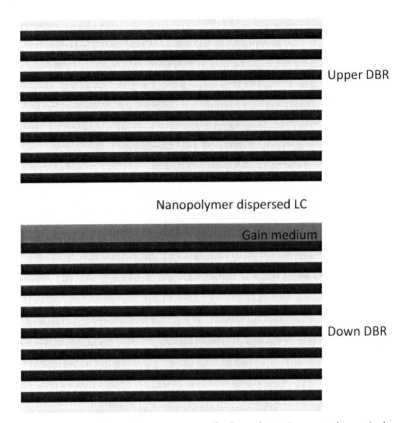

Upper DBR

Nanopolymer dispersed LC

Gain medium

Down DBR

Figure 6.11 Schematic structure of the photonic crystal vertical-cavity surface-emitting laser using an electro-optic index modulator.

Three-Dimensional Case: Jin *et al.* reported a SiO_2 opal photonic crystal microcavity laser containing a dye-doped polymer film [63,64]. The opal was fabricated by use of the vertical deposition method using SiO_2 nanospheres with a diameter of 253 nm. The photonic crystal microcavity consisted of a layer of gain medium with a thickness of 3 μm sandwiched between two SiO_2 opal. The gain medium was made of polymethyl methacrylate doped with organic dye allyl derivative of fluorescence (allyl-FL) and carbosilane dendrimer. A 355 nm, 8 ns laser was used as the pump light. When the excitation energy of the pump light was larger than the threshold pump energy of 15 μJ, remarkable laser emission with a wavelength of 589 nm could be obtained. The wavelength of the laser mode was in the photonic bandgap of the SiO_2 opal. Even when a layer of gain medium with irregular structure and strong scattering properties is sandwiched between two Bragg reflectors, remarkable laser emission could also be achieved due to the strong feedback of the photonic crystal microcavity [65].

Line defect waveguide microcavity

It is found that the group velocity of the guided electromagnetic modes of the line defect waveguide in the photonic bandedge is approximately zero, which means that a strong optical gain can be obtained at this frequency [66]. Checoury *et al.* reported a single-mode line defect waveguide laser in a square lattice InP photonic crystal under optical pump at the room temperature [42]. Calculation found that the group velocity of the electromagnetic modes at the second folding of the fundamental mode of the line defect waveguide was approximately zero at the Γ point, which means that a strong optical gain can be obtained at this frequency. They fabricated a two-dimensional InP photonic crystal containing InGaAsP quantum well material as the gain medium. A 1064 nm pulsed laser was used as the pump light. When the pump power was larger than the threshold pump power of about 1.3 W, remarkable laser emission could be obtained. The laser wavelength varied with the structural parameters of the photonic crystal. When the lattice constant increased from 460 to 520 nm, the laser wavelength changed from 1420 to 1580 nm.

The electric field distribution of the line defect waveguide will extend into the several, lateral surrounding photonic crystal lattices [41]. Sugitatsu *et al.* found that when a point defect was placed in the vicinity of the line defect waveguide, the point defect could act as a strong light transmitter radiating light from in plane to the vertical direction even though the resonant frequency of the point defect mode does not match the edge frequency of the waveguide mode of the line defect waveguide [67]. Therefore, the laser mode generated in the line defect waveguide can be trapped and emitted by the point defect placed in the vicinity of the line defect waveguide without satisfying the frequency resonant condition [66]. Sugitatsu *et al.* fabricated a triangular lattice semiconductor photonic crystal made of InGaAsP quantum well material, which was also used as the gain medium. The lattice constant and the distance between the waveguide and the point defect were both 420 nm. The resonant wavelength of the point defect was 1530 nm, while the wavelength of the laser oscillation generated in the line defect waveguide was 1560 nm when the gain medium was excited by a 980 nm intense pulse laser at the room temperature. The 1560 nm laser mode was emitted from the point defect due to the wide electric field distribution at the waveguide mode edge and it is distorted by the adjacent point defect [67].

When an electromagnetic wave is incident in a photonic crystal, the m-th order diffracted light will be reflected along the incident path if the vector component parallel to photonic crystal surface is $m\pi/a$, which is also called the Littrow diffraction [68]. When a wide waveguide is sandwiched between two photonic crystals, the forward-propagating light and the back-reflected light from the waveguide boundary forms a stationary resonant mode, also called the Littrow mode. So, a wide line defect waveguide could sustain the traditional Fabry–Perot-type resonant modes with their vector component parallel to waveguide direction is zero, and the unique Littrow resonant modes with their vector component parallel to waveguide direction is π/a, which corresponds to the Brillouin zone-edge modes [68]. Therefore, the Littrow diffraction could also be used to provide a strong feedback for the establishment of laser oscillation.

The guided modes of the photonic crystal line defect waveguide are confined strongly in the plane of the photonic crystal slab due to the photonic bandgap effect. The confinement of the guided modes in the vertical direction depends on the total internal reflection effect in the interface of photonic crystal and air, which limits the realization of high quality-factor microcavity mode. Gardin *et al.* found that when a cladding layer is covered on the surface of the line defect waveguide region, the effective refractive index of the line defect will increase, which will benefit for the enhancement of the photon confinement effect of the line defect waveguide [69]. They fabricated a two-dimensional InAsP photonic crystal line defect waveguide laser with the gain medium of InAsP quantum well materials. A 110 nm thick PMMA strip was covered in the center of the photonic crystal region. The lattice constant and the filling factor were 473 nm and 40%, respectively. The width of the PMMA strip was about 2.8 μm. The threshold pump power was 1.6 mW, which was much lower than that of the photonic crystal line defect waveguide without the PMMA cladding layer.

Extracting the laser emission from the line defect waveguide microcavity laser could be achieved by directly butt-jointing an output waveguide with the line defect waveguide laser, or coupling the laser mode to an output waveguide placed aside through the evanescent field [70,71]. Zheng *et al.* reported a butt joint line defect waveguide microcavity laser in a two-dimensional semiconductor photonic crystal slab made of InP slab containing the InGaAsP quantum well material, which was used as the gain medium [70]. The photonic crystal was composed of an output photonic crystal waveguide butt-jointed with a photonic crystal line defect waveguide microcavity laser. The other end of the line defect waveguide microcavity was terminated by the photonic crystal, as shown in Fig. 6.12. The laser emission could be coupled to the output waveguide due to the low microcavity quality factor in plane. The lattice constant and the diameter of the triangular lattice photonic crystal were 420 and 260 nm, respectively. To reduce the diffraction loss and the reflection loss at the interface of the output waveguide, the diameter of the inner row of air holes of the output waveguide was chirped from 260 to 190 nm. Under excitation of a 980 nm pump light, the laser emission with a wavelength of 1575 nm could be extracted from the line defect waveguide microcavity and propagate along the output waveguide. Halioua *et al.* reported an InGaAs/InGaAsP photonic crystal line defect waveguide laser vertically coupled with a silicon output waveguide through the evanescent field [71]. The silicon waveguide and the photonic crystal line defect waveguide, separated by a low refractive index polymer material benzocyclobutene, were aligned strictly. The laser mode generated in the photonic crystal line defect waveguide could be coupled into the silicon output waveguide and propagated in it.

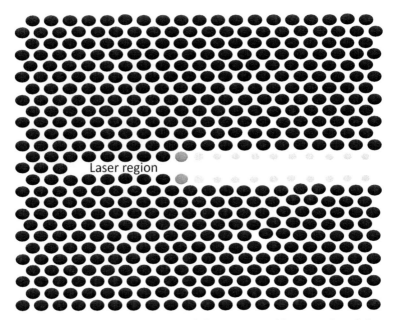

Figure 6.12 Schematic structure of photonic crystal butt joint line defect waveguide microcavity laser.

6.1.1.2 Based on distributed feedback mechanism

It has been pointed out that the spatially periodic modulation of refractive index could provide a strong frequency selective coupling between oppositely traveling waves and introduces the feedback needed for the laser oscillation, which is also called the distributed feedback mechanism [72]. The strongest coupling occurs when the electromagnetic modes in the gain medium satisfy the Bragg scattering condition [73]. Grating structures and one- and two-dimensional photonic crystals could be adopted to provide strong distributed feedback effect [74]. In practice, a film containing the active material was covered on the surface of the grating, one-, or two-dimensional photonic crystal structure [75]. When the excitation power of the pump light was larger than the threshold pump power, the laser oscillation could be sustained when the gain spectrum, i.e., the photoluminescence spectrum, overlapped with the photonic bandedge region [76,77]. By adopting different active materials, laser emission with a wavelength ranging from the visible to the infrared range could be obtained

[78]. Two-dimensional photonic crystal can provide more reciprocal lattice vectors than one-dimensional photonic crystal. So, there are various feedback mechanisms in two-dimensional photonic crystal. In the four lowest photonic bandedges of the M_1, K_1, M_2, Γ_1 of a two-dimensional hexagonal photonic crystal, the feedback conditions are as follows [79]:

$$\vec{k} + (\vec{k} - \vec{G}_i) \text{ in the } M_1 \text{ direction,} \tag{6.1}$$

$$\vec{k} + (\vec{k} - \vec{G}_i) + (\vec{k} - \vec{G}_j) \text{ in the } K_1 \text{ direction,} \tag{6.2}$$

$$\vec{k} + (\vec{k} - \vec{G}_i - \vec{G}_j) \text{ in the } M_2 \text{ direction,} \tag{6.3}$$

$$\vec{k} + (\vec{k} - 2\vec{G}_i) \text{ in the } \Gamma_1 \text{ direction,} \tag{6.4}$$

where $\vec{G}_i$ and $\vec{G}_j$ are unit reciprocal lattice vectors, $i \neq j$.

Meier *et al.* reported a multi-wavelength laser action in a two-dimensional SiO_2 photonic crystal covered by a layer of organic gain medium based on distributed feedback mechanism [80]. The two-dimensional photonic crystal consisted of triangular lattice of air holes embedded in the SiO_2 slab with a thickness of 40 nm. The lattice constant and the diameter of air holes were 400 and 200 nm, respectively. A film of organic gain medium made of 2-(4-biphenylyl)-5-(4-tertbutylphenyl)-1,3,4-oxadiazole-doped organic dye Coumarin 490 and DCM was deposited on the surface of the two-dimensional SiO_2 photonic crystal. The schematic structure of the photonic crystal is shown in Fig. 6.13. The doping concentration of Coumarin 490 and DCM was 1%. The thickness of the gain medium film was 150 nm, which only supported the lowest order TE and TM waveguide modes. A 337 nm pulsed laser with a pulse duration of 2 ns was used as the pump light. When the intensity of the pump light exceeded the threshold intensity of 50 kW/cm^2, two laser emissions with a wavelength of 596 nm for the TE mode and 580 nm for the TM mode were obtained. The wavelength of 596 nm was located in the edge of the lowest TE photonic bandgap in the *M* direction, while the wavelength of 580 nm was in the edge of the lowest TM photonic bandgap in the *M* direction. Owing to the strong Bragg scattering, the group velocity of these two electromagnetic waves was almost zero, which provides the strong feedback needed for the laser oscillation.

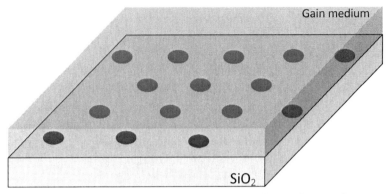

Figure 6.13 Schematic structure of distributed feedback SiO_2 photonic crystal laser.

Riechel *et al.* reported a surface emitting photonic crystal laser based on the distributed feedback mechanism [81]. The two-dimensional photonic crystal was composed of periodic patterns constructed by a superposition of two perpendicular sinusoidal height modulations in the acrylic film on the poly(ethylene terephthalate) substrate. The lattice constant was 300 nm. An active material film made of the conjugated polymer ladder type poly (p-phenylene) was deposited on the surface of the photonic crystal. The thickness of the active material film was 300 nm. A 400 nm pulsed laser with a pulse duration of 150 fs was used as the pump light. The feedback in the active material film waveguide sustains the laser oscillation. The laser mode could be coupled into the radiation modes in the free space perpendicular to the surface of the substrate through the first-order Bragg scattering effect [79]. When the energy of the pump light exceeded the threshold energy of 1.2 nJ, remarkable laser emission with a wavelength of 491 nm was achieved. The line width of the laser mode was 0.25 nm.

When the photonic crystal structure was directly etched on the top side of the gain medium, a distributed feedback laser could also be realized [82,83]. Harbers *et al.* reported an organic photonic crystal laser based on distributed feedback mechanism, fabricated by the laser interferometer lithography method [84].

A thick film of gain material made of photoresist UV II-HS doped with laser dye Coumarin 6H and DCM was fabricated. The periodic photonic crystal pattern was fabricated by use of two-step exposures along with a 90° rotation of the sample. The lattice constant and the depth of air holes were 400 and 100 nm, respectively. The schematic structure of the organic photonic crystal laser is shown in Fig. 6.14. A 355 nm, 8 ns intense pulsed laser was used as the pump light. When the excitation energy of the pump light was larger than the threshold pump energy of 10 μJ, remarkable laser emission with a wavelength of 585 nm could be obtained.

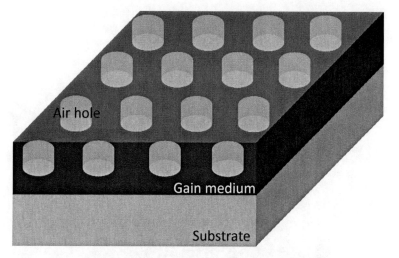

Figure 6.14   Schematic structure of distributed feedback organic photonic crystal laser etched in the gain medium directly.

6.1.2   Metallodielectric Photonic Crystal Laser

Metallodielectric photonic crystals can also provide remarkable photonic bandgap effect and can be used to realize photonic crystal laser devices. Stehr *et al.* reported a low-threshold laser emission based on a two-dimensional metallodielectric photonic crystal coated with conjugated polymer material based on the distributed feedback mechanism [85]. The two-dimensional metallodielectric photonic crystal was composed of square lattice of gold nanodisks embedded in the indium tin oxide matrix, as shown in Fig. 6.15. The lattice constant and the diameter of gold nanodisks were

300 nm and 110 nm, respectively. The thickness of the metallodielectric photonic crystal was 30 nm. A 460 nm-thick film of conjugated polymer methyl-substituted ladder-type poly(para phenylene) (LPPP) was coated on the metallodielectric photonic crystal, which was used as the gain medium. The metallodielectric photonic crystal was used to provide the distributed feedback effect. The second-order Bragg scattering was used to provide the feedback needed for the generation of laser modes. The first-order Bragg scattering was used to couple the generated laser modes out from the LPPP film in the vertical direction [20]. A 400 nm beam from a femtosecond laser system (with a pulse duration of 130 fs and a repetition rate of 1kHz) was used as the pump light. Strong laser emission with a wavelength of 492 nm was achieved when the pump energy was 2.7 nJ, which was much higher than the pump threshold of 1.9 nJ.

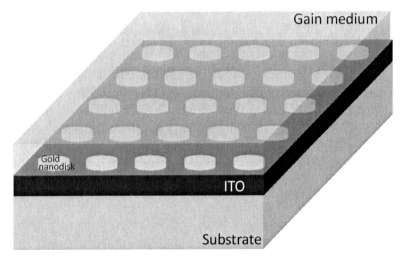

Figure 6.15   Schematic structure of metallodielectric-dielectric photonic crystal laser.

6.1.3   Liquid Crystal Photonic Crystal Laser

Cholesteric liquid crystals (CLCs) can be regarded as a kind of one-dimensional photonic crystals, because the rod-shaped liquid crystal molecules were arranged to form a helical structure due to the intrinsic twisting force of the CLC molecules [86]. In the direction along the helical axis, the dielectric constant of the liquid crystal is periodically modulated. The circular polarized light with the same handedness as the cholesteric medium can see the photonic bandgap. The center wavelength of the photonic bandgap λ_c can be calculated by [87]

$$\lambda_c = \sqrt{\frac{n_o^2 + n_e^2}{2}} \cdot p \qquad (6.5)$$

where n_o and n_e are the refractive indexes of the liquid crystal for the ordinary and extraordinary light, respectively. p is the helical pitch. So, laser emission can also be achieved in a CLC.

The CLC can provide perfect photonic bandgap structures, which is the important basis for the realization of photonic crystal laser devices. When organic fluorescence dye is doped in the CLC, laser emission with a wavelength located in the photonic bandgap can be obtained when the excitation power of the pump light exceeds the threshold pump power [88]. More often than not, the laser emission will appear in the position of the long-wavelength edge of the photonic bandgap of the CLC. The reason is that the fluorescence dye molecules are easy to be aligned in the direction parallel to the direction of the local liquid crystal molecules. Accordingly, the transition dipole moment of the fluorescence dye is parallel to the local director of the CLC [89]. Moreover, the optical eigen-modes in the long-wavelength photonic bandedge are linearly polarized and the electric filed vectors are parallel to the local director of the CLC. While in the short-wavelength photonic bandedge, the electric field vectors of the eigen-modes are perpendicular to the local director of the CLC [89]. However, under strong laser excitation, the periodic dielectric structure of the CLC can be distorted easily by the heat dissipation induced by the pump laser. Two methods can be adopted to consolidate the structural stability of the CLC. One method is to cool the molecules of the CLC below the glass transition temperature. The other method is to chemically cross-link the molecules of the CLC based on the photo-polymerization reactions [90].

Schmidtke *et al.* reported the laser emission properties in an organic dye doped in the molecules of the cholesteric polymer network [91]. The CLC was made of a mixture of the nematic p-pentylphenyl-2-chloro-4-(pentylbenzoyloxy)-benzoate   and   the cholesteric cholesteryl nonanoate. The CLC was fixed by a photon-polymerization of monomers. Organic dye 4-(dicyanomethylene)-2-methyl-6-(4-dimethylamino styryl)-4-H-pyran (DCM) was doped in the CLC. A 532 nm laser beam with a pulse duration of 7 ns was used as the pump light, as shown in Fig. 6.16. A white light source was used to measure the transmittance and the reflection spectrum of the CLC photonic crystal. A 532 nm continuous-wave laser was used to measure fluorescence spectrum. When the energy of the pump light was higher than the threshold energy of 0.1 µJ, remarkable laser emission with a wavelength of 601 nm could be achieved in the direction parallel to the cholesteric axis. The wavelength of 601 nm was located at the long-wavelength edge of the photonic bandgap. The laser emission reached the maximum

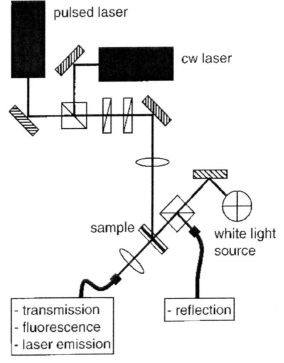

Figure 6.16 Experimental setup of a cholesteric liquid crystal laser. Reprinted with permission from *Adv. Mater.*, **14**, J. Schmidtke, *et al.*, Laser emission in a dye doped cholesteric polymer network, 746–479. Copyright © 2002 WILEY-VCH Verlag GmbH & Co. KGaA, Weinheim.

value of 0.14 µJ when the pump energy was 0.55 µJ. A high conversion efficiency was realized. When the pump energy was larger than 0.55 µJ, the heat dissipation gradually destroyed the periodic structure of the CLC, which resulted in the decrease in the energy of the laser emission, as shown in Fig. 6.17.

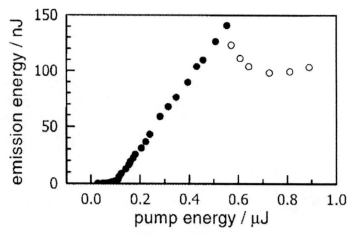

Figure 6.17   Changes of the laser emission energy as a function of the pump light energy of a cholesteric liquid crystal doped with DCM. Reprinted with permission from *Adv. Mater.*, **14**, J. Schmidtke, *et al.*, Laser emission in a dye doped cholesteric polymer network, 746–479. Copyright © 2002 WILEY-VCH Verlag GmbH & Co. KGaA, Weinheim.

Creating defect modes in the photonic bandgap is another approach to realizing laser emission in CLC photonic crystals. The defect structure can be introduced in the CLC by directly inserting a thin layer of isotropic material in the center of a CLC [92] by twisting one part of the CLC about its helical axis without separating two parts [93] or by inducing a phase jump in the cholesteric helix in a CLC photonic crystal heterostructure [94]. The schematic structure of the twist defect is shown in Fig. 6.18. Schmidtke *et al.* reported a defect mode laser emission in a dye-doped cholesteric polymer network, constructed by stacking two layers of CLC film fabricated by the ultraviolet polymerization of a cholesteric mixture made of diacrylate monomers doped with laser dye DCM [94]. In the interface of the heterostructure, the alignment of the director of the liquid crystal molecules in the bottom layer was perpendicular to that of the up layer, which led to a 90° phase jump of the cholesteric helix in the interface of the liquid crystal heterostructure. The wavelength of the defect mode was located at 590 nm. A 532 nm, 7 ns pulsed laser was used as the pump light. When the energy of the pump light was larger than the threshold pump energy of 50 nJ, strong laser emission with a wavelength of 590 nm could be obtained. Song *et al.* also reported the laser emission based on the defect mode in a 3.6 μm-thick polymeric CLC containing a 2 μm-thick defect layer made of nematic liquid crystal ZLI2293 doped with organic dye of triptycene poly (p-phenylenevinylene) (T-PPV) [95]. The doping concentration of the organic dye was 93%. The helical pitch was 510 nm. The defect mode in the photonic crystal was located at 511 nm. A 400 nm laser beam from an optical parametric oscillator system was used as the light source. When the energy density of the pump light was higher than the threshold energy density of 1.5 mJ/cm^2, a strong laser emission with the wavelength of 511 nm was achieved, as shown in Fig. 6.19. The line width of the laser mode was 3 nm.

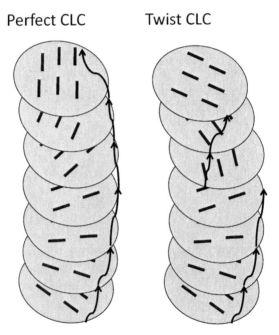

Perfect CLC   Twist CLC

Figure 6.18   Schematic structure of the twist defect of cholesteric liquid crystal photonic crystal.

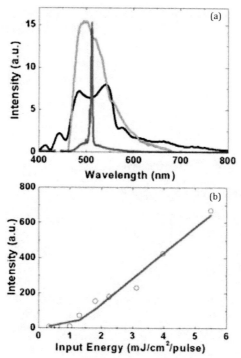

Figure 6.19 Polymeric cholesteric liquid crystal laser based on defect mode. (a) Spectrum of laser emission mode under excitation of exceeding the threshold pump energy. The red line is laser mode. The black line is the reflection spectrum of the polymeric cholesteric liquid crystal. The green line is the fluorescence spectrum of T-PPV. (b) Changes of the laser emission intensity as a function of the pump light energy. Reprinted with permission from *Adv. Mater.*, **16**, M. Song, *et al.*, Effect of phase retardation on defect-mode lasing in polymeric cholesteric liquid crystals, 779–783. Copyright © 2004 WILEY-VCH Verlag GmbH & Co. KGaA, Weinheim.

6.2 Tunable Photonic Crystal Laser

The frequencies of the laser radiation of a tunable photonic crystal laser vary with the external parameters. So, the tunable photonic crystal laser can find more flexible applications in practice. Various schemes have been proposed to demonstrate tunable photonic crystal laser, such as rotating the crystalline planes of three-dimensional photonic crystals, adjusting the refractive index contrast by use of light or electric field, changing the gain medium, and directly modulating the photonic crystal structure parameters [96]. Moreover, it is possible to achieve switchable lasing in a multimode photonic crystal microcavity laser by injecting seeds [97]. By injecting a weak laser pulse before and during the laser oscillation in the photonic crystal microcavity, the stimulated emission of a needed mode can be established from the seed laser field rather than from the random noise [97]. Therefore, the required laser mode can be obtained at will.

6.2.1 Tunable Dielectric Photonic Crystal Laser

6.2.1.1 Mechanical tuning method

Three-dimensional photonic crystals fabricated by the self-assembling method have many structural defects, which sustain resonant defect modes in the three-dimensional photonic crystal. These defect modes can provide a strong feedback for the realization of photonic crystal laser. According to the Bragg scattering condition, the central wavelength λ_{hkl} of the photonic bandgap in the (hkl) direction of the opal can be calculated by [98]

$$\lambda_{hkl} = 2n_{\text{eff}}d_{hkl} = 2\sqrt{n_{\text{sphere}}^2 \times 0.74 + n_{\text{i}}^2 \times 0.26} \cdot d_{hkl} \qquad (6.6)$$

where n_{eff} is the effective refractive index of the opal. n_{sphere} and n_{i} are the refractive index of microspheres and the infiltrated medium, respectively. d_{hkl} is the interplane spacing in the (hkl) direction. The interplane spacing varies with the crystalline direction. Therefore, it is possible to achieve different laser radiations by rotating the crystalline planes of the photonic crystal, modulating the lattice constant, or adjusting the refractive index contrast.

Shkunov *et al.* reported a tunable photonic crystal laser throughout the visible range by use of a SiO_2 opal infiltrated with laser dye [98]. They fabricated a SiO_2 opal with a face-centered cubic structure by slow gravity sedimentation of a colloidal suspension of SiO_2 microspheres with a diameter of 295 nm. There were many structure defects in the SiO_2 opal fabricated by the slow gravity sedimentation method, which sustains resonant defect modes in the three-dimensional photonic crystal. Strong laser radiation can be realized by use of these defect modes. The 355 nm light from a Nd:YAG laser system (with a pulse duration of 80 ps and a repetition rate of 100 Hz) was used as the excitation light, which was focused on the front surface of the opal by a cylindrical lens to from an excitation stripe with a width of 50 μm and a length of 800 μm, respectively. The laser dye whose photoluminescence band overlaps a particular photonic bandgap λ_{hkl} of the opal was used. A red organic dye Oxazine 725 in methanol was infiltrated into the SiO_2 opal. The axis of the excitation stripe was along the [111] direction of the three-dimensional photonic crystal. Remarkable laser emission at 687 nm can be obtained in the [111] direction when the intensity of the pump light exceeded the threshold value of 8 MW/cm^2, as shown in Fig. 6.20. The line width was only 0.2 nm for the emission light. The narrow line width, the highly directional radiation along the [111] direction, and the threshold effect ensure laser properties of the radiation. When a blue organic dye Stilbene 420 in a mixture of methanol and alcohol was infiltrated in the SiO_2 opal and the axis of the excitation stripe was along the [220] direction of the photonic crystal, strong

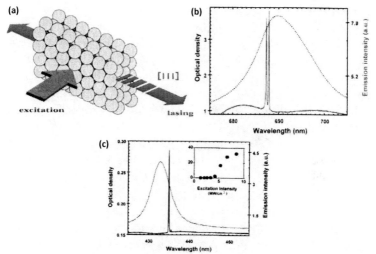

Figure 6.20 Tunable laser radiation in SiO_2 opal infiltrated with organic dyes. (a) Schematically experimental geometry for the laser radiation in the [111] direction. (b) 687 nm laser spectrum in the [111] direction (solid line), and the photoluminescence background (dashed line). (c) 432 nm laser spectrum in the [220] direction (solid line), and the photoluminescence background (dashed line). Insert is the I–I curve. Reprinted with permission from *Adv. Funct. Mater.*, **12**, M. N. Shkunov, *et al.*, Tunable, gap-state lasing in switchable directions for opal photonic crystals, 21–26. Copyright © 2002 WILEY-VCH Verlag GmbH & Co. KGaA, Weinheim.

laser emission at 432 nm was achieved in the [220] direction when the pump intensity was higher than the threshold intensity of 4 MW/cm^2. When both the red and blue organic dyes of Oxazine 725 and Stilbene 420 were infiltrated, and pumped at an excitation angle of 35.3° between the [111] and [220] direction, the simultaneous laser emission in red and blue light could be achieved. Therefore, large tunability of more than 250 nm in the wavelength of the laser emission could be reached by using photonic bandgaps in different crystalline directions (*hkl*). A wavelength tunability of the laser radiation of about 70 nm can be reached in a single photonic bandgap by changing the refractive index contrast of the photonic crystal through adopting different organic dye and solvent. A fine wavelength tunability of about 6 nm could be achieved in a single photonic bandgap by adopting a 20° rotation of the crystalline planes.

Lawrence *et al.* also reported a mechanically tunable three-dimensional colloidal photonic crystal laser by use of compressive strain force [99]. The three-dimensional colloidal photonic crystal was fabricated by electrostatically self-assemble monodispersed cross-linked polystyrene nanospheres with a diameter of 150 nm. Then the colloidal crystal was encapsulated with a photoinitiated free radical polymerized methacrylate functionalized poly(ethylene glycol) hydrogel, and subsequently the 2-methoxyethyl acrylate. Finally the colloidal crystal was photopolymerized to form a robust film. A layer of poly(methyl methacrylate) doped with organic dye Rhodamine B was used as the gain medium, which was sandwiched between the colloidal film and a dielectric stack mirror. The photoluminescence band of Rhodamine B covers the whole photonic bandgap of the colloidal photonic crystal. The dielectric mirror and the colloidal photonic crystal constructed a resonator for the operation of laser emission, as shown in Fig. 6.21. A 520 nm beam from a dye laser system (with a pulse duration of 0.5 ns and a repetition rate of 6 Hz) was used as the pump light. When the energy of the pump light was larger than the threshold energy of 3 μJ, remarkable laser emission could be obtained. The wavelength and the line width of the laser mode were 613 nm and 2 nm, respectively. When a compressive strain was exerted on the colloidal photonic crystal, the lattice constant was reduced, which makes the central wavelength of the photonic bandgap shift in the direction of short-wavelength. Accordingly, the wavelength of the laser mode shifts in the short-wavelength direction. The tuning range of the wavelength of the laser mode of 32 nm could be reached.

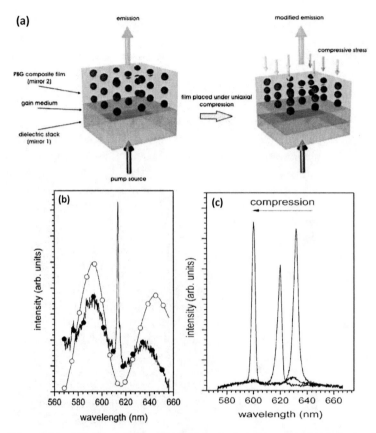

Figure 6.21   Mechanically tunable photonic crystal laser made of a layer of active medium sandwiched between a polystyrene opal and a dielectric mirror. (a) Schematic structure of the photonic crystal laser and the tuning mechanism. (b) Laser spectrum. The hollow circle is below the threshold energy. The filled circle is above the threshold energy. (c) Laser spectra with different degree of compressive strain. Reprinted with permission from *Adv. Mater.*, **18**, J. R. Lawrence, *et al.*, Dynamic tuning of organic lasers with colloidal crystals, 300–303. Copyright © 2006 WILEY-VCH Verlag GmbH & Co. KGaA, Weinheim.

6.2.1.2   Covering tuning method

The effective refractive index of the photonic crystal can be modulated through covering a thin cladding layer on the surface of the photonic crystal. Compared with the photonic crystal without the cladding layer, the effective refractive index of the photonic crystal will increase when covered with a cladding layer. This makes the photonic bandgap and defect modes shift in the direction of long wavelength [100]. As a result, the feedback properties of the photonic crystal and the laser emission wavelength of the photonic crystal laser can be tuned by use of the covering tuning method.

Arango *et al.* reported a tunable organic photonic crystal bandedge laser by covering the photonic crystal with fluids with different refractive indexes [101]. They fabricated an organic photonic crystal made of the photoresist SU-8 doped with organic laser dye pyrromethene 597 (P597) by use of the nanoimprint and photolithography method. The lattice constant and the diameter of air holes were 384 and 230.4 nm, respectively. The slab thickness of the SU-8 doped with P597 and the air hole were 450 and 100 nm, respectively. The schematic structure of the photonic crystal is shown in Fig. 6.22. A 532 nm, 5 ns pulsed laser was used as the pump light. When the excitation energy of the pump light was larger than the threshold pump energy, remarkable laser emission with a wavelength of 590 nm could be obtained. The wavelength of 590 nm was located at the photonic bandedge. When the photonic crystal was covered with fluids with different refractive indexes, the emission wavelength of the laser mode was changed. The laser wavelength increased with the increment of

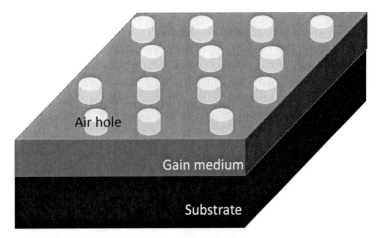

Figure 6.22   Schematic structure of tunable organic photonic crystal laser by use of covering fluid cladding.

the refractive index of the cladding fluids. When the refractive index of the fluid increased from 1.33 to 1.55, the laser wavelength changed from 590 to 599 nm.

6.2.1.3 Immersion tuning method

If a photonic crystal was immersed into a liquid material, the refractive index of the background material will change from air to the liquid. This leads to an increase of the effective refractive index of the photonic crystal. As a result, the photonic bandgap and the defect modes will shift in the long wavelength direction. Therefore, it is possible to achieve a tunable photonic crystal laser based on this method.

Kim *et al.* reported a tunable photonic crystal laser by immersing the photonic crystal into different liquid solution [102]. They fabricated a two-dimensional InGaAsP photonic crystal having a honeycomb lattice, as shown in Fig. 6.23. The lattice constant and the diameter of air holes are 400 and 160 nm, respectively. A 980 nm, 20 ns pulsed laser was used as the pump light. When the power of the pump light was larger than the threshold pump power of 2 μW, laser emission with a wavelength of 1540 nm could be obtained. When the photonic crystal was immersed into a liquid solution, the wavelength of the laser emission increased with the increment of the refractive index of the liquid. The laser changed from 1540 to 1572 nm when the refractive index of the background material changing from 1 to 1.373.

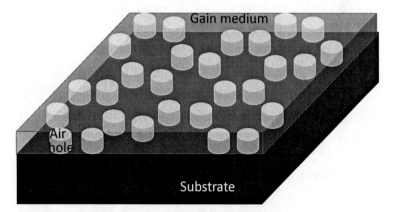

Figure 6.23 Schematic structure of tunable semiconductor photonic crystal laser by use of the immersion method.

6.2.1.4   Additional-material tuning method

The properties of the laser oscillation in a photonic crystal laser extremely depend on the photonic bandgap effect. The position and width of the photonic bandgap can be modulated by changing the refractive index contrast of the photonic crystal. The introduction of an additional material in the photonic crystal will change the refractive index contrast of the photonic crystal, which will change the laser wavelength.

Stroisch *et al.* reported a tunable one-dimensional photonic crystal distributed feedback laser by oblique angle deposition of an intermediate high refractive-index layer [103]. Stroisch *et al.* fabricated a line surface grating structure in an acrylic resist by the nanoimprint technology. The grating period and thickness were 400 and 95 nm, respectively. A thin layer of tantalum pentoxide was deposited only on one well of the grating grooves by use of the oblique angle electron beam evaporation method. Then a 350 nm-thick layer of *tris*-(8-hydroxyquinoline) aluminum (Alq$_3$) doped with laser dye 4-dicyanomethylene-2-methyl-6-(p-dimethylaminostyryl)-4H-pyran (DCM) was coated on the surface of the grating structure as the gain medium. The refractive index of tantalum pentoxide is much larger than that of the gain medium and the photoresist substrate. Therefore, the introduction of the tantalum pentoxide layer enlarges the refractive index contrast of the photonic crystal, which broadens the width of the photonic bandgap. When the thickness of the additional tantalum pentoxide layer increased from 0 to 28 nm, the width of the photonic bandgap was enlarged from 3 to 16 nm. Accordingly, a tunable wavelength of the laser emission ranging from 641 to 653 nm could be realized.

6.2.1.5   Optical tuning method

The orientation of the liquid crystal molecules can be changed when illuminated with a laser beam, which leads to a remarkable refractive index change of the liquid crystal. Therefore, when a photonic crystal laser is infiltrated with liquid crystal, the laser wavelength can be tuned through optically exciting the liquid crystal component [104].

6.2.2 Tunable Liquid Crystal Photonic Crystal Laser

The refractive index of liquid crystal can be changed by modulating the temperature or using an external applied voltage. Moreover, the structure of a liquid crystal photonic crystal can be changed by using the mechanical stress [105,106]. Therefore, an electrically, thermally, or mechanically tunable liquid crystal photonic crystal laser can be realized.

6.2.2.1 Thermal tuning method

Ozaki *et al.* reported a temperature tunable ferroelectric liquid crystal laser based on the photonic bandedge effect [105]. The ferroelectric liquid crystal is a kind of chiral smectic liquid crystal possessing ferroelectricity and a helical structure, which leads to the formation of photonic bandgap just like that of the CLC. The ferroelectric liquid crystal was doped with a laser dye Coumarin 500, which was used as the gain medium. A 400 nm laser beam (with a pulse duration of 150 fs and a repetition rate of 1 kHz) was used as the pump light. When the energy of the pump light was larger than the threshold pump energy of 3 µJ, strong laser emission with a wavelength of 475 nm could be obtained at a temperature of 20°C. The 475 nm laser wavelength was located at the photonic bandedge of the ferroelectric liquid crystal photonic crystal. The helical pitch of the ferroelectric liquid crystal increased with the increment of the temperature, which resulted in the shift of the photonic bandgap in the long wavelength direction. When the temperature increased from 20°C to 26°C, the central wavelength of the photonic bandgap shifted from 465 to 490 nm. Accordingly, the wavelength of the laser emission shifted to 483 nm.

Huang *et al.* reported a thermally tunable dye-doped CLC laser by use of a one-dimensional temperature gradient [107]. They fabricated a CLC made of a mixture of liquid crystal BL006 and a chiral material ZLI-811, which was doped with organic dye DCM. The concentration of ZLI-811 was 34%, which was much larger than the solubility of ZLI-811 in BL-006, 25%. Therefore, at the room temperature, a large amount of ZLI-811 did not dissolve. The solubility of ZLI-811 increases with the increment of the temperature, which results in a reduction of the cholesteric pitch. As a result, the photonic bandgap of the CLC shifts in the short-wavelength direction with the increase of the temperature. When the temperature increased from 25°C to 40°C, the central wavelength of the photonic bandgap shifted from 825 to 525 nm. To create a temperature controlling, one side of the sample was placed on a heating stage with the other side in the air. A 532 nm, 8 ns laser was used as the pump light. When the excitation energy was larger than the threshold pump energy of 4.7 μJ, remarkable laser emission with a wavelength at the photonic bandedge could be obtained. The wavelength of the laser mode could be tuned by changing the excitation regions with different temperature in the sample. The laser wavelength can be tuned from 577 nm to 670 nm. A wide tunability in the laser wavelength could be achieved.

6.2.2.2 Electric tuning method

Furumi *et al.* reported an electrically tunable CLC photonic crystal laser [106]. The CLC was formed by a mixture of liquid crystal R-1011 and RDP-60774 doped with a laser dye of 4-(dicyanomethylene)-2-methyl-6-(4-dimethlyaminostryl)-4H-pyran (DCM) with a doping concentration of 5.9%. The photonic bandgap ranged from 606 to 658 nm, which overlapped with the fluorescence band of DCM. A 532 nm laser (with a pulse duration of 6 ns and a repetition rate of 5 Hz) was used as the pump light. When the energy of the pump light was larger than the threshold pump energy of 0.9 μJ, a laser emission with a wavelength of 650 nm can be obtained. The wavelength of the laser mode was located at the photonic bandedge, which shows that the low group velocity and long photon lifetime of the electromagnetic modes at the photonic bandedge provide the strong feedback for the laser oscillation. Moreover, a 0.8 nm line width could be achieved for the laser mode when the pump energy exceeded the threshold energy. While under excitation of an applied voltage, the electric field–induced structural phase change from cholesteric phase to the nematic phase destroyed the photonic band structure. Accordingly, the laser oscillation depending on the photonic bandedge effect will be stopped gradually with the increase of the applied voltage. The insert shows the reflection spectrum of the liquid crystal photonic crystal. It is very clear that the peak reflectivity and the width of the reflection decreased with the increase of the applied voltage, which indicates that the perfect photonic bandgap structure disappeared gradually.

The interference of two laser beams in the mixture of liquid crystal, acrylate monomers, and photoinitiators can be used to fabricate the holographic polymer-dispersed liquid crystal based on the photopolymerization effect. The photopolymerization of acrylate monomers occurs in the high-intensity regions, which pushes the liquid crystal to the low-intensity regions [108]. Then a periodic structure made of alternative polymer-rich layers and liquid crystal-rich layers can be fabricated. The dielectric constant of the holographic polymer-dispersed liquid crystal is periodically modulated, which leads to the formation of photonic bandgap. So, the holographic polymer-dispersed liquid crystal can also be regarded as a kind of one-dimensional photonic crystal. The refractive index of liquid crystal can be modulated by use of an electric field. Therefore, the holographic polymer-dispersed liquid crystal and the CLC could be used to realize tunable one-dimensional photonic crystal laser. Jakubiak *et al.* fabricated a holographic polymer-dispersed liquid crystal made of the Bl045 liquid crystal doped with laser dye Coumarin 485, the photoinitiator of the rose bengal ethyl acetate, the coinitiator of N-phenylglycine, the chain extender of N-vinyl pyrrolidinone, the fatty acid of octanoic acid, and the monomer of the dipentaerythritol penta/hexa acrylate [109]. The lattice constant a and the effective refractive index n_{eff} of the polymer-dispersed liquid crystal were 163 nm and 1.54, respectively. The central wavelength of the photonic bandgap λ_c can be determined by

$$\lambda_c = 2n_{eff} \cdot a \qquad (6.7)$$

A 355 nm beam from a Nd:YAG laser system (with a pulse duration of 5 ns and a pulse repetition rate of 20 Hz) was used as the light source. Under excitation of the pump light, the line width of the 495 nm photoluminescence peak was decreased. When the pump intensity exceeded the threshold of 3.5 mJ/cm^2, the line width decreased rapidly. The application of a 30 V/μm static electric field made a remarkable reduction of the emission intensity and a 30 nm broadening of the line width.

Ozaki *et al.* reported an electrically tunable defect mode laser in a one-dimensional photonic crystal containing liquid crystal defect layer [110]. The one-dimensional photonic crystal was composed of a defect layer made of organic laser dye DCM-doped nematic liquid crystal E47 sandwiched between two dielectric multiplayer made of an alternating stacking of SiO_2 and TiO_2 layers. The thickness of the SiO_2 and TiO_2 layers was 103 and 64 nm, respectively. The schematic structure of the one-dimensional photonic crystal laser was shown in Fig. 6.24. A 532 nm laser (with a pulse duration of 8 ns and a repetition rate of 10 Hz) was used as the pump light. When the energy of the pump light was larger than the threshold pump energy of 3 µJ, a strong laser emission with a wavelength of 617 nm could be obtained in the defect mode. While under excitation of the applied voltage, the molecules of the liquid crystal were reoriented by the external electric field, which led to a reduction of the refractive index of the liquid crystal and a blue shift of the resonant wavelength of the defect mode. Accordingly, the laser mode shifted in the short wavelength. Under excitation of a very low applied voltage of 2 V, a 25 nm shift of the laser mode could be reached. Ozaki *et al.* also reported an electrically tunable laser based on defect mode in a one-dimensional photonic crystal containing organic conjugated polymer and liquid crystal defect layer [111]. The one-dimensional photonic crystal was composed of 200 nm-thick active material layer of poly(2-methoxy-5-dodecyl-oxy-1,4-phenylene vinylene) (MDDO-PPV) and 2 µm-thick nematic liquid crystal E7 layer sandwiched between two multilayers made of alternating stack of SiO_2 and TiO_2 layers. The thickness was 111 nm for SiO_2 layers and 69 nm for TiO_2 layers, respectively. A 532 nm laser (with a pulse duration of 8 ns and a repetition rate of 10 Hz) was used as the pump light. When the energy of the pump light was larger than the threshold pump energy of 26 nJ, the laser emission with a wavelength of 625 nm can be achieved. The wavelength of 625 nm corresponds to the resonant wavelength of the defect structure. Under excitation of the applied voltage, the molecules of the liquid crystal will be reoriented, which results in a reduction of the refractive index of the liquid crystal. Accordingly, the resonant wavelength of the defect mode shifts in the direction of the short wavelength. As a result, the wavelength of the laser mode will be blue shift with the increase of the applied voltage. When the applied voltage increased from 0 to 10 V, the wavelength of the laser mode decreased from 625 nm to 610 nm. A tuning range of 15 nm of the laser mode could be obtained.

Figure 6.24   Schemtic structure of electrically tunable one-dimensional photonic crystal laser containing a liquid crystal defect layer.

The dielectric constant of a negative dielectric anisotropy liquid crystal with its director parallel to the applied electric field is less than that perpendicular to the applied electric field [112]. Under excitation of an applied voltage, the refractive index of the negative dielectric anisotropy liquid crystal will decrease due to the electrohydrodynamical effect. Lin *et al.* fabricated a CLC with negative dielectric anisotropy, which was made of chiral material S811 and nematic liquid crystals MLC6608 and ZLI2293, which were negative dielectric anisotropy materials [112]. The

CLC was doped with organic laser dye DCM. Under excitation of an applied voltage, the photonic bandgap shifts in the direction of short wavelength due to the electrohydrodynamical effect. Under excitation of an intense 532 nm laser, tunable laser emission could be achieved by varying the applied voltage. The laser wavelength changed from 643 nm to 629 nm when the applied voltage increasing from 0 to 150 V.

6.2.2.3 Optical tuning method

Under excitation of an ultraviolet light, the light-induced structure transformation in the chiral dopant occurs in CLC. This results in the variation of the cholesteric pitch, and accordingly the shift of the photonic bandgap [113,114]. Chanishvili reported an optically tunable CLC photonic crystal laser [78]. The CLC was composed of a mixture of nematic liquid crystal ZLI-1695 and ZLI-811, which was doped with organic dye terphenyl derivative. A 355 nm beam from a Nd:YAG laser system (with a pulse duration of 6 ns and a repetition rate of 10 Hz) was used as the pump light. When the energy of the pump light was larger than the threshold pump energy of 1 μJ, laser emission with a wavelength of 382 nm could be obtained. The wavelength of 382 nm was located at the long-wavelength edge of the photonic bandgap. The line width was 2 nm. While under excitation of a 100 W mercury lamp, the central wavelength of the photonic bandgap increased due to the light-induced transformation of the structure chiral dopant. When the illuminating time changed from 0 to 48 minutes, the center wavelength of the photonic bandgap shifted from 370 to 410 nm. A tuning range of the wavelength of the laser mode from 385 to 415 nm could be realized, which indicates a large tuning range of 35 nm.

The cholesteryl iodide will undergo a photolysis action when illuminated by a deep ultraviolet 254 nm light, which leads to an increase of the refractive index of the cholesteryl iodide [115]. If a CLC was doped with cholesteryl iodide, the photonic bandgap of the CLC can be tuned by changing the excitation energy of a 254 nm deep ultraviolet light illuminating the CLC. Furumi *et al.* reported an optically tunable CLC laser doped with fluorescent dye (containing cholesteryl nonanoate and cholesteryl oleyl carbonate) and cholesteryl iodide. The photonic bandgap continuously shifted to the long-wavelength direction with the increase of excitation power of the 254 nm light. Under excitation of a 1 J/cm² ultraviolet light, the photonic bandgap shifted from 550 to 720 nm. The wavelength of the laser mode could be tuned from 610 to 700 nm under excitation of a 300 mJ/cm² ultraviolet light. A large tuning range of 90 nm could be obtained for the wavelength of the laser emission.

The light-sensitive chiral Azo molecules will undergo a configuration change from *trans*-form state to *cis*-form state under excitation of an ultraviolet light, which leads to a remarkable reduction in the refractive index of Azo molecules [116]. With the increase in the excitation time, the configuration change of the Azo molecules becomes more and more complete, which makes the refractive index of Azo molecules smaller and smaller. Therefore, when a CLC is doped with Azo material, the central wavelength of the photonic bandgap shifts in the direction of the short wavelength under excitation of an ultraviolet light. Lin *et al.* reported an optically tunable CLC laser with a wide tuning capability [117]. They fabricated a CLC made of a mixture of chiral material S811, a nematic liquid crystal ZLI2293, and a chiral AzoB, which was doped with organic laser dye 4-(dicyanomethylene)-2-methyl-6-(4-dimethlyaminostryl)-4H-pyran (DCM) and pyrromethene 580 (P580). Under illumination of a 350 nm ultraviolet light with an intensity of 7.74 mW/cm² for 15 minutes, the long-wavelength edge of the photonic bandgap of the dye-doped CLC shifted from 670 to 520 nm. Accordingly, when a 532 nm intense pulse light was used to excite the dye-doped CLC, the wavelength of the laser emission could be tuned from 667 to 563 nm. A large tunability in a range of about 100 nm in the laser emission wavelength could be achieved for the AzoB-doped CLC laser.

References

1. G. Guerrero, D. L. Boiko, and E. Kapon, "Photonic crystal hetero-structures implemented with vertical-cavity surface-emitting lasers," *Opt. Express* **12**, 4922–4928 (2004).

2. T. D. Happ, A. Markard, M. Kamp, A. Forchel, and S. Anand, "Single-mode operation of coupled-cavity lasers based on two-dimensional photonic crystals," *Appl. Phys. Lett.* **79**, 4091–4093 (2001).

3. S. Ahn, S. Kim, and H. Jeon, "Single-defect photonic crystal cavity laser fabricated by a combination of laser holography and focused ion beam lithography," *Appl. Phys. Lett.* **96**, 131101 (2010).

4. J. P. Dowling, M. Socalora, M. J. Bloemer, and C. M. Bowden, "The photonic band edge laser: a new approach to gain enhancement," *J. Appl. Phys.* **75**, 1896–1899 (1994).

5. A. Mekis, M. Meier, A. Dodabalapur, R. E. Slusher, and J. D. Joannopoulos, "Lasing mechanism in two-dimensional photonic crystal lasers," *Appl. Phys. A* **69**, 111–114 (1999).

6. R. Colombelli, K. Srinivasan, M. Troccoli, O. Painter, C. F. Gmachl, D. M. Tennant, A. M. Sergent, D. L. Sivco, A. Y. Cho, and F. Capasso, "Quantum cascade surface-emitting photonic crystal laser," *Science* **302**, 1374–1377 (2003).

7. K. Nozaki, H. Watanabe, and T. Baba, "Photonic crystal nanolaser monolithically integrated with passive waveguide for effective light extraction," *Appl. Phys. Lett.* **92**, 021108 (2008).

8. B. Ellis, L. Fushman, D. Englund, B. Y. Zhang, Y. Yamamoto, and J. Vuckovic, "Dynamics of quantum dot photonic crystal lasers," *Appl. Phys. Lett.* **90**, 151102 (2007).

9. C. S. Kim, M. Kim, W. W. Bewlwy, J. R. Lindle, C. L. Canedy, J. A. Nolde, D. C. Larrabee, I. Vurgaftman, and J. R. Mayer, "Broad-stripe, single-mode, mid-IR interband cascade laser with photonic-crystal distributed-feedback grating," *Appl. Phys. Lett.* **92**, 071110 (2008).

10. M. H. Kok, W. Lu, J. C. W. Lee, W. Y. Tam, G. K. L. Wong, and C. T. Chan, "Lasing from dye-doped photonic crystals with graded layers in dichromate gelatin emulsions," *Appl. Phys. Lett.* **92**, 151108 (2008).

11. A. M. Familia, A. Sarangan, and T. R. Nelson, "Optically pumped photonic crystal polymer lasers based on [2-methoxy-5-(2-ethylhexyloxy)-1,4-phenylenevinylenne]," *Opt. Express* **13**, 3136–3143 (2005).

12. Y. Chassagneux, R. Colombelli, W. Maineult, S. Barbier, H. E. Beere, D. A. Ritchie, S. P. Khanna, E. H. Linfield, and A. G. Davies, "Electrically pumped photonic-crystal terahertz lasers controlled by boundary conditions," *Nature* **457**, 174–178 (2009).

13. K. Yoshino, S. Tatsuhara, Y. Kawagishi, M. Ozaki, A. A. Zakhidov, and Z. V. Vardeny, "Amplified spontaneous emission and lasing in conducting polymers and fluorescent dyes in opals as photonic crystals," *Appl. Phys. Lett.* **74**, 2590–2592 (1999).

14. S. Nojima, "Optical-gain enhancement in two-dimensional active photonic crystals," *J. Appl. Phys.* **90**, 545–551 (2001).

15. R. C. Polson, A. Chipouline, and Z. V. Vardeny, "Random lasing in π-conjugated films and infiltrated opals," *Adv. Mater.* **13**, 760–764 (2001).

16. M. Loncar, T. Yoshie, A. Scherer, P. Gogna, and Y. Qiu, "Low-threshold photonic crystal laser," *Appl. Phys. Lett.* **81**, 2680–2682 (2002).

17. M. H. Kok, W. Lu, W. Y. Tam, and G. K. L. Wong, "Lasing from dye-doped icosahedral quasicrystals in dichromate gelatin emulsions," *Opt. Express* **17**, 7275–7284 (2009).

18. C. Monat, C. Seassai, X. Letartre, P. Regreny, P. R. Romeo, P. Viktorovitch, M. L. V. Dyerville, D. Vassagne, J. P. Albert, E. Jalaguier, S. Pocas, and B. Aspar, "InP-based two-dimensional photonic crystal on silicon: in-plane Bloch mode laser," *Appl. Phys. Lett.* **81**, 5102–5104 (2002).

19. D. S. Song, Y. J. Lee, H. W. Choi, and Y. H. Lee, "Polarization-controlled, single-transverse-mode, photonic-crystal, vertical-cavity, surface-emitting lasers," *Appl. Phys. Lett.* **82**, 3182–3184 (2003).

20. N. Yokouchi, A. J. Danner, and K. D. Choquette, "Vertical-cavity surface-emitting laser operating with photonic crystal seven-point defect structure," *Appl. Phys. Lett.* **82**, 3608–3610 (2003).

21. S. H. Kwon, H. Y. Ryu, G. H. Kim, Y. H. Lee, and S. B. Kim, "Photonic bandedge lasers in two-dimensional square lattice photonic crystal slabs," *Appl. Phys. Lett.* **83**, 3870–3872 (2003).

22. C. O. Cho, J. Jeong, J. Lee, H. Jeon, I. Kim, D. H. Jang, Y. S. Park, and J. C. Woo, "Photonic crystal band edge laser array with a holographically generated square-lattice pattern," *Appl. Phys. Lett.* **87**, 161102 (2005).

23. S. Yokoyama, T. Nakahama, S. Mashiko, and M. Nakao, "Photonic crystal templates for organic solid-state lasers," *Appl. Phys. Lett.* **87**, 191101 (2005).

24. M. Scharrer, A. Yamilov, X. Wu, H. Cao, and R. P. H. Chang, "Ultraviolet lasing in high-order bands of three-dimensional ZnO photonic crystals," *Appl. Phys. Lett.* **88**, 201103 (2006).

25. M. B. Christiansen, A. Kristensen, S. S. Xiao, and N. A. Mortensen, "Photonic integrated in k-space: enhancing the performance of photonic crystal dye lasers," *Appl. Phys. Lett.* **93**, 231101 (2008).

26. S. W. Chen, T. C. Lu, Y. J. Hou, T. C. Liu, H. C. Kuo, and S. C. Wang, "Lasing characteristics at different band edges in GaN photonic crystal surface emitting lasers," *Appl. Phys. Lett.* **96**, 071108 (2010).

27. D. S. Song, S. H. Kim, H. G. Park, C. K. Kim, and Y. H. Lee, "Single-fundamental-mode photonic-crystal vertical-cavity surface-emitting lasers," *Appl. Phys. Lett.* **80**, 3901–3903 (2002).

28. H. Masuda, M. Yamada, F. Matsumoto, S. Yokoyama, S. Mashiko, M. Nakao, and K. Nishio, "Lasing from two-dimensional photonic crystals using anodic porous alumina," *Adv. Mater.* **18**, 213–216 (2006).

29. H. Y. Ryu, S. H. Kwon, Y. J. Lee, Y. H. Lee, and J. S. Kim, "Very-low-threshold photonic band-edge lasers from free-standing triangular photonic crystal slabs," *Appl. Phys. Lett.* **80**, 3476–3478 (2002).

30. M. Notomi, H. Suzuki, T. Tamamura, and K. Edagawa, "Lasing action due to the two-dimensional quasiperiodicity of photonic quasicrystal with a Penrose lattice" *Phys. Rev. Lett.* **92**, 123906 (2004).

31. Y. Yamada, T. Nakamura, and K. Yano, "Optical response of mesoporous synthetic opals to the adsorption of chemical species," *Langmuir* **24**, 2779–2784 (2008).

32. H. Yamada, T. Nakamura, Y. Yamada, and K. Yano, "Colloidal-crystal laser using monodispersed mesoporous silica spheres," *Adv. Mater.* **21**, 1–5 (2009).

33. B. Liu, A. Yamilov, and H. Cao, "Effect of Kerr nonlinearity on defect lasing modes in weakly disordered photonic crystals," *Appl. Phys. Lett.* **83**, 1092–1094 (2003).

34. D. Englund, H. Altug, and J. Vuckovic, "Time-resolved lasing action from single and coupled photonic crystal nanocavity array lasers emitting in the telecom band," *J. Appl. Phys.* **105**, 093110 (2009).

35. L. M. Chang, C. H. Hou, Y. C. Ting, C. C. Chen, C. L. Hsu, J. Y. Chang, C. C. Lee, G. T. Chen, and J. I. Chyi, "Laser emission from GaN photonic crystals," *Appl. Phys. Lett.* **89**, 071116 (2006).

36. M. Nomura, S. Iwamoto, M. Nishioka, S. Ishida, and Y. Arakawa, "Highly efficient optical pumping of photonic crystal nanocavity lasers using cavity resonant excitation," *Appl. Phys. Lett.* **89**, 161111 (2006).

37. K. Nozaki, S. Kita, Y. Arita, and T. Baba, "Resonantly photopumped lasing and its switching behavior in a photonic crystal nanolaser," *Appl. Phys. Lett.* **92**, 021501 (2008).

38. K. Nozaki and T. Baba, "Laser characteristics with ultimate-small modal volume in photonic crystal slab point shift nanolasers," *Appl. Phys. Lett.* **88**, 211101 (2006).

39. Z. Zhang, T. Yoshie, X. Zhu, J. Xu, and A. Scherer, "Visible two-dimensional photonic crystal slab laser," *Appl. Phys. Lett.* **89**, 071102 (2006).

40. Y. S. Choi, M. T. Rakher, K. Hennessy, S. Strauf, A. Badolato, P. M. Petroff, D. Bouwmeester, and E. L. Hu, "Evolution of the onset of coherent

in a family of photonic crystal nanolasers," *Appl. Phys. Lett.* **91**, 031108 (2007).

41. H. G. Park, S. H. Kim, S. H. Kwon, Y. G. Ju, J. K. Yang, J. H. Baek, S. B. Kim, and Y. H. Lee, "Electrically driven single-cell photonic crystal laser," *Science* **305**, 1444–1447 (2004).

42. A. Sugitatsu, T. Asano, and S. Noda, "Characterization of line-defect-waveguide lasers in two-dimensional photonic-crystal slabs," *Appl. Phys. Lett.* **84**, 5395–5397 (2004).

43. X. Checoury, P. Boucaud, J. M. Lourtioz, O. G. Lafaye, S. Bonnefont, D. Mulin, J. Valentin, F. L. Dupuy, F. Pommereau, C. Cuisin, E. Derouin, O. Drisse, L. Legouezigou, F. Lelarge, F. Poingt, G. H. Duan, and A. Talneau, "1.5 μm room-temperature emission of square-lattice photonic crystal waveguide lasers with a single line defect," *Appl. Phys. Lett.* **86**, 151111 (2005).

44. H. G. Park, J. K. Hwang, J. Huh, H. Y. Ryu, Y. H. Lee, and J. S. Kim, "Nondegenerate monopole-mode two-dimensional photonic band gap laser," *Appl. Phys. Lett.* **79**, 3032–3034 (2001).

45. O. Painter, R. K. Lee, A. Scherer, A. Yraiv, J. D. Obrien, P. D. Dapkus, and I. Kim, "Two-dimensional photonic bandgap defect mode laser," *Science* **284**, 1819–1821 (1999).

46. H. S. Ee, K. Y. Jeong, M. K. Seo, Y. H. Lee, and H. G. Park, "Ultrasmall square-lattice zero-cell photonic crystal laser," *Appl. Phys. Lett.* **93**, 011104 (2008).

47. S. H. Kim, H. Y. Ryu, H. G. Park, G. H. Kim, Y. S. Choi, and J. S. Kim, "Two-dimensional photonic crystal hexagonal waveguide ring laser," *Appl. Phys. Lett.* **81**, 2499–2501 (2002).

48. A. R. Alija, L. J. Martinez, P. A. Postigo, C. Seassal, and P. Viktorovitch, "Coupled-cavity two-dimensional photonic crystal waveguide ring laser," *Appl. Phys. Lett.* **89**, 101102 (2006).

49. M. H. Shih, W. Kuang, A. Mock, M. Bagheri, E. H. Hwang, J. D. Obrien, and P. D. Dapkus, "High-quality-factor photonic crystal heterostructure laser," *Appl. Phys. Lett.* **89**, 101104 (2006).

50. P. T. Lee, T. W. Lu, F. M. Tsai, T. C. Lu, and H. C. Kuo, "Whispering gallery mode of modified octagonal quasiperiodic photonic crystal single-defect microcavity and its side-mode reduction," *Appl. Phys. Lett.* **88**, 201104 (2006).

51. P. T. Lee, T. W. Lu, F. M. Tsai, and T. C. Lu, "Investigation of whispering gallery mode dependence on cavity geometry of quasiperiodic

photonic crystal microcavity lasers," *Appl. Phys. Lett.* **89**, 231111 (2006).

52. K. Nozaki and T. Baba, "Quasiperiodic photonic crystal microcavity lasers," *Appl. Phys. Lett.* **84**, 4875–4877 (2004).

53. J. Yoon, W. Lee, J. M. Caruge, M. Bawendi, E. L. Thomas, S. Kooi, and P. N. Prasad, "Defect-mode mirrorless lasing in dye-doped organic/inorganic hybrid one-dimensional photonic crystal," *Appl. Phys. Lett.* **88**, 091102 (2006).

54. M. Dabbicco, T. Maggipinto, and M. Brambilla, "Optical bistability and stationary patterns in photonic-crystal vertical-cavity surface-emitting lasers," *Appl. Phys. Lett.* **86**, 021116 (2005).

55. A. J. Danner, J. J. R. Jr, N. Yokouchi, and K. D. Choquette, "Transverse modes of photonic crystal vertical-cavity lasers," *Appl. Phys. Lett.* **84**, 1031–1033 (2004).

56. P. B. Dayal, N. Kitabayashi, T. Miyamoto, F. Koyama, T. Kawashima, and S. Kawakami, "Polarization control of 1.5 µm vertical-cavity surface-emitting lasers using autocloned photonic crystal polarizer," *Appl. Phys. Lett.* **91**, 041110 (2007).

57. S. Boutami, B. Benbakir, J. L. Leclercq, and P. Viktorovtich, "Compact and polarization controlled 1.55 µm vertical-cavity surface-emitting laser using single-layer photonic crystal mirror," *Appl. Phys. Lett.* **91**, 071105 (2007).

58. A. J. Liu, M. X. Xing, H. W. Qiu, W. Chen, W. J. Zhou, and W. H. Zheng, "Reduced divergence angle of photonic crystal vertical-cavity surface-emitting laser," *Appl. Phys. Lett.* **94**, 191105 (2009).

59. A. J. Liu, W. Chen, M. X. Xing, W. J. Zhou, H. W. Qu, and W. H. Zheng, "Phase-locked ring-defect photonic crystal vertical-cavity surface-emitting laser," *Appl. Phys. Lett.* **96**, 151103 (2010).

60. J. J. Raftery, J. A. J. Danner, J. C. Lee, and K. D. Choquette, "Coherent coupling of two-dimensional arrays of defect cavities in photonic crystal vertical cavity surface-emitting lasers," *Appl. Phys. Lett.* **86**, 201104 (2005).

61. L. D. A. Lundeberg, D. L. Boiko, and E. Kapon, "Coupled islands of photonic crystal heterostructures implemented with vertical-cavity surface-emitting lasers," *Appl. Phys. Lett.* **87**, 241120 (2005).

62. C. Levallois, B. Caillaud, J.-L. de Bougrenet de la Tocnaye, L. Dupont, A. L. Corre, H. Folliot, O. Dehaese, and S. Loualiche, "Long-wavelength vertical-cavity surface-emitting laser using an electro-optic

index modulator with 10 nm tuning range," *Appl. Phys. Lett.* **89**, 011102 (2006).

63. F. Jin, C. F. Li, X. Z. Dong, W. Q. Chen, and X. M. Duan, "Laser emission from dye-doped polymer film in opal photonic crystal cavity," *Appl. Phys. Lett.* **89**, 241101 (2006).

64. F. Jin, Y. Song, X. Z. Dong, W. Q. Chen, and X. M. Duan, "Amplified spontaneous emission from dye-doped polymer film sandwiched by two opal photonic crystals," *Appl. Phys. Lett.* **91**, 031109 (2007).

65. X. T. Long, Q. F. Dai, H. H. Fan, Z. C. Wei, M. M. Wu, and H. Z. Wang, "Dominant mode in closed photonic crystal microcavity filled with high scattering irregular gain medium". *Appl. Phys. Lett.* **89**, 251105 (2006).

66. A. Sugitatsu and S. Noda, "Room temperature operation of 2D photonic crystal slab defect-waveguide laser with optical pump," *Electron. Lett.* **39**, 213–215 (2003).

67. A. Sugitatsu, T. Asano, and A. Noda, "Line-defect-waveguide laser integrated with a point defect in a two-dimensional photonic crystal slab," *Appl. Phys. Lett.* **86**, 171106 (2005).

68. O. Khayam, C. Camboumac, H. Benisty, M. Ayre, R. Brenot, G. H. Duan, and W. Pernice, "In-plane Littrow lasing of broad photonic crystal waveguides," *Appl. Phys. Lett.* **91**, 041111 (2007).

69. S. Gardin, F. Bordas, X. Letartre, C. Seassal, A. Rahmant, R. Bozio, and P. Viktorovitch, "Microlasers based on effective index confined slow light modes in photonic crystal waveguides," *Opt. Express* **16**, 6331–6339 (2008).

70. W. H. Zheng, G. Ren, M. X. Xing, W. Chen, A. J. Liu, W. J. Zhou, T. Baba, K. Nozaki, and L. H. Chen, "High efficiency operation of butt joint line-defect-waveguide microlaser in two-dimensional photonic crystal slab," *Appl. Phys. Lett.* **93**, 081109 (2008).

71. Y. Halioua, T. J. Karle, F. Raineri, P. Monnier, I. Sagnes, G. Roelkens, D. V. Thourhout, and R. Raj, "Hybrid In-P-based photonic crystal lasers on silicon on insulator wires," *Appl. Phys. Lett.* **95**, 201119 (2009).

72. C. V. Shank, J. E. Bjorkholm, and H. Kogelnik, "Tunable distributed-feedback dye laser," *Appl. Phys. Lett.* **18**, 395–396 (1971).

73. H. Kogelnik and C. V. Shank, "Stimulated emission in a periodic structure," *Appl. Phys. Lett.* **18**, 152–154 (1971).

74. M. Z. Rossi, S. Perissinotto, G. Lanzani, M. Salerno, and G. Gigli, "Laser dynamics in organic distributed feedback lasers," *Appl. Phys. Lett.* **89**, 181105 (2006).

75. D. Hofstetter, R. Theron, A. H. E. Shaer, A. Bakin, and A. Waag, "Demonstration of a ZnO/MgZnO-based one-dimensional photonic crystal multiquantum well laser," *Appl. Phys. Lett.* **93**, 101109 (2008).

76. M. Imada, S. Noda, A. Chutinan, T. Tokuda, M. Murata, and G. Sadaki, "Coherent two-dimensional lasing action in surface-emitting laser with triangular-lattice photonic crystal structure," *Appl. Phys. Lett.* **75**, 316–318 (1999).

77. W. W. Bewley, C. S Kim, M. Kim, C. L. Canedy, J. R. Lindle, I. Vurgaftman, J. R. Meyer, R. E. Muller, P. M. Echternach, and R. Kaspi, "Broad-stripe midinfrared photonic-crystal distributed-feedback lasers with laser-ablation confinement," *Appl. Phys. Lett.* **83**, 5383–5385 (2003).

78. M. Kim, C. S. Kin, W. W. Bewley, J. R. Lindle, C. L. Canedy, I. Vurgaftman, and J. R. Meyer, "Surface-emitting photonic-crystal distributed-feedback laser for the midinfrared," *Appl. Phys. Lett.* **88**, 191105 (2006).

79. M. Notomi, H. Suzuki, and T. Tamamura, "Directional lasing oscillation of two-dimensional organic photonic crystal lasers at several photonic band gaps," *Appl. Phys. Lett.* **78**, 1325–1327 (2001).

80. M. Meier, A. Mekis, A. Dodabalapur, A. Timko, R. E. Slusher, J. D. Joannopoulos, and O. Nalamasu, "Laser action from two-dimensional distributed feedback in photonic crystals," *Appl. Phys. Lett.* **74**, 7–9 (1999).

81. S. Riechel, C. Kallinger, U. Lemmer, J. Feldmann, A. Gombert, V. Wittwer, and U. Scherf, "A nearly diffraction limited surface emitting conjugated polymer laser utilizing a two-dimensional photonic band structure," *Appl. Phys. Lett.* **77**, 2310–2312 (2000).

82. C. F. Lai, P. Yu, T. C. Wang, H. C. Kuo, T. C. Lu, S. C. Wang, and C. K. Lee, "Lasing characteristics of a GaN photonic crystal nanocavity light source," *Appl. Phys. Lett.* **91**, 041101 (2007).

83. M. M. Jorgensen, S. R. Petersen, M. B. Christiansen, T. Bub, C. L. C. Smith, and A. Kristensen, "Influence of index contrast in two dimensional photonic crystal lasers," *Appl. Phys. Lett.* **96**, 231115 (2010).

84. R. Harbers, J. A. Hoffnagle, W. D. Hinsberg, R. F. Mahrt, N. Moll, D. Erni, and W. Bachtold, "Lasing in interferometrically structured organic materials," *Appl. Phys. Lett.* **87**, 241124 (2005).

85. J. Stehr, J. Crewett, F. Schindler, R. Sperling, G. V. Plessen, U. Lemmer, J. M. Lupton, T. A. Klar, J. Feldmann, A. W. Holleitner, M. Forster, and U. Scherf, "A low threshold polymer laser based on metallic nanoparticle gratings," *Adv. Mater.* **15**, 1726–1729 (2003).

86. H. Finkelmann, S. T. Kim, A, Munoz, P. P. Muhoray, and B. Taheri, "Tunable mirrorless lasing in cholesteric liquid crystalline elastomers," *Adv. Mater.* **13**, 1069–1072 (2001).

87. S. Furumi and Y. Sakka, "Chiroptical properties induced in chiral photonic-bandgap liquid crystals leading to a highly efficient laser-feedback effect," *Adv. Mater.* **18**, 775–780 (2006).

88. L. M. Blinov, G. Cipparrone, V. V. Lazarev, and B. A. Umanskii, "Planar amplifier for a microlaser on a cholesteric liquid crystal," *Appl. Phys. Lett.* **91**, 061102 (2007).

89. F. Araoka, K. C. Shin, Y. Takanishi, K. Ishikawa, H. Takezoe, Z. G. Zhu, and T. M. Swager, "How doping a cholesteric liquid crystal with polymeric dye improves an order parameter and makes possible low threshold lasing," *J. Appl. Phys.* **94**, 279–283 (2003).

90. V. I. Kopp, B. Fan, H. K. M. Vithana, and A. Z. Genack, "Low-threshold lasing at the edge of a photonic stop band in cholesteric liquid crystals," *Opt. Lett.* **23**, 1707–1709 (1998).

91. J. Schmidtke, W. Stille, H. Finkelmonn, and S. T. Kim, "Laser emission in a dye doped cholesteric polymer network," *Adv. Mater.* **14**, 746–749 (2002).

92. Y. C. Yang, C. S. Kee, J. E. Kee, J. E. Kim, H. Y. Park, J. C. Lee, and Y. J. Jeon, "Photonic defect modes of cholesteric liquid crystals," *Phys. Rev. E* **60**, 6852–6854 (1999).

93. V. I. Kopp and A. Z. Genack, "Twist defect in chiral photonic structures," *Phys. Rev. Lett.* **89**, 033901 (2002).

94. J. Schmidtke, W. Stille, and H. Finkelmann, "Defect mode emission of a dye doped cholesteric polymer network," *Phys. Rev. Lett.* **90**, 083902 (2003).

95. M. H. Song, B. Park, K. C. Shin, T. Ohta, Y. Tsunoda, H. Hoshi, Y. Takanishi, K. Ishikawa, J. Watanabe, S. Nishimura, T. Toyooka, Z. Zhu, and T. M. Swager, "Effect of phase retardation on defect mode lasing in polymeric cholesteric liquid crystals," *Adv. Mater.* **16**, 779–783 (2004).

96. S. Noda, M. Yokoyama, M. Imada, A. Chutinan, and M. Mochizuki, "Polarization mode control of two-dimensional photonic crystal laser by unit cell structure design," *Science* **293**, 1123–1125 (2001).

97. S. V. Zhukovsky, D. N. Chigrin, A. V. Lavrinenko, and J. Kroha, "Switchable lasing in multimode microcavities," *Phys. Rev. Lett.* **99**, 073902 (2007).

98. M. N. Shkunov, Z. V. Vardeny, M. C. Delong, R. C. Polson, A. A. Zakhidov, and R. H. Baughman, "Tunable, gap-state lasering in switchable directions for opal photonic crystals," *Adv. Funct. Mater.* **12**, 21–26 (2002).

99. J. R. Lawrence, Y. Ying, P. Jiang, and S. H. Foulger, "Dynamic tuning of organic lasers with colloidal crystals," *Adv. Mater.* **18**, 300–303 (2006).

100. K. Rivoire, A. Kinkhabwala, F. Hatami, W. T. Masselink, Y. Avlasevich, K. Mullen, W. E. Moerner, and J. Vuckovic, "Lithographic positioning of fluorescent molecules on high-Q photonic crystal cavities," *Appl. Phys. Lett.* **95**, 123113 (2009).

101. F. B. Arango, M. B. Christiansen, M. G. Hansen, and A. Kristensen, "Optofluidic tuning of photonic crystal band edge lasers," *Appl. Phys. Lett.* **91**, 223503 (2007).

102. S. Kim, J. Lee, H. Jeon, and H. J. Kim, "Fiber-coupled surface-emitting photonic crystal band edge laser for biochemical sensor applications," *Appl. Phys. Lett.* **94**, 133503 (2009).

103. M. Stroisch, C. T. Morin, T. Woggon, M. Gerken, U. Lemmer, K. Forberich, and A. Gombert, "Photonic stopband tuning of organic semiconductor distributed feedback lasers by oblique angle deposition of an intermediate high index layer," *Appl. Phys. Lett.* **95**, 021112 (2009).

104. B. Maune, J. Witzens, T. B. Jones, M. Kolodrubetz, H. Atwater, A. Scherer, R. Hagen, and Y. Qiu, "Optically triggered Q-switched photonic crystal laser," *Opt. Express* **13**, 4699–4707 (2005).

105. M. Ozaki, M. Kasano, D. Ganzke, W. Haase, and K. Yoshino, "Mirrorless lasing in a dye-doped ferroelectric liquid crystal," *Adv. Mater.* **14**, 306–309 (2002).

106. S. Furumi, S. Yokoyama, A. Otomo, and S. Mashiko, "Electrical control of the structure and lasing in chiral photonic band-gap liquid crystals," *Appl. Phys. Lett.* **82**, 16–18 (2003).

107. T. Huang, Y. Zhou, and S. T. Wu, "Spatially tunable laser emission in dye-doped photonic liquid crystals," *Appl. Phys. Lett.* **88**, 011107 (2006).

108. R. L. Sutherland, L. V. Natarajan, V. P. Tondiglia, and T. J. Bunning, "Bragg gratings in an acrylate polymer consisting of periodic polymer-dispersed liquid-crystal planes," *Chem. Mater.* **5**, 1533–1538 (1993).

109. R. Jakubiak, T. J. Bunning, R. A. Vaia, and L. V. P. Tondiglia, "Electrically switchable, one-dimensional polymeric resonators from holographic

photopolymerization: a new approach for active photonic bandgap materials," *Adv. Mater.* **15**, 241–244 (2003).

110. R. Ozaki, T. Matsui, M. Ozaki, and K. Yoshino, "Electrically color-tunable defect mode lasing in one-dimensional photonic-band-gap system containing liquid crystal," *Appl. Phys. Lett.* **82**, 3593–3595 (2003).

111. R. Ozaki, Y. Matsuhisa, M. Ozaki, and K. Yoshino, "Electrically tunable lasing based on defect mode in one-dimensional photonic crystal with conducting polymer and liquid crystal defect layer," *Appl. Phys. Lett.* **84**, 1844–1846 (2004).

112. T. H. Lin, H. C. Jau, C. H. Chen, Y. J. Chen, T. H. Wei, C. W. Chen, and A. Y. G. Fuh, "Electrically controllable laser based on cholesteric liquid crystal with negative dielectric anisotropy," *Appl. Phys. Lett.* **88**, 061122 (2006).

113. A. Chanishvili, G. Chilaya, G. Petriashvili, R. Barberi, R. Bartolina, G. Cipparrone, A. Mazzulla, and L. Oriol, "Phototunable lasing in dye-doped cholesteric liquid crystals," *Appl. Phys. Lett.* **83**, 5353–5355 (2003).

114. A. Y. G. Fuh, T. H. Lin, J. H. Liu, and F. C. Wu, "Lasing in chiral photonic liquid crystals and associated frequency tuning," *Opt. Express* **12**, 1857–1863 (2004).

115. S. Furumi, S. Yokoyama, A. Otomo, and S. Mashiko, "Phototunable photonic bandgap in a chiral liquid crystal laser device," *Appl. Phys. Lett.* **84**, 2491–2493 (2004).

116. P. V. Shibaev, R. L. Sanford, D. Chiappetta, V. Milner, A. Genack, and A. Bobrovsky, "Light controllable tuning and switching of lasing in chiral liquid crystals," *Opt. Express* **13**, 2358–2363 (2005).

117. T. H. Lin, Y. J. Chen, C. H. Wu, A. Y. G. Fuh, J. H. Liu, and P. C. Yang, "Cholesteric liquid crystal laser with wide tuning capability," *Appl. Phys. Lett.* **86**, 161120 (2005).

Questions

6.1 What is the physical mechanism of the photonic bandedge laser, photonic crystal microcavity laser, and the photonic crystal waveguide laser?

6.2 How can a tunable photonic crystal laser be realized?

Chapter 7

Photonic Crystal Logic Devices

All-optical logic devices play a very important role in the fields of all-optical computing system, ultrahigh speed information processing, and optical communication. Adopting photons instead of electrons as the information carriers can achieve higher data transmission rate, lower energy losses, and really parallel processing ability. The central processing unit (CPU) is the arithmetic core and the controlling core of an electronic computer. The central processing unit consists of a large number of electronic logic devices, such as various logic gates, adder, multiplier, divider, and so on. The photon is adopted as the information carrier in the all-optical computing system. Therefore, it is necessary to achieve all-optical logic devices. The realization of all-optical logic devices mainly depends on the interactions of photon and matter. The photonic crystal possesses unique properties of controlling the propagation states of photons due to the photonic bandgap effect. Moreover, two-dimensional semiconductor photonic crystal can be easily integrated into the photonic integrated circuits. Great efforts have been made to realize photonic crystal all-optical logic devices. However, up to now, there has been little progress in the research of photonic crystal logic devices.

Photonic Crystals: Principles and Applications
Qihuang Gong and Xiaoyong Hu
Copyright © 2014 Pan Stanford Publishing Pte. Ltd.
ISBN 978-981-4267-30-4 (Hardcover), 978-981-4364-83-6 (eBook)
www.panstanford.com

7.1 All-Optical Logic Gates

All-optical logic gates are the essential components for the construction of various logic arithmetic units. Photonic crystal all-optical logic gates include "AND" gate, "OR" gate, "NOT" gate, and so on. More often than not, photonic crystal waveguide and microcavity are adopted to realize the functions of all-optical logic gates.

7.1.1 "AND" Gate

The "AND" gate can perform the following logic operations:

    0 AND 0 = 0;
    1 AND 0 = 0;
    0 AND 1 = 0;
    1 AND 1 = 1;

Photonic crystal all-optical "AND" gate can be realized by use of nonlinear frequency conversion, tunable microcavity, nonlinear optical waveguide, and so on.

The "AND" gate can be realized by use of the nonlinear frequency conversion in a nonlinear photonic crystal. Mccutcheon *et al.* reported an all-optical conditional "AND" gate in a two-dimensional nonlinear InP photonic crystal [1]. The photonic crystal consisted of two microcavities, cavity 1 and 2, and four linear waveguide, as shown in Fig. 7.1. Microcavity 1 was used to trap the picosecond control light A with a frequency of ω_1 from the bus waveguide and transferred it to the input waveguide of microcavity 2. Microcavity 2 could sustain two resonant modes with the frequency of ω_4 and ω_5, respectively. The signal light $\omega_A = \omega_1$ was not in resonance with the resonant modes of microcavity 2, i.e., $\omega_A \neq \omega_4$, and $\omega_A \neq \omega_5$. After signal light ω_A enters the microcavity 2, a new signal light ω_B with the frequency of $\omega_B = \omega_A + \omega_4$ or $\omega_A + \omega_5$ could be generated through a nonlinear sum frequency process in the microcavity 2 due to the greatly enhanced optical nonlinearity. When the microcavity mode ω_5 was selectively resonant excited by another femtosecond signal laser $\omega_C = \omega_5$, the signal $\omega_B = \omega_A + \omega_5$ could be obtained. This is the operation "A (1) AND C (1) = B (1)." When the frequency of ω_A was shifted to $\omega_A + \Delta\omega$, the signal $\omega_B = \omega_A + \omega_5$ did not appear. This is the operation "A (0)

AND C (1) = B (0)." When the microcavity mode ω_5 was not excited, the signal $\omega_B = \omega_A + \omega_5$ also did not appear. This is the operation "A (1) AND C (0) = B (0)." Therefore, the logic gate of "AND" operation could be achieved.

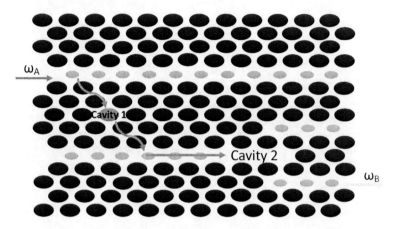

Figure 7.1 Schematic structure of photonic crystal "AND" logic gate based on nonlinear frequency conversion.

The photonic crystal "AND" gate can also be realized by use of a dynamically tunable photonic crystal microcavity. Soljacic and Joannopoulos designed a photonic bandgap microcavity that could sustain two dipole-type resonant modes: One is odd with respect to the x axis and the other is even in a two-dimensional square lattice photonic crystal [2]. The schematic structure of the photonic crystal logic gate is shown in Fig. 7.2. The center of the microcavity was placed an elliptical rod, which was used to break the degeneracy of the two resonant modes. Because of this, the resonant frequencies of two resonant modes are different, i.e., $\omega_x \neq \omega_y$. As a consequence, the signal light that propagates in the waveguide in the x direction cannot be transferred into the waveguide in the y direction, because crosstalk between the two waveguides in the x- and y directions does not exist. When no signal light propagates in the input waveguide A and B, there is no signal transmission in the output waveguide C. This is the logic operation "0 AND 0 = 0." When a signal light with a frequency of $\omega_A = \omega_x + \Delta\omega$ propagates in the input waveguide A, no signal can transmit from the output waveguide C because the signal light

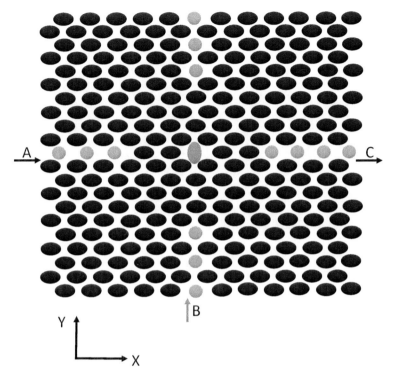

Figure 7.2 Schematic structure of photonic crystal "AND" logic gate based on dynamically tunable microcavity.

ω_A is not in resonance with the microcavity mode ω_x. This is the logic operation "1 AND 0 = 0." Similarly, when another signal light with a frequency of $\omega_B = \omega_y + \Delta\omega$ propagates in the input waveguide B, no signal can transmit from the output waveguide C because crosstalk between the two waveguides in the x- and y- directions does not exist. This is the logic operation "0 AND 1 = 0." When the signal lights ω_A and ω_B appear simultaneously, the signal light ω_A is in resonance with the microcavity mode ω_x due to the signal light ω_B induced variation of the effective refractive index of the photonic bandgap microcavity. So, the signal light ω_A can transmit through the output waveguide C. This is the logic operation "1 AND 1 = 1". Therefore, the "AND" logic gate could be realized.

Nonlinear photonic crystal waveguide can also be adopted to realize the functions of all-optical "AND" gate. Zhu *et al.* presented a scheme to achieve high-contrast "AND" gate based on two-dimensional nonlinear photonic crystal optical waveguide [3]. The two-dimensional photonic crystal consisted of square lattice dielectric rods with a length of infinite embedded in the background medium of air. The optical waveguide was composed of two input waveguides A and B, an output waveguide F, and an interference waveguide, as shown in Fig. 7.3. There are three dielectric rods with third-order optical nonlinearity in the end of the interference waveguide. The interference waveguide was properly designed so that the signal light entering the waveguide A or B will be reflected in the terminal of the interference waveguide. The interference of the incident light and the reflected light leads to the formation of a standing wave in the interference region, and as a result, no signal light can reach the output waveguide F. Therefore, when the signal light was input from the waveguide A, the signal light output from the waveguide F was almost zero.

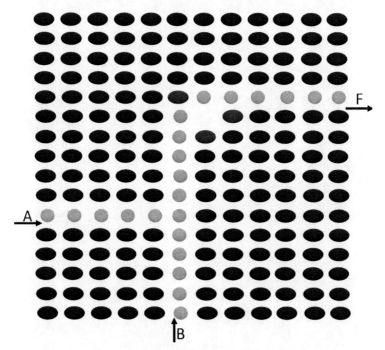

Figure 7.3   Schematic structure of photonic crystal "AND" logic gate based on bend optical waveguide.

This corresponds to the logic operation of "1 AND 0 = 0." Similarly, when the signal light was input from the waveguide B, the signal light output from the waveguide F was also zero. This corresponds to the logic operation of "0 AND 1 = 0." When the signal light was input from the waveguide A and B simultaneously, the intensity of light exciting the three nonlinear rods in the bend region was enlarged remarkably. This results in a great change in the refractive index of the three nonlinear dielectric rods. Perfect reflection cannot be maintained at the bend region. The signal light can propagate through the interference waveguide and the bend region. Accordingly, there is strong output signal from the waveguide F. This corresponds to the logic operation of "1 AND 1 = 1." Liu *et al.* also presented a schematic to realize the logic operation of "AND" gate based on photonic crystal waveguide, as shown in Fig. 7.4 [4]. Three nonlinear dielectric rods were inserted in the two-dimensional square lattice dielectric rod–type

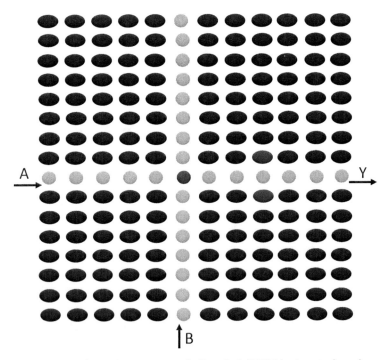

Figure 7.4 Schematic structure of all-optical "AND" logic gate based on photonic crystal cross waveguides.

photonic crystal waveguide. The nonlinear rod in the cross region of two waveguide was used to diffract the signal light. Another two nonlinear rods in the two lateral side of the waveguide were used to introduce a reflection for the signal light. The photonic crystal waveguide was carefully designed so that when the signal light was input only from the port A, the transmittance of the port Y was zero owing to the destructive interference of the diffracted and the reflected light in the output waveguide. This corresponds to the logic operation of "1 AND 0 = 0." Similarly, when the signal light was input only from the port B, the transmittance of the port Y was also zero because of the destructive interference of the diffracted and the reflected light in the output waveguide. This corresponds to the logic operation of "0 AND 1 = 0." When the signal light was input from ports A and B simultaneously, the intensity of light in the nonlinear rods was enhanced greatly, and as a result, the destructive interference condition was no longer maintained. Strong signal light is output from the port Y. This corresponds to the logic operation of "1 AND 1 = 1."

7.1.2 "OR" Gate

The "OR" gate can perform the following logic operations:

0 AND 0 = 0;
1 AND 0 = 1;
0 AND 1 = 1;
1 AND 1 = 1;

The self-collimation effect of the photonic crystal can be adopted to realize all-optical "OR" gate. The self-collimation effect originates from the strong anisotropic dispersion properties of the photonic crystal. The propagation direction of an electromagnetic wave in the photonic crystal, i.e., the direction of the group velocity, is perpendicular to the equal-frequency contours. If photonic crystal has an ultraflat equal-frequency contour within certain angular and frequency range, the electromagnetic wave with a certain frequency will propagate in the photonic crystal without any diffraction [5]. The self-collimated light can be split into a reflected beam and a diffracted beam when it encounters a line defect with a low effective refractive index. There will be a phase shift of $\pi/2$ between the reflected and the diffracted beams when the rod diameter of the line defect is smaller than that of the photonic crystal [6,7]. Therefore, an "OR" logic gate can be realized on the basis of the constructive or destructive interference of the reflected and diffracted beams of two self-collimated signal lights.

Zhang *et al.* presented a scheme to demonstrate all-optical "OR" gate based on self-collimated beams in a two-dimensional photonic crystal composed of square lattice of dielectric rods embedded in air [8]. A line defect made of a row of dielectric rods with a reduced diameter was induced in the photonic crystal along the Γ–X direction, as shown in Fig. 7.5. The self-collimation phenomenon occurs in the frequency of 0.194 (a/λ) along the Γ–M direction. When the signal light I_1 enters the photonic crystal, the reflected beam can reach the output port O_1. This corresponds to the logic operation of "1 OR 0 = 1." Similarly, when the signal light I_2 enters the photonic crystal, the diffracted beam can reach

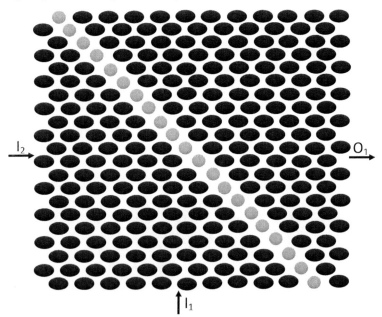

Figure 7.5 Schematic structure photonic crystal "OR" logic gate based on self-collimation effect.

the output port O_1. This corresponds to the logic operation of "0 OR 1 = 1." When the signal lights I_1 and I_2, possessing an initial phase difference of $\pi/2$ enter the photonic crystal simultaneously, the reflected beam of the signal light I_1 and the diffracted beam of the signal light I_2 constructively interfere, which leads to a strong output signal in the port O_1. This corresponds to the logic operation of "1 OR 1 = 1."

7.1.3 "XOR" Gate

The "XOR" gate can perform the following logic operations:

> 0 AND 0 = 0;
>
> 1 AND 0 = 1;
>
> 0 AND 1 = 1;
>
> 1 AND 1 = 0;

The all-optical "XOR" logic gate can be realized on the basis of photonic crystal cross waveguides. Liu *et al.* presented a scheme to achieve the function of all-optical "XOR" logic gate based on dielectric rod–type photonic crystal cross waveguides with four single-defects [4]. The schematic structure of the photonic crystal cross waveguides is shown in Fig. 7.6. The dielectric rod in the center of the cross region is used to diffract the signal light. The photonic crystal cross waveguide was carefully designed so that when the signal light was input from port A, a phase shift of π was induced in the diffracted signal light to the port Y. When the signal light was input from port B, a zero phase shift was introduced in the diffracted signal light to the port Y. Therefore, when the signal light was input from the port A, the diffracted light could output from the port Y. This corresponds to the logic operation of "1 XOR 0 = 1." Similarly, when the signal light was input from the port B, the diffracted light could also output from the port Y. This corresponds to the logic operation of "0 XOR 1 = 1." When the signal light was input from the ports A and B simultaneously, the destructive interference of the diffracted light in the output waveguide resulted in no signal light transmission from the port Y. This corresponds to the logic operation of "1 XOR 1 = 0."

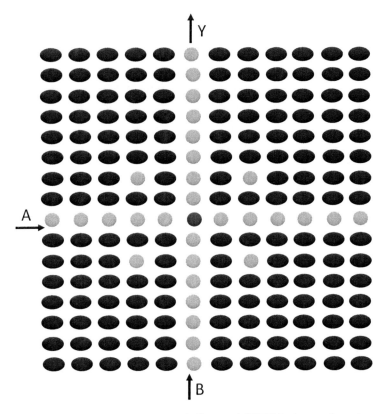

Figure 7.6   Schematic structure of all-optical "XOR" logic gate based on photonic crystal cross waveguides.

7.2   All-Optical Logic Flip-Flop

The electronic logic set-reset flip-flop is a key component of electronic integrated circuits and information processing systems. Similarly, the all-optical logic flip-flop is also an essential and indispensable component of the photonic integrated circuits, all-optical computing, and ultrahigh-speed information processing. The photonic crystal can also be adopted to realize the set-reset flip-flop function [9]. The set-reset flip-flop can perform the following logic operations:

"Output R = 1" and "Output S = 0" for "Input R = 1"
"Output R = 0" and "Output S = 1" for "Input S = 1"

Notomi *et al.* proposed a scheme to realize the set-rest flip-flop function based on dynamically tunable photonic crystal cavities [10]. The schematic structure of the photonic crystal microcavity is shown in Fig. 7.7. There exist two photonic crystal microcavities, Cv_R and Cv_S, and each have two resonant modes. The structure of the microcavities is carefully defined so that the resonant frequency of the low-order mode of two microcavities is equal to each other, while that of the high-order mode is different. The low-order mode was resonant with the single continuous-wave bias input B, which results in the cross feedback between the microcavity Cv_R and Cv_S, i.e., when one of the microcavity is set to the "ON" state, the bias input for the other microcavity is reduced

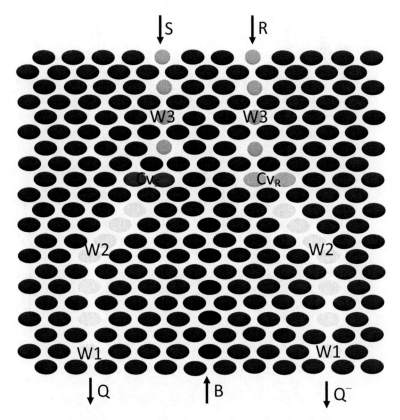

Figure 7.7 Schematic structure of the all-optical logic flip-flop based on photonic crystal microcavities.

greatly, making the other microcavity in the state of "OFF." The WG1 waveguide can transmit the signal light in resonance with all the microcavity modes. The WG2, and 3 waveguides only transmit the single light in resonance with the low- and high-order microcavity modes, respectively. Therefore, when an input signal S enters the waveguide WG3, there are signal transmitted from the Output S port. While the transmittance of the Output R port is zero. This corresponds to the logic operation "Output R = 0" and "Output S = 1" for "Input S = 1." Similarly, when an input signal R enters the waveguide WG3, there are signal transmitted from the Output R port. While the transmittance of the Output S port is zero. This corresponds to the logic operation "Output S = 0" and "Output R = 1" for "Input R = 1."

References

1. M. W. McCutcheon, G. W. Rieger, J. F. Young, D. Dalacu, P. J. Poole, and R. L. Williams, "All-optical conditional logic with a nonlinear photonic crystal nanocavity," *Appl. Phys. Lett.* **95**, 221102 (2009).

2. M. Soljacic and J. D. Joannopoulos, "Enhancement of nonlinear effects using photonic crystals," *Nat. Mater.* **3**, 211–219 (2004).

3. Z. H. Zhu, W. M. Ye, J. R. Ji, X. D. Yuan, and C. Zen, "High-contrast light-by-light switching and AND gate based on nonlinear photonic crystals," *Opt. Express* **14**, 1783–1788 (2006).

4. Q. Liu, Z. B. Ouyang, C. J. Wu, C. P. Liu, and J. C. Wang, "All-optical half adder based on cross structures in two-dimensional photonic crystals," *Opt. Express* **16**, 18992–19000 (2008).

5. D. Y. Zhao, J. Zhang, P. J. Yao, X. Y. Jiang, and X. Y. Chen, "Photonic crystal Mach-Zehnder interferometer based on self-collimation," *Appl. Phys. Lett.* **90**, 231114 (2007).

6. X. F. Yu and S. H. Fan, "Bends and splitters for self-collimated beams in photonic crystals," *Appl. Phys. Lett.* **83**, 3251–3253 (2003).

7. S. G. Lee, S. S. Oh, J. E. Kim, H. Y. Park, and C. S. Kee, "Line-defect-induced bending and splitting of self-collimated beams in two-dimensional photonic crystals," *Appl. Phys. Lett.* **87**, 181106 (2005).

8. Y. L. Zhang, Y. Zhang, and B. J. Li, "Optical switches and logic gates based on self-collimated beams in two-dimensional photonic crystals," *Opt. Express* **15**, 9287–9292 (2007).

9. K. Asakawa, Y. Sugimoto, Y. Watanabe, N. Ozaki, A. Mizutani, Y. Takata, Y. Kitagawa, H. Ishikawa, N. Ikeda, K. Awazu, X. Wang, A. Watanabe, S. Nakamura, S. Ohkouchi, K. Inoue, M. Kristensen, O. Sigmund, P. I. Borel, and R. Baets, "Photonic crystal and quantum dot technologies for all-optical switch and logic devices," *New J. Phys.* **8**, 208–235 (2006).

10. M. Notomi, T. Tanabe, A. Shinya, E. Kuramochi, H. Taniyama, S. Mitsugi, M. Morita, "Nonlinear and adiabatic control of high-Q photonic crystal nanocavities," *Opt. Express* **15**, 17458–17481 (2001).

Question

7.1 Please discuss the different realization mechanisms of the photonic crystal "AND" logic gate.

Chapter 8

Photonic Crystal Sensors

Sensors are essential devices that have very important applications in various fields of modern science and technology. Many kinds of sensors have been designed and fabricated for the control and detection of gases, fluids, cells, and so on. The microscale size and the photonic bandgap properties make photonic crystals the promising basis for the realization of label-free and integrated sensors. Owing to the strong photon confinement effect of photonic crystal microcavity and line defect, ultrahigh resolution of single-molecule level or ultrasmall volume of femtoliter level can be achieved for photonic crystal sensors. Therefore, photonic crystal sensors make possible the realization of the dream of "laboratory on a chip" [1]. Even though great progress has been made in the field of photonic crystal sensors, great efforts and penetrating research are still needed for the practical applications of photonic crystal sensors.

8.1 Fluid Sensor

Fluid sensors can be realized by utilizing the photonic crystal microcavity or line defect waveguide. When the fluid penetrates into the air hole of the microcavity region, the effective refractive

Photonic Crystals: Principles and Applications
Qihuang Gong and Xiaoyong Hu
Copyright © 2014 Pan Stanford Publishing Pte. Ltd.
ISBN 978-981-4267-30-4 (Hardcover), 978-981-4364-83-6 (eBook)
www.panstanford.com

index of the microcavity increases, which leads to the blue shift of the resonant frequency of the microcavity mode. Similarly, a line defect waveguide can be formed by a row of air holes with reduced diameter. When the air holes of the line defect waveguide are filled with the fluid, the effective refractive index of the waveguide increases, which makes the blue shift of the frequency of the guided mode. According to the shift magnitude of the resonant frequency of the microcavity and guided modes, the changes of the refractive index can be determined. Accordingly, the fluid can be distinguished [2,3].

Hasek *et al.* reported a photonic crystal fluid sensor based on the transmittance changes of the photonic crystal waveguide in the microwave range [4]. They fabricated a 10 mm-thick two-dimensional triangular lattice photonic crystal waveguide made of high-density polyethylene. The photonic crystal waveguide was constructed by a row of air holes with a reduced diameter. The lattice constant, diameter of normal, and reduced air holes were 1.346, 0.7, and 0.51 mm, respectively. The SEM image of the photonic crystal waveguide is shown in Fig. 8.1. The photonic bandgap was located in the frequency range 97–109 GHz. The transmittance of the photonic crystal waveguide was 23 dB. When the air holes of the waveguide region were filled with cyclohexane, the transmittance of the photonic crystal waveguide increased to 20 dB. When the air holes of the waveguide region were filled with CCl_4, the transmittance of the photonic crystal waveguide increased to 17 dB. The introduction of fluid in the air holes of the photonic crystal waveguide region increase the effective refractive index of the photonic crystal waveguide, which leads to the appearance of more guided modes. This results in an increase of the transmittance of the photonic crystal waveguide. Based on the transmittance changes of the photonic crystal waveguide, different fluids could be detected and distinguished.

Loncar *et al.* reported a fluid sensor based on a triangular lattice photonic crystal microcavity laser with the InGaAsP quantum well material used as the gain medium [5]. The photonic crystal microcavity was formed by a reduced air hole in the center of a two-dimensional triangular lattice photonic crystal and a fractional edge dislocation by extending the length of a row of air holes to break the symmetry of the microcavity, as shown in Fig. 8.2. As a result, the electric field distribution of the microcavity mode was mainly confined in the central defect hole. The lattice constant, the diameter of the normal air holes, and the diameter of the defect air holes were 446, 268, and 67 nm, respectively. An 830 nm, 30 ns pulsed laser was used as the pump light. A 67 nm red shift of the laser wavelength could be reached when the photonic crystal microcavity laser was moved from air to the isopropyl alcohol solution with a refractive index of 1.377. A 1 nm wavelength shift could be achieved for a refractive index change of 0.0056.

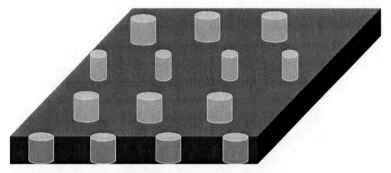

Figure 8.1 Schematic structure of the fluid sensor based on photonic crystal waveguide.

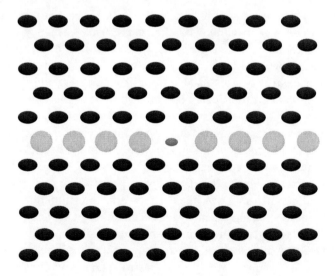

Figure 8.2 Schematic structure of the fluid sensor based on photonic crystal microcavity laser.

Photonic crystal channel-drop filter can also be adopted to perform the function of fluid sensing. When the photonic crystal channel-drop filter is immersed in a fluid solution, the effective refractive index of the photonic crystal increases. This makes the resonant wavelength of the microcavity mode shift in the direction of the long wavelength. Different fluids results in different microcavity resonant frequencies. Thus, a fluid sensor can be obtained. Dorfner *et al.* reported a silicon triangular lattice photonic crystal channel-drop filter structure for fluid sensing applications [6]. The photonic crystal microcavity was constructed by removing three air holes. The lattice constant and the diameter of air holes were 380 and 237 nm, respectively. The microcavity resonant wavelength was 1468.5 nm. The microcavity resonant wavelength changed to 1511.7 nm when the photonic crystal was immersed in isopropanol. The refractive index sensitivity of $\Delta n/n = 0.006$ could be reached.

8.2 Biochemical Sensor

The biochemical sensing technique with an ultrafast response, ultrasmall analyte requirement, and ultrahigh sensitivity plays a very important role in the fields of life science, medical diagnostics, and environmental detection systems. The electric field distribution of a photonic crystal microcavity is mainly confined in the center of the defect structure. Owing to the strong photon confinement effect, the interactions of light and matter are enhanced greatly. This makes it possible to detect and measure a single biological molecule very correctly in the real time. When a biological molecule, such as a DNA molecule, protein, and antibody, is placed on the surface of a photonic crystal microcavity, the resonant frequency of the microcavity mode blue-shifts owing to the increase of the effect of the refractive index of the microcavity structure [7]. More often than not, different biological molecules have different refractive indexes. The shift magnitude is different for different biological molecules. Therefore, biological molecules can be distinguished by different shift magnitudes of the microcavity mode. The biochemical sensor can be realized by use of the photonic crystal microcavity due to strong photon confinement and high sensitivity to the refractive index change of the surrounding medium.

8.2.1 Using Photonic Crystal Microcavity

Chow fabricated a biochemical sensor based on a two-dimensional silicon photonic crystal microcavity [8]. The microcavity was formed by reducing the diameter of an air hole in the center of a triangular lattice photonic crystal with a lattice constant and diameter of air holes of 440 and 255 nm, respectively. The diameter of the air hole constructing the single defect was 176 nm. The schematic structure of the photonic crystal microcavity is shown in Fig. 8.3. The resonant wavelength of the microcavity was around 1550 nm. When chemical solution was dropped on the surface of the photonic crystal microcavity, the effective refractive index of the microcavity region increased, which made the resonant wavelength of the microcavity mode shift in the long-wavelength direction. When the environmental refractive index changed 0.002, the resonant wavelength of the microcavity mode shifted 0.4 nm, which indicates a high detection sensitivity of 200 nm/RIU. The detection resolution of the refractive index was better than 0.001 for the photonic crystal biochemical sensor.

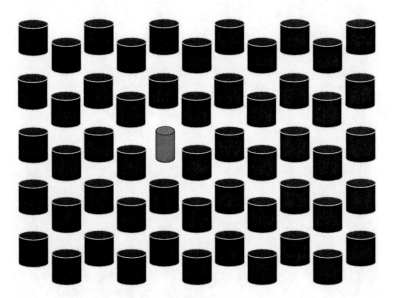

Figure 8.3   Schematic structure of biochemical sensor based on single-defect photonic crystal microcavity.

The coupling of several identical photonic crystal microcavities can be used to construct a coupled-resonator optical waveguide. The sensitivity can be extremely enhanced by use of coupled-resonator optical waveguide for the biochemical sensing because of enhanced light–matter interaction originating from low group velocity and strong confinement of electric field at the site where the analyte is loaded [9]. It has been pointed out that the detection of nanoliter volumes of biochemical analyte and the refractive index resolution down to 10^{-6} refractive index units (RIU) are realistic for photonic crystal biochemical sensors [10,11].

The slot waveguide, consisting of a slot (or a layer of low-dielectric material) sandwiched between two layers of high-dielectric material, have strong photon confinement effect. Owing to the electric filed discontinuity at the interface of the high and low-dielectric materials, the electric field distribution of the guided mode was mainly localized in the slot region [12]. Small refractive index change in the order of 10^{-4} RIU could be detected by use of the slot waveguide due to the strong interactions of light and matter in the slot region [13]. Similarly, the slot photonic crystal waveguide could be obtained by etching a narrow slot in the center of the photonic crystal waveguide. The detection sensitivity $\Delta\lambda/\Delta n$ of 500 nm/RIU could be reached. Moreover, a slot photonic crystal microcavity can be formed by use of a heterostructure configuration, in which a local compression of the lattice constant is required. Based on the slot photonic crystal microcavity, the detection sensitivity $\Delta\lambda/\Delta n$ in the order of 10^3 nm/RIU could be obtained [14].

The surface wave, having the resonant frequency dropping in the photonic bandgap, can propagate in the interface of the dielectric material and air of a truncated photonic crystal due to the strong photonic bandgap effect. The electric field distribution of the surface wave is confined around the surface of the truncated photonic crystal. Therefore, the surface wave is very sensitive to the refractive index change of the region surrounding the truncated region of the photonic crystal. Therefore, photonic crystal surface wave can also be adopted to realize biochemical sensors. Lu *et al*. fabricated a two-dimensional InGaAsP photonic crystal surface wave microcavity by use of the electron beam lithography and reactive ion etching technique [15]. The lattice constant and the diameter of air holes were 510 and 234 nm, respectively. The distance from the slab edge to the center of the nearest air hole was 71.4 nm. The surface wave microcavity was constructed by shrinking and shifting the air holes of the barrier region at the slab edge, as shown in Fig. 8.4. The barrier region acted as the Bragg mirrors for the surface wave in the microcavity region based on the mode gap effect [16]. So the surface wave was confined in the surface of the truncated photonic crystal by the photonic bandgap effect and the mode gap effect. When the environmental refractive index changed from 1 to 1.02, the resonant wavelength of the microcavity mode shifted 12.5 nm, which indicates a large sensitivity $\Delta\lambda/\Delta n$ of 625 nm/RIU. The detectable variation of the environmental refractive index achieved 3.6×10^{-6} theoretically.

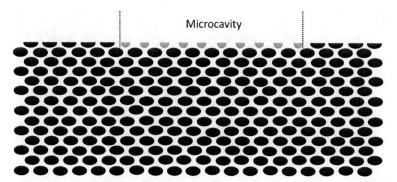

Figure 8.4 Schematic structure of the photonic crystal biochemical sensor based on surface wave microcavity.

8.2.2 Using Guided Resonance Effect

The guided resonance effect could be adopted to realize a photonic crystal biochemical sensor. Ganesh *et al.* reported a photonic crystal biochemical sensor with a high detection resolution based on the guided resonance effect [17]. The schematic structure of the photonic crystal biochemical sensor is shown in Fig. 8.5. The photonic crystal biochemical sensor was constructed by periodic grooves with a lattice period of 250 nm and a thickness of 100 nm, respectively, etched in a porous glass, which was coated with a

15 nm-thick TiO_2 film. A photonic bandgap was formed in the direction parallel to the substrate surface, while the zero-order diffraction mode was transmitted in the direction normal to the substrate surface. When the biochemical analyte was adhered to the surface of the one-dimensional photonic crystal, the guided resonance condition was changed due to the increase of the effective refractive index of the photonic crystal. This resulted in the red shift of the wavelength of the guided resonance. The wavelength of the guided resonance shifted 19 nm when a drop of isopropyl alcohol was adhered on the surface of the photonic crystal biochemical sensor.

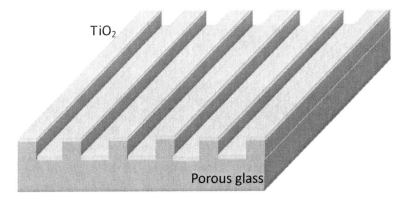

Figure 8.5   Schematic structure of the photonic crystal biochemical sensor based on guided resonance effect.

The porous silicon possesses a large number of pores, which leads to a very large surface-bulk ratio of the order of 10^6 cm^2/cm^3 [18,19]. The large pore networks of porous silicon result in a larger filling capacity of the porous silicon photonic crystals compared with the photonic crystals made of conditional dielectric materials. This makes a larger change of the effective refractive index of the porous silicon photonic crystal. A larger shift magnitude of the reflected or transmitted wavelength can be obtained for the porous silicon photonic crystal. Therefore, much higher sensitivity can be expected for porous silicon photonic crystal biochemical sensors [20].

8.3 Gas Sensor

The gas sensor plays a very important role in the modern industry, scientific research, and environmental protection. Based on the modern advanced microfabrication technology, high quality factor of 10^6 order can be reached in a photonic crystal microcavity. Small refractive index change of the ambient gas will modify the resonant wavelength of the microcavity modes. Therefore, high-sensitivity gas sensor can be realized based on photonic crystal microcavity. On the other hand, the porous materials have very large surface-bulk ratio, which is very suitable for the applications of gas sensors [21].

8.3.1 Using Photonic Crystal Microcavity

Sunner *et al.* reported a gas sensor based on high quality factor photonic crystal microcavity [22]. They fabricated a GaAs triangular lattice photonic crystal microcavity through modifying the air holes along a photonic crystal waveguide, as shown in Fig. 8.6. The lattice constant and the diameter of the normal air holes were 415 and 208 nm, respectively. The region A, having enlarged air holes with a diameter of 224 nm, was used as a mirror. The diameter of air holes in region B was tapered, which was used to reduce the reflection. The quality factor of the photonic crystal microcavity reached 40,000. When the refractive index of the ambient gas increases, the effective refractive index of the photonic crystal microcavity increases, which makes the resonant wavelength of the microcavity mode-shift in the long-wavelength direction. The resonant wavelength shifted 8 pm when a refractive index change of the order of 10^{-4} of the ambient gas was introduced. The detection sensitivity $\Delta\lambda/\Delta n$ can achieve 80 nm/RIU. For the same kind of ambient gas, the pressure change can also influence the refractive index of the gas. Therefore, the gas sensor can also be adopted to detect the change of the ambient gas pressure. With the increase of the gas pressure, the density of the ambient gas will be enlarged. This results in the increase in the refractive index of the ambient gas. As a result, the resonant wavelength of the microcavity mode shifts in the direction of the long wavelength. When the pressure of the ambient SF_6 gas increased from 0.1×10^4 to 5×10^4 Pa, a 30 pm shift magnitude in the resonant wavelength could be obtained for the photonic crystal microcavity.

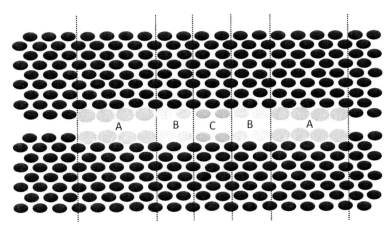

Figure 8.6   Schematic structure of the gas sensor based on photonic crystal microcavity through modifying the air holes along a photonic crystal waveguide.

8.3.2   Using Porous Materials

Silicon nanowire arrays have a spatially periodic distribution of dielectric constant, which can be regarded as a kind of two-dimensional photonic crystal. The interactions of silicon nanowire with NO molecules result in a decrease of the electrical resistance of the silicon nanowire by increasing the density of the free carrier inside the silicon nanowire [23]. This indicates that silicon nanowire arrays can be adopted as a potential candidate for gas sensor devices. Peng *et al.* reported a NO gas sensor based on silicon nanowire arrays [23]. They fabricated high-quality vertically aligned silicon nanowire on the substrate of n-silicon wafer by use of the metal assisted chemical etching method. Two gold electrodes were deposited on the surface of the silicon nanowire arrays. The electrical resistance was measured with different NO concentration in air at the room temperature. The sensitivity of the electrical resistance response can be described by

$$\frac{\Delta R}{R_0} = \frac{R_0 - R}{R_0}$$

(8.1)

where R_0 and R are the electrical resistance without and with the NO gas, respectively. With the NO concentration increased from 10 to 100 ppm, the electrical resistance response $\Delta R/R$ was enlarged from 0.05 to 0.5. High detection sensitivity of less than 10 ppm could be achieved. Carbon nanotubes can also be used to realize the NO_2 gas sensor based on the fact that the interactions of NO_2 molecules and carbon atoms results in a transition from n-type to p-type semiconductor of carbon nanotubes [24]. Jeong *et al.* reported a flexible NO_2 gas sensor by use of carbon nanotubes/reduced graphene film [25]. The NO_2 gas sensor consists of carbon nanotubes/reduced graphene film on the polyimide substrate, gold electrodes, and microheaters. Owing to the n-type to p-type semiconductor transition of carbon nanotubes under exposure of NO_2 gas, the electrical resistance of the carbon nanotubes increased. The sensitivity of the electrical resistance response can be estimated by

$$\frac{\Delta R}{R_0} = \frac{R - R_0}{R_0}$$

(8.2)

where R_0 and R are the electrical resistance without and with the NO_2 gas, respectively. With the increment of the NO_2 gas concentration, the value of the $\Delta R/R_0$ increased. The gas sensor has a very high sensitivity of 0.5 ppm. Owing to the excellent flexibility of the reduced graphene film, a flexible NO_2 gas sensor could be reached.

Jia *et al.* reported an ethanol gas sensor based on metallic ion doped SnO_2 porous films, which was fabricated by the sol-gel and colloidal nanoparticle template technique [26]. The mechanism of the ethanol gas sensing of the SnO_2 porous film originates from the strong interaction of ambient ethanol gas molecules with the O_2 molecules absorbed in the surface of the SnO_2 porous material, which leads to a remarkable reduction of the electric resistance of the SnO_2 porous material. When the metallic ion Cr^{3+} was doped in the SnO_2 porous film, the interactions of the metallic ion Cr^{3+} with the SnO_2 resulted in a chemical shift of the binding energy of SnO_2. Accordingly, the Fermi energy level of SnO_2 was changed. On the other hand, the doping of chromium ions induced a strong absorption of ethanol molecules due to the *d-d* electron transition related to Cr^{3+} [27]. These two factors lead to a high detection sensitivity of ethanol gas. The sensitivity can be defined as R/R_0, where R and R_0 are the electrical resistance of SnO_2 porous material in air and in the target gas, respectively. The ethanol concentration was 1000 ppm. High sensitivity R/R_0 of 50 could be reached when the doping concentration of chromium ions was 1%.

References

1. H. Kurt and D. S. Citrin, "Photonic crystals for biochemical sensing in the terahertz," *Appl. Phys. Lett.* **87**, 041108 (2005).

2. J. Wu, D. Day, and M. Gu, "A microfluidic refractive index sensor based on an integrated three-dimensional photonic crystal," *Appl. Phys. Lett.* **92**, 071108 (2008).

3. J. Topolancik, P. Bhattachatya, J. Sabarinathan, and P. C. Yu, "Fluid detection with photonic crystal-based multichannel waveguides," *Appl. Phys. Lett.* **82**, 1143–1145 (2003).

4. T. Hasek, H. Kurt, D. S. Citrin, and M. Koch, "Photonic crystals for fluid sensing in the subterahertz range," *Appl. Phys. Lett.* **89**, 173508 (2006).

5. M. Loncar, A. Scherer, Y. M. Qiu, "Photonic crystal laser sources for chemical detection," *Appl. Phys. Lett.* **82**, 4648–4650 (2003).

6. D. F. Dorfner, T. Hurlimann, T. Zabel, L. H. Frandsen, G. Abstreiter, and J. J. Finley, "Silicon photonic crystal nanostructures for refractive index sensing," *Appl. Phys. Lett.* **93**, 181103 (2008).

7. L. Martiradonna, F. Pisanello, T. Stomeo, A. Qualtieri, G. Vecchio, S. Sabella, R. Cingolani, M. D. Vittorio, and P. P. Pompa, "Spectral tagging by integrated photonic crystal resonators for highly sensitive and parallel detection in biochips," *Appl. Phys. Lett.* **96**, 113702 (2010).

8. E. Chow, A. Grot, L. W. Mirkarimi, M. Sigalas, and G. Girolami, "Ultracompact biochemical sensor built with two-dimensional photonic crystal microcavity," *Opt. Lett.* **29**, 1093–1095 (2004).

9. H. Kurt and D. S. Citrin, "Coupled-resonator optical waveguides for biochemical sensing of nanoliter volumes of analyte in the terahertz region," *Appl. Phys. Lett.* **87**, 241119 (2005).

10. B. Cunningham, P. Li, B. Lin, and J. Pepper, "Colorimetric resonant reflection as a direct biochemical assay technique," *Sens. Actuators B* **81**, 316–328 (2002).

11. H. Altug and J. Vuckovic, "Polarization control and sensing with two-dimensional coupled photonic crystal microcavity arrays," *Opt. Lett.* **30**, 982–984 (2005).

12. V. R. Almeida, Q. F. Xu, C. A. Barrios, and M. Lipson, "Guiding and confining light in void nanostructure," *Opt. Lett.* **29**, 1209–1211 (2004).

13. C. A. Barrios, K. B. Gylfason, B. Sanchez, A. Griol, H. Sohlstrom, M. Holgado, and R. Casquel, "Slot-waveguide biochemical sensor," *Opt. Lett.* **32**, 3080–3082 (2007).

14. A. D. Falco, L. Ofaolain, and T. F. Krauss, "Chemical sensing in slotted photonic crystal heterostructure cavities," *Appl. Phys. Lett.* **94**, 063503 (2009).

15. T. W. Lu, Y. H. Hsiao, W. D. Ho, and P. T. Lee, "Photonic crystal heteroslab-edge microcavity with high quality factor surface mode for index sensing," *Appl. Phys. Lett.* **94**, 141110 (2009).

16. B. S. Song, S. Noda, T. Asano, and Y. Akahane, "Ultra-high-Q photonic double-heterostructure nanocavity," *Nat. Mater.* **4**, 207–210 (2005).

17. N. Ganesh, I. D. Block, and B. T. Cunningham, "Near ultraviolet-wavelength photonic-crystal biosensor with enhanced surface-to-bulk sensitivity ratio," *Appl. Phys. Lett.* **89**, 023901 (2006).

18. A. Motohashi, M. Kawakami, H. Aoyagi, A. Kinoshita, and A. Satou, "Rapid, reversible, sensitive porous silicon gas sensor," *J. Appl. Phys.* **91**, 2519–2523 (2002).

19. G. G. Rong, J. D. Ryckman, R. L. Mernaugh, and S. M. Weiss, "Label-free porous silicon membrane waveguide for DNA sensing," *Appl. Phys. Lett.* **93**, 161109 (2008).

20. L. Moretti, L. Rea, L. D. Stefano, and I. Rendina, "Periodic versus aperiodic: enhancing the sensitivity of porous silicon based optical sensors," *Appl. Phys. Lett.* **90**, 191112 (2007).

21. G. Barillaro, G. M. Lazzerini, and L. M. Strambini, "Modeling of porous silicon junction field effect transistor gas sensor: insight into NO_2 interaction," *Appl. Phys. Lett.* **96**, 162105 (2010).

22. T. Sunner, T. Stichel, S. H. Kwon, T. W. Schlereth, S. Hofling, M. Kamp, and A. Forchel, "Photonic crystal cavity based on gas sensor," *Appl. Phys. Lett.* **92**, 261112 (2008).

23. K. Q. Peng, X. Wang, and S. T. Lee, "Gas sensing properties of single crystalline porous silicon nanowires," *Appl. Phys. Lett.* **95**, 243112 (2009).

24. J. X. Wang, X. W. Sun, Y. Yang, and C. M. L. Wu, "N-P transition sensing behaviors of ZnO nanotubes exposed to NO_2 gas," *Nanotechnology* **20**, 465501 (2009).

25. H. Y. Jeong, D. S. Lee, H. K. Choi, D. H. Lee, J. E. Kim, J. Y. Lee, W. J. Lee, S. O. Kim, and S. Y. Choi, "Flexible room-temperature NO_2 gas sensors based on carbon nanotubes/reduced graphene hybrid films," *Appl. Phys. Lett.* **96**, 213105 (2010).

26. L. Jia, W. Cai, and H. Q. Wang, "Metal ion-doped SnO_2 ordered porous films and their strong gas sensing selectivity," *Appl. Phys. Lett.* **96**, 103115 (2010).

27. S. Morandi, E. Comini, G. Faglia, and G. Ghiotti, "Cr–Sn oxide thin films: electrical and spectroscopic characterization with CO, NO_2, NH_3 and ethanol," *Sens. Actuators B* **118**, 142–148 (2006).

Questions

8.1 What is the general mechanism on which photonic crystal sensors are based?

8.2 For the purpose of the integration applications, what kind of gas sensor is more suitable?

8.3 Could you design a photonic crystal structure that can perform the function of biochemical sensor?

Index